The
Fluorescent Protein
Revolution

Series in Cellular and Clinical Imaging

Series Editor
Ammasi Periasamy

PUBLISHED

Coherent Raman Scattering Microscopy
edited by Ji-Xin Cheng and Xiaoliang Sunney Xie

Imaging in Cellular and Tissue Engineering
edited by Hanry Yu and Nur Aida Abdul Rahim

Second Harmonic Generation Imaging
edited by Francesco S. Pavone and Paul J. Campagnola

The Fluorescent Protein Revolution
edited by Richard N. Day and Michael W. Davidson

FORTHCOMING

Natural Biomarkers for Cellular Metabolism:
Biology, Techniques, and Applications
edited by Vladimir V. Gukassyan and Ahmed A. Heikal

Optical Probes in Biology
edited by Jin Zhang and Carsten Schultz

SERIES IN CELLULAR AND CLINICAL IMAGING
AMMASI PERIASAMY, SERIES EDITOR

The Fluorescent Protein Revolution

Edited by

Richard N. Day
Michael W. Davidson

CRC Press
Taylor & Francis Group
Boca Raton London New York

CRC Press is an imprint of the
Taylor & Francis Group, an **informa** business

CRC Press
Taylor & Francis Group
6000 Broken Sound Parkway NW, Suite 300
Boca Raton, FL 33487-2742

First issued in paperback 2019

© 2014 by Taylor & Francis Group, LLC
CRC Press is an imprint of Taylor & Francis Group, an Informa business

No claim to original U.S. Government works

ISBN-13: 978-1-4398-7508-7 (hbk)
ISBN-13: 978-0-367-37870-7 (pbk)

Library of Congress Cataloging-in-Publication Data

The fluorescent protein revolution / editors, Richard N. Day, Michael W. Davidson.
    p. ; cm. -- (Series in cellular and clinical imaging)
    Includes bibliographical references and index.
    ISBN 978-1-4398-7508-7 (hardcover : alk. paper)
    I. Day, Richard N., editor of compilation. II. Davidson, Michael W. (Michael Wesley), 1950- editor of compilation. III. Series: Series in cellular and clinical imaging.
    [DNLM: 1. Luminescent Proteins. 2. Microscopy, Fluorescence--methods. QU 55]

RB43
616.07'58--dc23   2013041602

Visit the Taylor & Francis Web site at
http://www.taylorandfrancis.com

and the CRC Press Web site at
http://www.crcpress.com

Contents

Series Preface

A picture is worth a thousand words.

This proverb says everything. Imaging began in 1021 with use of a pinhole lens in a camera in Iraq; later in 1550, the pinhole was replaced by a biconvex lens developed in Italy. This mechanical imaging technology migrated to chemical-based photography in 1826 with the first successful sunlight picture made in France. Today, digital technology counts the number of light photons falling directly on a chip to produce an image at the focal plane; this image may then be manipulated in countless ways using additional algorithms and software. The process of taking pictures ("imaging") now includes a multitude of options—it may be either invasive or noninvasive, and the target and details may include monitoring signals in two, three, or four dimensions.

Microscopes are an essential tool in imaging used to observe and describe protozoa, bacteria, spermatozoa, and any kind of cell, tissue, or whole organism. Pioneered by Antoni van Leeuwenhoek in the 1670s and later commercialized by Carl Zeiss in 1846 in Jena, Germany, microscopes have enabled scientists to better grasp the often misunderstood relationship between microscopic and macroscopic behavior by allowing to study the development, organization, and function of unicellular and higher organisms, as well as structures and mechanisms at the microscopic level. Further, the imaging function preserves temporal and spatial relationships that are frequently lost in traditional biochemical techniques and gives two- or three-dimensional resolution that other laboratory methods cannot. For example, the inherent specificity and sensitivity of fluorescence, the high temporal, spatial, and three-dimensional resolution that is possible, and the enhancement of contrast resulting from the detection of an absolute rather than relative signal (i.e., unlabeled features do not emit) are several advantages of fluorescence techniques. Additionally, the plethora of well-described spectroscopic techniques providing different types of information, and the commercial availability of fluorescent probes such as visible fluorescent proteins (many of which exhibit an environment- or analytic-sensitive response), increase the range of possible applications, such as the development of biosensors for basic and clinical research. Recent advancements in optics, light sources, digital imaging systems, data acquisition methods, and image enhancement, analysis, and display methods have further broadened the applications in which fluorescence microscopy can be applied successfully.

Another development has been the establishment of multiphoton microscopy as a three-dimensional imaging method of choice for studying biomedical specimens from single cells to whole animals with submicron resolution. Multiphoton microscopy methods utilize naturally available endogenous fluorophores—including nicotinamide adenine dinucleotide (NAD), flavin adenine dinucleotide (FAD), tryptophan (TRP), and so on—whose autofluorescent properties provide a label-free approach. Researchers may then image various functions and organelles at molecular levels using two-photon and fluorescence lifetime imaging (FLIM) microscopy to distinguish normal conditions from cancerous ones. Other widely used nonlabeled imaging methods are coherent anti-Stokes Raman scattering spectroscopy (CARS) and stimulated Raman scattering (SRS) microscopy, which allow the imaging of molecular function using the molecular vibrations in cells, tissues, and whole organisms. These techniques have been widely used in gene therapy, single-molecule imaging, tissue engineering, and stem cell research. Another nonlabeled method is harmonic generation, which is also widely used in clinical imaging, tissue engineering, and stem cell research. There are many more advanced technologies developed for cellular and clinical imaging, including multiphoton tomography, thermal imaging in animals, ion imaging (calcium, pH) in cells, etc.

The goal of this series is to highlight these seminal advances and the wide range of approaches currently used in cellular and clinical imaging. Its purpose is to promote education and new research across a broad spectrum of disciplines. The series emphasizes practical aspects, with each volume focusing on a particular theme that may cross various imaging modalities. Each title covers basic to advanced imaging methods, as well as detailed discussions dealing with interpretations of these studies. The series also provides cohesive, complete, state-of-the-art, cross-modality overviews of the most important and timely areas within cellular and clinical imaging.

Since my graduate student days, I have been involved and interested in multimodal imaging techniques applied to cellular and clinical imaging. I have pioneered and developed many imaging modalities throughout my research career. The series manager, Luna Han, recognized my genuine enthusiasm and interest to develop a new book series on cellular and clinical imaging. This project would not have been possible without the support of Luna. I am sure that all the volume editors, chapter authors, and myself have benefited greatly from her continuous input and guidance to make this series a success.

Equally important, I personally would like to thank the volume editors and the chapter authors. This has been an incredible experience working with colleagues who demonstrate such a high level of interest in educational projects, even though they are all fully occupied with their own academic activities. Their work and intellectual contributions based on their deep knowledge of the subject matter will be appreciated by everyone who reads this book series.

SERIES EDITOR

Ammasi Periasamy PhD – Professor and Center Director, W.M. Keck Center for Cellular Imaging, University of Virginia, Charlottesville, Virginia

Preface

It has been over four decades since a small group of investigators began making annual pilgrimages to the University of Washington's Friday Harbor Marine Laboratories to embark on studies that would ultimately elucidate the photochemistry responsible for the green light emitted by the crystal jellyfish (*Aequorea victoria* or *Aequorea aequorea*). The group included Frank Johnson (Princeton University), Osamu Shimomura (Princeton University and Woods Hole Oceanographic Institute), Milton Cormier (University of Georgia), and William Ward (then a postdoc with Milton Cormier, now at Rutgers University). Recently, William Ward reminisced to us about his summers spent collecting jellyfish:

> Milt Cormier had great scientific instincts, and before any of the rest of us had the vision, he was convinced of great biochemical similarities between bioluminescent components of the sea pansy and the jellyfish. Soon after joining the Cormier lab in 1973, I was assigned the project of purifying and characterizing green-fluorescent protein (GFP) from the sea pansy *Renilla*. It was to validate Milt's intuitions that I was sent to Friday Harbor in 1975 to collect jellyfish.

> Every day, at sun-up, the team "hit the floating docks" with their collection of pool skimming nets and dozens of 6-gallon utility pails. The pails were immediately filled with seawater and placed at intervals along the floating docks. No time was wasted as the team quickly began scooping up the jellies and plopping them into buckets—up to 10,000 jellyfish collected each day. The parade of workers carrying pairs of buckets from the docks to the lab always reminded me of the Disney cartoon version of the *Sorcerer's Apprentice*, starring Mickey Mouse. Little did I know, back in 1975, that for 16 additional seasons, I would return to jellyfish "heaven". So much did I enjoy working with bioluminescence, that I chose to make this field of biochemistry my life's work.

Together they caught over a million jellyfish and spent decades investigating the biochemical mechanisms responsible for jellyfish bioluminescence. A crucial event occurred in the 1980s when Douglas Prasher, also working with Milton Cormier, prepared a cDNA library from *Aequorea victoria* mRNA. He used this to obtain a single full-length clone that encoded the complete *Aequorea* GFP sequence. Since no information was available at the time regarding the biosynthetic pathway leading to the chromophore formation necessary for fluorescence, it was not possible to predict whether the jellyfish protein could be produced in other biological systems. Fortunately, it turned out that the GFP chromophore is part of the primary amino acid sequence and forms spontaneously without the need for additional cofactors (described in Parts I and II of this book).

With the demonstration that GFP still fluoresced when produced in other organisms, the astonishing potential of this genetically encoded protein as a tool for studies in cell biology, medicine, and physiology was realized. It is now recognized that the identification of GFP from *Aequorea* was the first step in what has often been described as a "revolution" in cell biology and was ultimately recognized by the 2008 Nobel Prize in Chemistry. What is more, again fulfilling Milton Cormier's intuitions, many other marine organisms were identified in the late 1990s, which produce proteins with amino acid similar to the

Aequorea GFP. Some of these GFP-like proteins have been cloned and engineered for live-cell imaging applications, providing the rainbow of FPs that are described in Part II of this book.

In Part III of this book, the authors present some of the unique applications of the FPs to cell biology. These applications include superresolution microscopy of cellular fine structures and Förster resonance energy transfer (FRET) microscopy to visualize protein interactions and cell-signaling activities inside living cells. The application photobleaching and photoactivation techniques to visualize protein behaviors are discussed, and techniques that exploit the plant and algal photoreceptors to enable light-regulated control of enzymatic activities are described. Finally, the application of FPs to the noninvasive imaging of tumor–host interactions in living animals is presented. We realize the field has grown rapidly and that it is impossible to do justice to all who have contributed to the development and application of many novel FPs. Most importantly, we thank all the authors who have contributed chapters to this book.

EDITORS

Richard N. Day
Michael W. Davidson

Editors

Richard N. Day is a professor in the Department of Cellular and Integrative Physiology at the Indiana University School of Medicine, Indianapolis, Indiana. His research focuses on understanding the network of regulatory protein interactions that function to control cell-type-specific gene expression. His laboratory group uses biochemical and molecular approaches to define networks of protein interactions that are coordinated by specific transcription factors. These *in vitro* approaches are then complemented by noninvasive live-cell imaging techniques using the many different color variants of the marine invertebrate fluorescent proteins. He first published studies using green fluorescent protein (GFP) in 1996 and has continued to pioneer the use of the fluorescent proteins (FPs) for cellular imaging. Recent studies from the laboratory have used Förster resonance energy transfer (FRET)-based microscopy approaches to begin to define networks of protein interactions in living cells.

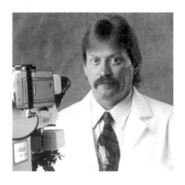

Michael W. Davidson is a scholar/scientist affiliated with the Department of Biological Science and the National High Magnetic Field Laboratory at Florida State University, Tallahassee, Florida.

Davidson's laboratory is involved in the development of educational websites that address all phases of optical microscopy, including brightfield, phase contrast, differential interference contrast (DIC), fluorescence, confocal, total internal reflection fluorescence (TIRF), multiphoton, and superresolution techniques. In addition to his interest in educational activities, Davidson is also involved in research dealing with the performance of traditional fluorescent proteins and optical highlighters in fusions for targeting and dynamics studies in live cells. Davidson's digital images and photomicrographs have graced the covers of over 2000 publications in the past two decades, and he has licensed images to numerous commercial partners. The proceeds of these licensing fees support Davidson's research.

Contributors

John R. Allen
National High Magnetic Field Laboratory
and
Department of Biological Science
Florida State University
Tallahassee, Florida

John M. Christie
College of Veterinary and Life Sciences
University of Glasgow
Glasgow, United Kingdom

Jun Chu
Department of Pediatrics
Stanford University School of Medicine
Palo Alto, California

Michael W. Davidson
National High Magnetic Field Laboratory
and
Department of Biological Science
Florida State University
Tallahassee, Florida

Richard N. Day
Department of Cellular and Integrative
    Physiology
Indiana University School of Medicine
Indianapolis, Indiana

Robert M. Hoffman
AntiCancer, Inc.
and
Department of Surgery
University of California, San Diego
San Diego, California

Michael Z. Lin
Department of Pediatrics
and
Department of Bioengineering
Stanford University School of Medicine
Palo Alto, California

Jennifer Lippincott-Schwartz
Eunice Kennedy Shriver Institute
    of Child Health and Human Development
National Institutes of Health
Bethesda, Maryland

Davide Mazza
National Cancer Institute
National Institutes of Health
Bethesda, Maryland

James G. McNally
National Cancer Institute
National Institutes of Health
Bethesda, Maryland

G. Ulrich Nienhaus
Institute of Applied Physics
Karlsruhe Institute of Technology
Karlsruhe, Germany

George H. Patterson
National Institute of Biomedical Imaging
    and Bioengineering
National Institutes of Health
Bethesda, Maryland

S. James Remington
Department of Physics
and
Institute of Molecular Biology
University of Oregon
Eugene, Oregon

Malte Renz
Eunice Kennedy Shriver Institute
    of Child Health and Human Development
National Institutes of Health
Bethesda, Maryland

Mark A. Rizzo
Department of Physiology
University of Maryland School
    of Medicine
Baltimore, Maryland

Stephen T. Ross
Nikon Instruments, Inc.
Melville, New York

Nathan C. Shaner
Department of Photobiology
    and Bioimaging
The Scintillon Institute
San Diego, California

Jörg Wiedenmann
National Oceanography Centre
University of Southampton
Southampton, United Kingdom

Yan Xing
Department of Pediatrics
Stanford University School of Medicine
Palo Alto, California

Marc Zimmer
Department of Chemistry
Connecticut College
New London, Connecticut

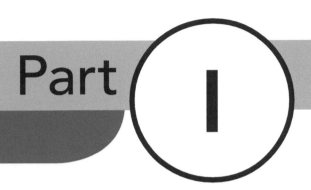

Part I

History and perspective

Introduction to fluorescent proteins

Marc Zimmer
Connecticut College

Contents

1.1 INTRODUCTION

The development of fluorescent proteins (FPs) and their subsequent use in all areas of science is an excellent example of basic science leading to practical biotechnological and medical applications. More than 40 years ago, a small band of researchers that included Frank Johnson (Princeton University), Osamu Shimomura (Princeton University and Woods Hole Oceanographic Institute), Milton Cormier (University of Georgia), and William Ward (Rutgers University) were interested in elucidating the photochemistry responsible for the green light emitted by the crystal jellyfish (*Aequorea victoria* or *Aequorea aequorea*). Together they caught over a million jellyfish and spent more than three quarters of a century studying jellyfish bioluminescence. Although the initial impetus for the research was pure basic science, they wanted to gain a better understanding of the chemistry responsible for the light emission from *Aequorea*; it did not take long before practical uses for this knowledge became apparent. By 1967, Ridgway (Ridgway and Ashley 1967) used aequorin luminescence to monitor the *in vivo* calcium release and reabsorption during stimulation of a single barnacle muscle fiber. It was these types of applications of aequorin, as an *in vivo* calcium monitor, that at least partly motivated Osamu Shimomura to continue his research on aequorin and, by association, green fluorescent protein (GFP) (see Section 1.4). This work ultimately led to the 2008 Nobel Prize, the rainbow of FPs (many of which are now commercially available), and the myriad of applications described in this book. However, I think it is safe to say that it is the interest in the photophysics of bioluminescent organisms, and not their application, which led to the discovery of all the FPs and techniques described in this book.

1.2 BIOLUMINESCENCE

A. victoria, the crystal jellyfish, has hundreds of photocytes positioned along the edge of its umbrella. Each photocyte gives off tiny pinpricks of green light—no one knows why the jellyfish needs this capability. It was from the photocytes that Osamu Shimomura first isolated the GFP. More than 80% of bioluminescent organisms are found in the world's oceans; very few occur in freshwaters or are terrestrial. They are found in most marine phyla from bacteria to fish and can be found throughout the oceans from the tropics to the poles and from the surface to the deepest ocean trenches (Haddock et al. 2010). Most of the marine bioluminescent organisms emit blue light ($\lambda_{max} \sim 475$ nm), the wavelength that travels the furthest through water. It has been estimated that bioluminescence has independently evolved at least 40 times. Bioluminescent flashes can be seen as from tens to hundreds of meters away (Warrant and Locket 2004). The main uses (Widder 2010) for bioluminescence are

1. To attract or communicate with a mate, for example, the light flashes emitted by fireflies
2. To defend against predators, for example, dinoflagellate colonies lighting up encroaching fish
3. To attract or find food, for example, the lure of the angler fish

1.2.1 LUCIFERIN/LUCIFERASE MODEL

In 1887, Raphaël Dubois reported that the bioluminescence of the common piddock involved the reaction of a substrate, called a luciferin, with an enzyme named luciferase. The luciferins are highly conserved—just four different luciferins (see Figure 1.1a through d) are responsible for the bioluminescence of all light-emitting marine organisms (Haddock et al. 2010). In contrast, the luciferases are much less conserved and little sequence similarity is seen across species. In most bioluminescent organisms, oxidation of luciferin, catalyzed by a luciferase, generates the observed light emission.

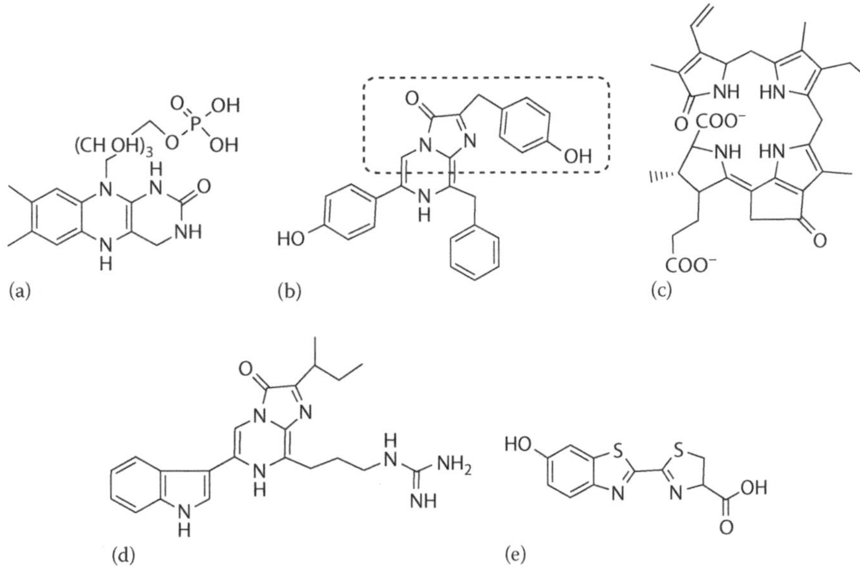

Figure 1.1 Only four different marine luciferins are found. They are (a) the bacterial luciferins, which occur in free living and symbiotic bacteria and require an aldehyde in addition to luciferin and luciferase; (b) coelenterazine, which are the most widely distributed luciferins and are found in at least nine phyla, for example, fish, ostracods, vampire squid, and cnidarians (compare the structure of the GFP chromophore, shown in Figure 1.3, with the substructure of coelenterazine encircled above); (c) dinoflagellate luciferins, which are found in dinoflagellates and euphausiids; and (d) cypridina luciferin that are also found in ostracods. There are a few more terrestrial luciferins. Firefly luciferin (e) is the most studied of the terrestrial luciferins.

1.2.2 BIOLUMINESCENCE IN *A. victoria*

There are bioluminescent organisms that have the luciferase, luciferin, and oxygen combined into one system, resembling the luciferase trapped in an intermediate state. These systems are known as photoproteins. Photoproteins require Ca^{2+} or Mg^{2+} to trigger a conformational change that results in a photon-emitting intramolecular reaction. The first photoprotein was discovered by Shimomura in *A. victoria* and it was named after the jellyfish—aequorin. To characterize the chromophore in aequorin, 100–200 mg of pure aequorin was needed. This meant that Shimomura and his coworkers had to catch 50,000 of the jellyfish that measure between 7.5 and 10 cm in diameter and weigh about 50 g each (Shimomura 1995). In 1973, the structure of the chromophore in aequorin was reported, demonstrating that it is a coelenterazine that is bound to the aequorin (Figure 1.1b), and this is responsible for the blue light emission from the photoprotein. However, the jellyfish produces green light, an additional protein is involved in *Aequorea* bioluminescence—the GFP. In the absence of GFP, activation of aequorin with Ca^{2+} results in a blue light emission (λ_{max} 470 nm). In the jellyfish, radiationless energy transfer occurs from the aequorin to GFP resulting in its characteristic green light emission (λ_{max} 509 nm; see Figure 1.2). Biochemical analysis, as well as UV and

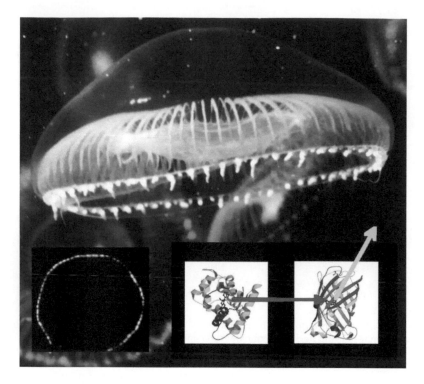

Figure 1.2 In the photocytes (bottom left) of *A. victoria* (top), a photoprotein called aequorin (center) generates an electronically excited product that undergoes radiationless energy transfer (blue arrow) to the GFP (right). GFP, in turn, gives off the green light (509 nm). This process is initiated by calcium ions binding the aequorin.

mass spectroscopy, and the synthesis of model compounds were used to determine the structure of the chromophore in GFP (Shimomura 1979), and this will be discussed in detail in Section 1.4.

1.3 FIREFLY LUCIFERASE: LIGHTING THE WAY FOR FPs

The development of the firefly luciferase (luc) from the North American firefly *Photinus pyralis* as an analytic tool for biological research preceded that of FPs, driven by the pioneering work of Bill McElroy and Marlene DeLuca. Firefly luciferases catalyze the production of an excited oxyluciferin species from ATP, oxygen, and luciferin. In 1976, luciferase was immobilized onto arylamine glass beads (Jablonski and DeLuca 1976), leading to the development of numerous rapid microbiological tests based on bioluminescent ATP assays. In 1985, firefly luciferase was successfully cloned and the resulting cDNA encoding was expressed in *Escherichia coli* (Dewet et al. 1985). This led to the use of the luciferase as a reporter of cellular functions. It is most commonly used as a "reporter gene," where the promoter of a gene of interest controls the transcription of luciferase. This allows *in vivo* bioluminescence associated with the luciferase expression to be used to measure transcriptional activity. The use of luciferase as a reporter of cellular activities presaged the use of FPs. The drawback of using firefly luciferase as a reporter, however, is that it always requires the addition of luciferin for bioluminescence.

A photograph of a luciferase-expressing tobacco plant luminescing upon watering with a luciferin-containing nutrient medium in the 1986 science paper entitled "Transient and stable expression of the firefly luciferase gene in plant cells and transgenic plants" was the first published image of a transgenic multicellular organism expressing bioluminescence or fluorescence (Ow et al. 1986). It was reprinted in many magazines, journals, and newspapers and is still to be seen in many textbooks. The luciferase-modified tobacco plant was the forerunner of Alba, the fluorescent rabbit, and GloFish, the FP zebra fish sold in pet shops all over the United States, except California.

1.4 *A. victoria* GREEN FLUORESCENT PROTEIN

The first person to conceive of the idea of using GFP as a genetic marker was Douglas Prasher. He was already using aequorin as a genetically incorporated calcium sensor and this seemed a logical progression. Prasher sequenced and cloned the cDNA for GFP. It has 238 amino acids in a configuration where residues S^{65}, Y^{66}, and G^{67} combine to form the chromophore (Prasher et al. 1992). However, when he attempted to express GFP in *E. coli*, no fluorescence was observed, and he concluded that "These results will enable us to construct an expression vector for the preparation of *non-fluorescent apoGFP*" (Prasher et al. 1992). This view that GFP would not be the genetically encoded fluorophore envisioned by Prasher was reinforced in a follow-up paper on the structure of the GFP chromophore in which Bill Ward wrote, "The posttranslational events required for chromophore formation are not yet understood. It is very unlikely that the chromophore forms spontaneously, but its formation probably requires enzymatic machinery" (Cody et al. 1993).

Fortunately, Martin Chalfie was not dissuaded, and in September 1992, he requested a copy of the GFP gene from Doug Prasher. One month after receiving the GFP cDNA, Ghia Euskirchen, a student in the Chalfie laboratory, succeeded in creating green fluorescent *E. coli*. There were two reasons she was successful. First, she used only the GFP coding region, and second, she had experience with and had access to fluorescence microscopes, which allowed her to distinguish between the inherent green autofluorescence of the bacteria and GFP fluorescence. This marked the start of the FP revolution. Unlike most other bioluminescent reporters (e.g., firefly luciferase), GFP fluoresces in the absence of any other proteins, substrates, or cofactors; it autocatalytically forms its own chromophore. Therefore, it is able to function as a genetic marker in any organism, as shown by Chalfie when he used known promoters to express GFP in the touch neurons of *Caenorhabditis elegans* (Chalfie et al. 1994).

The *A. victoria* green fluorescent protein (avGFP) chromophore is formed by an autocatalytic internal cyclization of the tripeptide $S^{65}Y^{66}G^{67}$ (numbers correspond to the amino acid position in the wild-type GFP) and subsequent oxidation of the intrinsically formed structure. The GFP fluorescence is not observed for up to 4 h after protein synthesis because while the protein folds quickly, the subsequent fluorophore formation and oxidation is relatively slow (Reid and Flynn 1997). GFP refolding from an acid-, base-, or guanidine HCl-denatured state (chromophore containing but nonfluorescent)

History and perspective

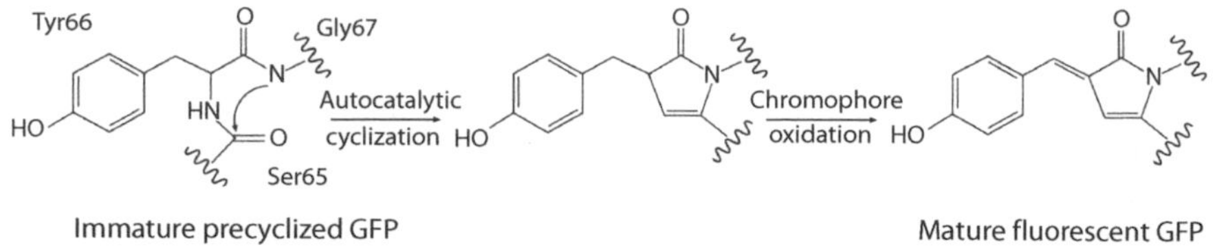

Figure 1.3 The immature precyclized protein (left) undergoes a rapid autocatalyzed cyclization, followed by a slower oxidation to form the mature fluorescent chromophore (right).

occurs with a half-life of between 24 s and 5 min, and the recovered fluorescence is indistinguishable from that of native GFP (Bokman and Ward 1981). In 1994, Roger Tsien proposed that the chromophore is formed by nucleophilic attack of the nitrogen of Gly67 on the carbonyl carbon of Ser65 (see Figure 1.3).

1.4.1 avAGP MUTANTS WITH DIFFERENT COLORS

Knowing that the GFP chromophore is formed from the $S^{65}Y^{66}G^{67}$ residues, Figure 1.3, made these residues obvious candidates for mutation. Therefore, even before the crystal structure of GFP was known, Roger Tsien created the Y66H mutant. The resultant blue fluorescent protein (BFP) was the first wavelength mutation of GFP (Heim et al. 1994). BFP has a much lower fluorescence quantum yield than GFP ($\Phi_{fl} = 0.20$ vs. 0.80). It has been suggested that this is due to the fact that His66 (BFP chromophore) forms fewer hydrogen bonds with the surrounding protein than Tyr66 (GFP chromophore) does and that the smaller imidazole ring (His66) in BFP may have more conformational freedom than the larger phenol (Tyr66), which leads to more intersystem crossing (Wachter et al. 1997). Azurite, a brighter BFP mutant, was created by decreasing the conformational freedom of the histidine containing chromophore (Mena et al. 2006). While the original BFP has a low fluorescence quantum yield, the new BFPs such as azurite, EBFP, SBFP2, and EBFP2 are as bright and stable as the other often used FPs.

Wild-type avGFP is rarely used in imaging because of its dim fluorescence, and peak excitation at near UV wavelengths made it difficult to distinguish GFP fluorescence from background fluorescence. Further, the avGFP has two excitation peaks due to the neutral and anionic forms of the chromophore. These drawbacks led to the search and development of new GFP mutants as *in vivo* imaging agents with higher intrinsic brightness (intrinsic brightness = fluorescence quantum yield × peak molar extinction coefficient) with more red-shifted excitation. For example, Tsien found that a chromophore mutation, the S65T GFP mutant, has only one excitation peak, a sixfold increased brightness and a fourfold increase in the rate of oxidation of chromophore. The avGFPS^{65}T variant is the basis of the most commonly used FP, enhanced green fluorescent protein (EGFP) (Cubitt et al. 1995; Heim et al. 1995).

Another early chromophore mutant of avGFP identified in the Tsien laboratory was cyan fluorescent protein (CFP) with blue-green emission. The CFP resulted from the substitution of Tyr^{66} with tryptophan (T^{66}W), producing a FP with two excitation peaks at 433 and 445 nm, and a broad emission profile with maxima at 475 and 503 nm (Cubitt et al. 1995; Heim et al. 1994). The continued engineering of CFP resulted in variants with improved characteristics including ECFP, cerulean, mTurquoise and TagCFP. These improved CFPs are often used in FRET and dual-color experiments. Other BFP and CFP, such as TagBFP and mTFP1, have been developed that have tyrosine-containing GFP-like chromophore.

1.4.2 avGFP CRYSTAL STRUCTURES

In the late eighties, diffraction quality crystals of GFP were grown by Ward (Perozzo et al. 1988), but it wasn't until 1996 that the crystal structure of GFP was solved simultaneously for wild-type GFP (Yang et al. 1996) and EGFP (Ormoe et al. 1996). GFP has an 11-stranded β-barrel with an α-helix running through the β-barrel. The chromophore is located in the center of the barrel and is protected from bulk solvent. The barrel has a diameter of about 24 Å and a height of 42 Å.

In the first of the rationally designed mutants based on the crystal structure of GFP, Tsien and coworkers decided to mutate T203 into a tyrosine so that it could π stack with the phenolic group of the tyrosine in the chromophore (Ormoe et al. 1996). The resultant yellow fluorescent protein (YFP) is red shifted by 16 nm relative to GFPS^{65}T and does indeed have a π stacking interaction between the chromophore and Tyr^{203} (Wachter et al. 1998). It is significantly brighter than GFP. The rationally designed chromophore of YFP is remarkably similar to that of a naturally occurring FP, phiYFP (Shagin et al. 2004). The continued engineering of YFP yielded the much improved citrine, Venus, topaz, TagYFP, and YPet.

In general, mutations of the chromophore-forming residues lead to large wavelength shifts, while changing the protein matrix around the chromophore corresponds to spectral fine-tuning. The side chains of residues 148, 165, 167, and 203 (GFP numbering) are in close contact with the chromophore. They affect the polarization, protonation state, and dihedral freedom of the chromophore and therefore tweak the spectral properties of the FPs.

1.5 SEARCH FOR RED FLUORESCENT PROTEINS

The green, blue, cyan, and yellow colors obtained by mutating avGFP were the first colors in the rainbow palette of FPs, but a very important color, red, was still missing. Since infrared wavelengths are minimally absorbed by hemoglobin, water- and lipid-imaging agents that emit in the 650–900 nm range would be ideally suited for *in vivo* imaging of deep tissue in live animals, and a search for red bioluminescent (Branchini et al. 2007; Caysa et al. 2009; Fischer and Lagarias 2004) or FPs (Kredel et al. 2009; Shcherbo et al. 2009; Suto et al. 2009) has been continuing unabated for the last 15 years.

Many research groups tried a variety of different mutagenesis schemes to engineer the avGFP into a red fluorescing form. This strategy was not very successful and we would have to wait until 2008 before a red mutant of *A. victoria* GFP was created (Mishin et al. 2008). Other groups took to oceans and its environs to look for red bioluminescent organisms. They were no more successful. However, in 1999, Lukyanov and Labas made a conceptual shift in the search for red fluorescent proteins (RFPs) (Matz et al. 1999). Realizing that the association between aequorin and GFP, Figure 1.2, might be an exception and that FPs did not necessarily have to be associated with other chemiluminescent proteins, they decided to look for organisms that were red fluorescent but were not bioluminescent. In aquarium shops in Moscow, they found corals containing the first RFP, DsRed.

History and perspective

Unfortunately, DsRed was not ideal; it is more orange than red, tetrameric, and slow to mature and goes through an intermediate green state before the red fluorescent form is obtained. Using mass spectroscopy, theoretical calculations and other methods Tsien showed that the DsRed chromophore has undergone a second oxidation to form an acylimine that extends the conjugation of the chromophore—hence, the red shift (see Figure 1.9). Their efforts to improve the characteristics of DsRed for *in vivo* imaging focused on first breaking the tetramer and then optimizing the monomeric RFP. It would take 33 mutations to DsRed to create the first monomeric RFP (mRFP1) (Campbell et al. 2002). Unfortunately, mRFP1 photobleaches quickly and has a significantly reduced intrinsic brightness. Using mRFP1 as a starting point, the Tsien laboratory applied a variety of directed evolution strategies, and in 2004, they introduced a palette of new FPs, the mFruits (Shaner et al. 2004; Wang et al. 2004). Today, there are more than 15 different red-emitting FPs ($\lambda_{emiss} > 600$ nm), most are variants of the RFPs obtained from corals and other organisms (Day and Davidson 2009).

1.6 BIOLOGICAL, CHEMICAL, AND ECOLOGICAL DIVERSITY OF FPs AND OTHER GFP-LIKE PROTEINS

FPs are members of a larger superfamily that encompasses both the FPs reviewed in this book and the globular-2 fragment (G2F) domains of nidogens (Chudakov et al. 2010). The G2F domains have the characteristic beta-barrel shape with an alpha-helix running through them, described in Section 1.4.2, but they do not undergo the chromophore-forming autocatalytic, cyclization described in Section 1.9.

Most GFP-like proteins are found in nonbioluminescent organisms (Alieva et al. 2008). Although most are fluorescent, some are nonfluorescent chromoproteins (e.g., the purple chromoprotein from *Anemonia sulcata*). Parsimony analysis (Shagin et al. 2004) and ancestral reconstruction experiments (Alieva et al. 2008; Chang et al. 2005) suggest that all but one of the nongreen colors arose from an ancestor with a canonical green chromophore. The exception is the yellow protein from *Zoanthus* sp., which is likely to have evolved from a DsRed-like red ancestor (Alieva et al. 2008). Figure 1.9 shows the variety of naturally occurring chromophores found in FPs and their maturation pathways. A single GFP-like gene is found in most organisms, except *Zoanthus* and *Discosoma* both with two distinctly different GFP-like proteins, *Montastraea cavernosa* with four different GFP-like genes and finally *Branchiostoma floridae* with a family of 16 GFP-like proteins (Bomati et al. 2009).

Over 150 distinct GFP-like proteins are currently known. GFP-like proteins have been found in marine organisms ranging from chordates (e.g., amphioxus) to cnidarians (e.g., corals and sea pansies). However, FP genes have only been found in two genomes of organisms that have been fully sequenced, perhaps due to the fact that most of the organisms sequenced are freshwater or terrestrial organisms. Both organisms, the lancelet and the starlet sea anemone, have both G2F and FP genes (Chudakov et al. 2010).

FPs are commonly found in corals, they are often responsible for the colors of the corals, and they can constitute up to 14% of the total protein content of corals. Not all FPs in corals, however, are fluorescent-non-FPs are often referred to as pocilloporins or GFP-like proteins (Roth et al. 2010). All naturally occurring pocilloporins have a chromophore in the *trans* and/or nonplanar conformation (see Section 1.9). In this chapter, GFP-like proteins refer to all fluorescent and non-FPs with a GFP fold (11 beta sheets in a beta-barrel with a central alpha-helix and tripeptide cyclization occurring at a central kink in the alpha-helix).

1.7 FUNCTION OF FPs

A number of functions have been proposed for the FPs. Since the FPs provide color, they might be involved in camouflage (Matz et al. 2006). Further, most FPs in corals are located near the photosynthetic apparatus, so it has been proposed that they could serve a photoprotective role (Dove 2004; Dove et al. 2001; Miyawaki 2002; Roth et al. 2010). Additionally, the exposure to high amounts of superoxide leads to an increase in SOD-like activity, and a change in the FP's protein structure without a significant change in fluorescence could potentially provide antioxidant protection (Bou-Abdallah et al. 2006; Mazel et al. 2003; Palmer et al. 2009). In this regard, when the tyrosine in the center of the chromophore was replaced with an alanine residue, the resulting non-FPs were better antioxidants than their parent FPs (Palmer et al. 2009). Finally, the FPs may function as a primitive proton pump (Agmon 2005) or potentially as light-induced electron donors (Bogdanov et al. 2009a,b).

If all naturally occurring FPs have a common or ancestral function that is not related to their color and/or fluorescent properties, one would find that the residues involved in these functions would be highly conserved. In contrast, these residues would be less conserved in artificially created mutant FPs—after all these were created for their enhanced fluorescence properties, not to improve the electron donation, photoprotection, or antioxidant properties (Ong et al. 2011). The most remarkable difference between the naturally occurring GFP-like proteins and those created for their improved imaging abilities is the complete conservation of the chromophoric tyrosine in all naturally occurring GFP-like proteins.

The identity of the central chromophore residue at position 66 does not affect cyclization. Although an aromatic residue is required for the chromophore to be fluorescent, chromophore formation still occurs for the Ser^{66} (Barondeau et al. 2005), Leu^{66} (Rosenow et al. 2004), and Gly^{66} (Barondeau et al. 2003) mutants, but they are nonfluorescent. There is no obvious reason why Tyr^{66} is conserved at this position in all wild-type structures. A His^{66} substitution results in a blue FP, while Trp^{66} results in a CFP. Yet all species sequenced to date have a tyrosine as the central amino acid in the chromophore-forming tripeptide. The Tyr codons (TAT and TAC) only differ from the His (CAT and CAC) and Phe (TTC) codons by one nucleotide base; with all the known diversity among the FPs, one would expect some species to have a His or Phe in their chromophores. Clearly, Tyr^{66} is of vital importance in the function of GFP-like proteins, and its function is not related to its color, since some species with fluorescent and nonfluorescent colors can be obtained with other residues in the central position of the chromophore.

Lukyanov et al. (Bogdanov et al. 2009a,b) have suggested that light-induced electron transfer should be considered a primary function of GFP and that "for both oxidative and anaerobic redding, the chromophore should be based on tyrosine to be converted into the red state" (Bogdanov et al. 2009a,b). Perhaps Tyr^{66} is completely conserved in all wild-type GFP-like proteins because only it can donate an electron to an electron acceptor forming a short-lived intermediate.

1.8 PHOTOPHYSICS OF FPs

The fluorescence emission of GFP occurs with high efficiency (quantum yield $\Phi_{fl} = 0.8$) and with a mean fluorescence lifetime (τ) of $\approx$ 3 ns (Striker et al. 1999). Upon denaturation, the fluorescence yield decreases by at least three orders of magnitude (Niwa et al. 1996). Model compounds of the chromophore do not fluoresce in solution (quantum yield $\Phi_{fl} < 10^{-3}$), unless the rotation of the aryl–alkene bond is restrained (Wu and Burgess 2008). Fluorescence can be obtained by lowering the temperature to 77 K, which freezes the solution and constrains the chromophore. It has been suggested that twisting between the phenolate and imidazolidinone groups of the chromophore is the mechanism for a ultrafast fluorescence quenching internal conversion process (Litvinenko et al. 2001, 2002, 2003; Litvinenko and Meech 2004; Webber et al. 2001).

According to quantum mechanical calculations, the ground and excited states for the τ one bond flip (OBF) and hula twist (HT) in the neutral form and the ϕ OBF in the zwitterionic form come very close to each other. It has been proposed that in the absence of the protein matrix, which surrounds the chromophore and prevents twisting, this process can lead to fluorescence quenching internal crossing (Weber et al. 1999), see Figure 1.4.

The photophysical behavior of avGFP is summarized in Figure 1.5 (Bell et al. 2000; Brejc et al. 1997). The electronic absorption spectrum of wild-type GFP has two peaks, which have been assigned to a dominant neutral form (the A state λ_{max} 395 nm)

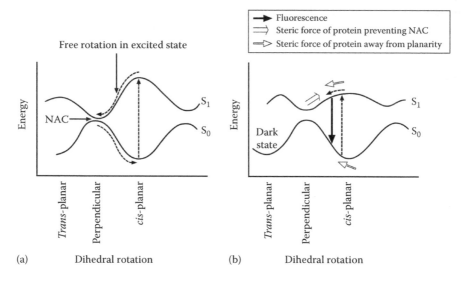

Figure 1.4 (a) Model compounds of the GFP chromophore in the ground state (S_0) can be excited to the first singlet state (S_1) in which an HT or OBF can freely occur (free rotation in excited state). Upon reaching the perpendicularly twisted conformation, fluorescence quenching nonadiabatic crossing (NAC) occurs. (b) In the ground state of GFP (S_0), the residues surrounding the chromophore exert a twisting force on the chromophore ($\Leftrightarrow$). Upon excitation, the conjugation across the ethylenic bridge of the chromophore is reduced and it will twist; however, the protein matrix prevents the chromophore from reaching the perpendicularly twisted conformation ($\Rightarrow$) and fluorescence quenching internal crossing is prevented. (Reprinted with permission from Zimmer, M., Photophysics and dihedral freedom of the chromophore in yellow, blue and green fluorescent protein, *J. Phys. Chem. B*, 113, 304, 2009. Copyright 2009 American Chemical Society.)

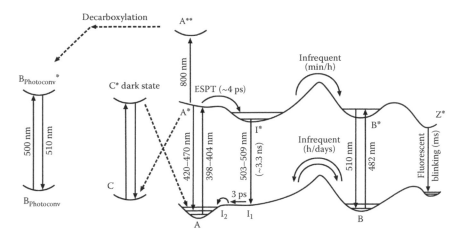

Figure 1.5 The neutral form (A) of the chromophore, with the phenolic oxygen protonated, can convert to the anionic species (B) by going through an intermediate state (I). The change from forms A to I is solely a protonation change, while the change from I to B is a conformational change with most changes occurring at Thr203. Upon excitation of the A state, an excited-state proton transfer (ESPT) occurs in which the proton is transferred from the chromophore to Glu222 in a timescale of the order of picoseconds. Following radiative relaxation from the excited state intermediate (I*) the systems returns to the ground state A through the ground state intermediates I_1 and I_2 (Kennis 2004). Excitation of the anionic B state results in direct emission from the excited state (B*) at 482 nm. A nonfluorescent dark state, state C, has been observed that is distinct from states A and B and absorbs at higher energies (Nifosi et al. 2003). The C state, perhaps the neutral *trans* form of the chromophore, may be populated by nonradiative decay from A* and it may be depopulated by excitation to the excited C* state with *trans–cis* isomerization to repopulate state A. In some mutants such as T203V GFP it was found that further excitation of A* with 800 nm leads to a higher excited A**, which is responsible for Glu222 decarboxylation under femtosecond excitation conditions (dotted arrow) (Langhojer et al. 2009). Fluorescent blinking has been ascribed to NAC and conversions between the neutral, anionic, and dark (Z) zwitterionic state (Chirico et al. 2005). (Reprinted with permission from Zimmer, M., Photophysics and dihedral freedom of the chromophore in yellow, blue and green fluorescent protein, *J. Phys. Chem. B.*, 113, 303. Copyright 2009 American Chemical Society.)

and the less populated anionic form (the B state λ_{max} 482 nm). The excited neutral form undergoes an ESPT reaction through a proton wire, resulting in an excited intermediate form (I*), which emits at 508 nm. The excited anionic (B* state) form emits at 510 nm.

1.9 CHROMOPHORE FORMATION IN FPs

The common denominator between all the FPs is their size (220–240 residues), their shape (beta-barrel), and the centrally located chromophore. It is the fact that this chromophore is an intrinsic part of the protein backbone that is formed autocatalytically that makes the FPs so useful as genetically encoded *in vivo* imaging tools. The most obvious conformational prerequisite for the autocatalytic cyclization step, shown in Figure 1.3, is that the nitrogen of Gly[67] and the carbonyl carbon of Ser[65] have to be in close proximity to each other. This conformation, which preorganizes Ser[65]–Gly[67] for cyclization, is termed the tight-turn conformation. In order to form such a tight turn,

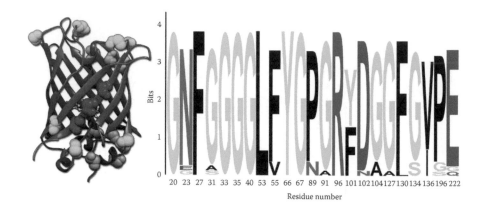

Figure 1.6 Left: GFP is composed of an 11 sheet β-barrel. A chromophore (tube in center of barrel) containing helix connects the third and fourth sheets. All highly conserved residues located in the lids are depicted as CPK atoms (left). Gly31, Gly33, and Gly35 are highly conserved in all species; they are colored red and are located on the second β-sheet. Conserved residues involved in chromophore formation Tyr66, Gly67, Arg96, and Glu222 are not shown. Right: A WebLogo representation (Crooks et al. 2004) of amino acid diversity among the most conserved residues of the wild-type GFP-like structures in all 85 species in the GFP family (Pfam 01353). Residues are numbered relative to avGFP. See figure left for location of some of these residues. A high bit score (y-axis) on the logo plot reflects invariant residues. (From Zimmer, M., Function and structure of GFP-like proteins in the protein data bank, *Mol. Biosyst.*, 7, 987, 2011. Reproduced by permission of The Royal Society of Chemistry.)

a conformationally flexible glycine residue is required as the third amino acid in the chromophore-forming tripeptide (Branchini et al. 1998). Therefore, it is not surprising that all GFP and GFP-like proteins with mature chromophores in the protein database (pdb) also have a Gly[67] (GFP numbering) (see Figure 1.6 [right]).

Before the crystal structure of GFP or any of its mutants was solved, it was also suggested that a tight-turn conformation in immature GFP was not the only requirement for an autocatalytic cyclization. It was proposed that an arginine residue would be found in close proximity to the chromophore-forming region and that it had a catalytic function in the chromophore formation (Zimmer et al. 1996). The subsequently released crystal structures of GFP (Ormoe et al. 1996; Yang et al. 1996) revealed that an arginine, Arg[96], was indeed in the position predicted. Although Arg[96] is not absolutely necessary for chromophore formation, it is highly conserved in all FPs (Matz et al. 1999). Figure 1.7 shows that all the arginines in the crystal structures of FPs with mature chromophores are in the same area relative to the cyclized five-membered ring. Since all the mature GFPs, except the four slow-maturing GFPs, have an arginine in the same position, it is safe to assume that Arg[96] plays an important role in catalyzing the cyclization of the GFP chromophore.

The autocatalytic cyclization and oxidation responsible for the chromophore in GFP-like proteins occurs by either a cyclization–oxidation–dehydration mechanism (Rosenow et al. 2004) or a cyclization–dehydration–oxidation mechanism (Barondeau et al. 2003) (see Figure 1.8). Crystal structures of immature GFP mutants have shown that the GFP protein matrix creates a dramatic bend at the chromophore-forming region of the central helix (Barondeau et al. 2003). This kink removes main-chain

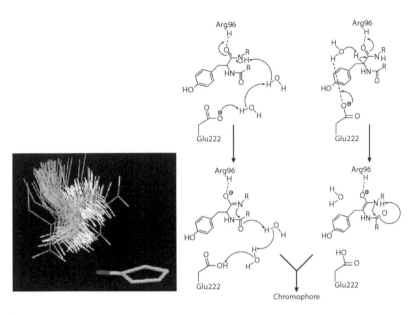

Figure 1.7 An Isostar overlay plot of all the crystal structures of GFP and GFP-like proteins with fully cyclized chromophores found in the pdb. The imidazolone rings of all the structures were overlapped in order to show the orientation of the arginine residue (Arg[96] in GFP numbering) relative to the position of the imidazolone ring. Two proposed roles (Barondeau et al. 2003; Wachter 2007) for Arg[96] and Glu222 are shown on the right. (Reprinted from *Chem. Phys.*, 348, Zimmer, M., The role of the tight-turn, broken hydrogen bonding, Glu22 and Arg96 in the post-transitional green fluorescent protein chromophore formation, 156, Copyright 2008, with permission from Elsevier.)

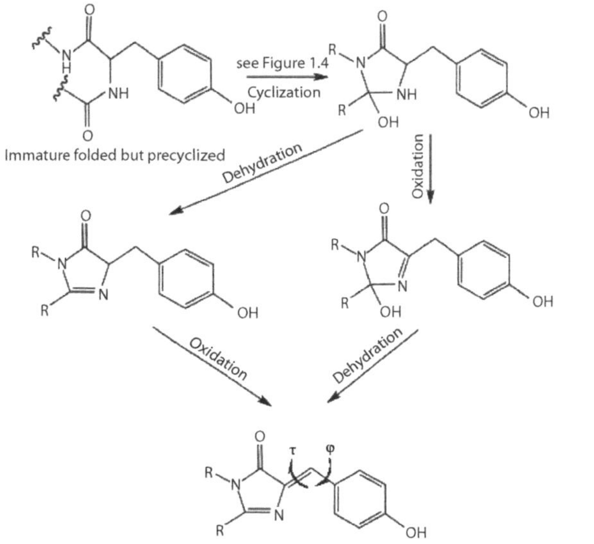

Figure 1.8 The cyclization–oxidation–dehydration mechanism proposed by Wachter (Rosenow et al. 2004) (right) and the cyclization–dehydration–oxidation mechanism proposed by Getzoff (Barondeau et al. 2003) (left).

hydrogen bonds that are commonly associated with an alpha-helix. It is presumed that this aids in chromophore formation because the hydrogen bonds would otherwise be broken during maturation.

Autocatalytic protein cyclizations are extremely rare, in fact those found in the FPs and the histidine ammonia lyase (HAL) are the only ones known to date. It is not known whether the chromophore-forming cyclization evolved independently or has a common origin. However, since Arg[96] is structurally and positionally conserved in all FPs (see Figures 1.6 and 1.7), it is safe to assume that the same initial autocatalytic Arg[96]-catalyzed chromophore-forming reaction is present in all FPs and that this forms an initial cyclized intermediate in all species. The diversity of colors and chromophores is subsequently created through a series of different pathways forming a number of different chromophores from the common cyclized intermediate (see Figure 1.9). For example, the RFPs have an extended chromophore that is formed by an additional oxidation that proceeds through either a green

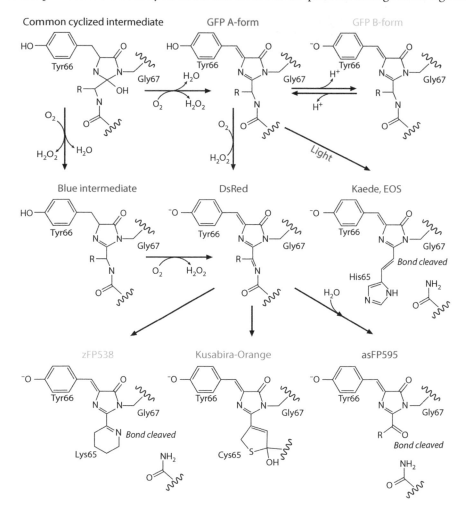

Figure 1.9 After the initial common Arg[96] catalyzed cyclization a variety of pathways are utilized by different organisms to generate six main types of chromophores.

intermediate that resembles the GFP chromophore (Gross et al. 2000) or alternatively a blue intermediate that has an acylimine group where the methylene group linking the two aromatic rings has yet to be oxidized (Pletnev et al. 2010; Strack et al. 2010).

1.10 CONFORMATIONAL SPACE AVAILABLE TO THE CHROMOPHORE IN FPs

A number of groups have noted that most light-emitting states of FPs have planar *cis* chromophore conformations and that the nonplanar *trans* forms tend to be nonfluorescent (Day and Davidson 2009; Seward and Bagshaw 2009). Figure 1.10 shows the τ and φ dihedral angles of all the 266 GFP-like structures in the protein databank in July 2010—very few of the chromophores are planar (i.e., τ and φ = 0.0° or 180°). The GFP-like protein matrix is not complementary with a planar chromophore, and the protein matrix exerts a twisting force on the chromophores in GFP-like proteins (Maddalo and Zimmer 2006; Megley et al. 2009). In order to accommodate the twisting force, the chromophore can undergo a tilting deformation around the τ dihedral angle (y-axis in Figure 1.10) or a twisting deformation around the φ dihedral (x-axis in Figure 1.10; Piatkevich et al. 2010).

Due to leverage effects, rotations around the τ dihedral angle require more free volume than those around the φ dihedral, and a concerted rotation of the τ and

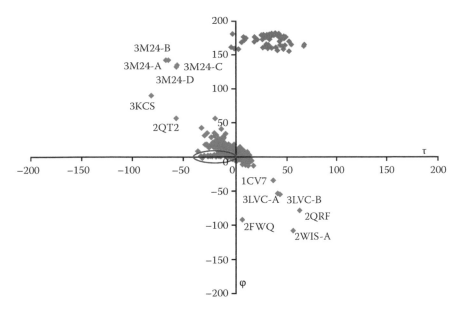

Figure 1.10 A plot of the τ versus φ dihedral angles (see Figure 1.1 for definition of dihedrals) of all GFP-like structures with an aromatic residue in position 66. *Trans* chromophores (orange) undergo mainly twisting deformations (φ), while GFP-like proteins with *cis* chromophores (blue) undergo a combination of τ and φ deformations, except the 46 structures encircled in red which undergo only a twisting deformation. All structures labeled with their pdb codes have an immature chromophore in the cyclized dehydrated intermediate stage. (From Zimmer, M., Function and structure of GFP-like proteins in the protein data bank, *Mol. Biosyst.*, 7, 989, 2011. Reproduced by permission of The Royal Society of Chemistry.)

φ dihedrals in opposite directions (an HT) requires the least volume. According to Figure 1.10, most structures are nonplanar with a concerted combination of twisting and tilting. The slope of the τ versus φ plot for all GFP-like structures with a *cis* conformation is −0.85 (a perfectly negatively correlated HT would have a slope of −1.00). No structures have undergone a pure tilting deformation, while 46 structures have a twisted chromophore (τ between −5.0° and 5.0° and φ greater than 5.0°). Recently, the Mathies laboratory (Fang et al. 2009) used femtosecond-stimulated Raman spectroscopy to examine the structural evolution of the excited state in GFP. They found that the dominant motion is a phenolic wag of the chromophore that is directed to a transition state perfectly aligned for ESPT by the aforementioned twisting force (Section 1.7).

This mechanism is supported by studies of the photoswitching FPs (see Chapters 6 and 9). These naturally occurring FPs can be switched back and forth between a green state and a dark state by 405 nm irradiation (e.g., Dronpa, mTFP0.7, KFP1). A *cis–trans* isomerization of the chromophore has been proposed as the structural basis for the photoswitching observed in Dronpa (Andresen et al. 2007). This mechanism was verified by mutating either Val[157] or Met[159] to less bulky residues, accelerating the photoswitching—presumably by decreasing steric hindrance to *cis–trans* isomerization (Stiel et al. 2007). The M159T and V157G mutations also decrease the quantum yield of Dronpa from 0.85 to quantum yields of 0.23 and 0.77, respectively (Stiel et al. 2007). Recently, it was suggested that adoption of a *trans* configuration cannot solely be responsible for the nonfluorescent form (Mizuno et al. 2008). Based on NMR analyses, Miyawaki et al. propose that "the fluorescence of the protein is regulated by the degree of flexibility of the chromophore but is not necessarily accompanied by *cis–trans* isomerization" (Mizuno et al. 2008).

1.11 THE FUTURE

In order to look into the future, I would like to take us back into the past. If we go back 5–10 years and review the major breakthroughs in FP technology, were some of the developments predictable or was their appearance unexpected and unpredictable? I think it would have been very easy to predict that new FPs would be found in marine environments, particularly since we still know so little about life in the oceans. It would be just as easy to predict that random and site-specific mutagenesis would produce brighter, red-shifted, faster-maturing, and/or monomeric versions of known FPs. But would we have been able to predict that FPs could be used as delivery vehicles, to monitor the cell cycle, as molecular assassins or as stopwatches?

1.11.1 HIGHLIGHTERS

In 2002, a GFP-like protein that undergoes a green to red photoconversion was found completely serendipitously. In the paper describing the isolation of the protein and its use, Atsushi Miyawaki writes: "We happened to leave one of the protein aliquots on the laboratory bench overnight. The next day, we found that the protein sample on the bench had turned red, whereas the others that were kept in a paper box remained green. Although the sky had been partly cloudy, the red sample had been exposed to sunlight through the south-facing windows.... To verify this serendipitous observation we put a green sample in a cuvette over a UV illuminator emitting 365-nm light and found that the sample turned red within several minutes…. At this point the protein was renamed Kaede, which means maple leaf in Japanese" (Ando et al. 2002).

EosFP, named after the goddess of dawn, is another photoconvertible FP discovered around the same time. Initially, it is green fluorescent, but irradiation with UV light (390 ± 30 nm) induces cleavage between the amide nitrogen and the α-carbon atom in the histidine adjacent to the chromophore resulting in a red fluorescent form (Wiedenmann et al. 2004). EosFP and Kaede are excellent examples of a group of FPs that have been found and created that change their emission upon irradiation. They are known as optical highlighters and have been subdivided into three groups: photoactivatable, photoconvertible, and photoswitchable FPs. Their discovery has led to new superresolution microscopy techniques such as fluorescence-photoactivated localization microscopy (FPALM), photoactivated localization microscopy (PALM), interferometric photoactivated localization microscopy (iPALM), and stochastic optical reconstruction microscopy (STORM) (see Chapters 9 and 12 in this book). These methods have in turn led to the development of a new generation of highlighter proteins, for example, IrisFP, mEos2, and PAmCherry.

1.11.2 KILLERRED AND THE TIMERS

In most FPs, the β-barrel protects the chromophore and limits the diffusion of oxygen and reactive oxygen species (ROS) in and out of the protein. KillerRed is known to kill cells that surround it by ROS whenever it is exposed to ~550 nm of light. It has a water channel not found in wild-type GFP, which could be the entrance for oxygen into the KillerRed beta-barrel. It was created by directed evolution from the non-FP anm2CP (Pletnev et al. 2009). In 2000, Lukyanov reported the creation of a DsRed mutant that changed from green to red over a period of 18 h. The color change was unaffected by pH, ionic strength, or protein concentration. The main drawback of this fluorescent timer (FT) was the fact that it is a tetramer. Vladislav Verkhusha took mCherry and very cleverly mutated it to obtain three new monomeric FTs. They change from blue to red with slow, medium, and fast rates. The blue intermediate has a $C\alpha2-C\beta2$ single bond that is oxidized to form the final red form of the timer FP (Pletnev et al. 2010). However, even the fast FT cannot be used to monitor cellular processes that occur on timescales of minutes rather than hours.

1.11.3 DNA DELIVERY

Harvard's David Liu has created a GFP mutant with an approximate charge of 36+. The supercharged protein folds correctly, doesn't aggregate, and binds to negative macromolecules such as siRNA and DNA. This is incredibly useful as the megapositive GFP penetrates mammalian cells with ease. By complexing the GFP36+ with nucleic acids, without forming any covalent bonds, the researchers were able to deliver their cargo into four different mammalian cell lines (McNaughton et al. 2009).

1.11.4 THE FUTURE REVISITED

Although one cannot page through an issue of *Science* or *Nature* without seeing an image showing the fluorescence of an FP and although more than 1.2 million FP experiments were performed in the United Kingdom in 2009, we still don't know its function in nature, many FP-containing species are yet to be discovered, we don't understand the chromophore-forming autocatalytic cyclization, and we can't yet

History and perspective

rationally fine-tune its emission spectrum. Would we have been able to predict that nature had a selection of highlighter proteins hidden up her sleeves or that FPs could be used as molecular assassins? I don't think so, but the more basic science we do and the more we know about bioluminescent organisms, the more likely we are to find new and unexpected FP derived applications.

In looking to the future of FPs, I think one needs to go no further than the last line of Roger Tsien's banquet speech at the ceremony for the 2008 Nobel Prizes in which he said, "While environmentally friendly or so-called 'green' chemistry has become all the rage in the chemical community, no human chemist can yet match what a single jellyfish gene directs: 238 ordered condensations + 1 cyclization + 1 oxidation, all done in a few minutes in aerated water with no protecting groups, only one slightly toxic byproduct, and essentially 100% yield of an extremely useful product that literally glows green. Corals produce yellow and red FPs with the same chemistry plus one extra oxidation. Yet coral reefs are also under world-wide jeopardy, due to acidification and warming of the oceans. So my final thanks are to both the jellyfish and corals: long may they have intact habitats in which to shine!"

REFERENCES

Agmon, N. 2005. Proton pathways in green fluorescence protein. *Biophysical Journal* 88 (4):2452–2461.

Alieva, N.O., K.A. Konzen, S.F. Field, E.A. Meleshkevitch, V. Beltran-Ramirez, D.J. Miller, A. Salih, J. Wiedenmann, and M.V. Matz. 2008. Diversity and evolution of coral fluorescent proteins. *PLoS One* 3 (7):e2680.

Ando, R., H. Hama, M. Yamamoto-Hino, H. Mizuno, and A. Miyawaki. 2002. An optical marker based on the UV-induced green-to-red photoconversion of a fluorescent protein. *Proceedings of the National Academy of Sciences of the United States of America* 99 (20):12651–12656.

Andresen, M., A.C. Stiel, S. Trowitzsch, G. Weber, C. Eggeling, M.C. Wahl, S.W. Hell, and S. Jakobs. 2007. Structural basis for reversible photoswitching in Dronpa. *Proceedings of the National Academy of Sciences of the United States of America* 104 (32):13005–13009.

Barondeau, D.P., C.J. Kassmann, J.A. Tainer, and E.D. Getzoff. 2005. Understanding GFP chromophore biosynthesis: Controlling backbone cyclization and modifying post-translational chemistry. *Biochemistry* 44 (6):1960–1970.

Barondeau, D.P., C.D. Putnam, C.J. Kassmann, J.A. Tainer, and E.D. Getzoff. 2003. Mechanism and energetics of green fluorescent protein chromophore synthesis revealed by trapped intermediate structures. *Proceedings of the National Academy of Sciences of the United States of America* 100 (21):12111–12116.

Bell, A.F., X. He, R.M. Wachter, and P.J. Tonge. 2000. Probing the ground state structure of the green fluorescent protein chromophore using Raman spectroscopy. *Biochemistry* 39:4423–4431.

Bogdanov, A.M., A.S. Mishin, I.V. Yampolsky, V.V. Belousov, D.M. Chudakov, F.V. Subach, V.V. Verkhusha, S. Lukyanov, and K.A. Lukyanov. 2009a. Green fluorescent proteins are light-induced electron donors. *Nature Chemical Biology* 5 (7):459–461.

Bogdanov, A.M., A.S. Mishin, I.V. Yampolsky, V.V. Belousov, D.M. Chudakov, F.V. Subach, V.V. Verkhusha, S. Lukyanov, and K.A. Lukyanov. 2009b. Green fluorescent proteins are light-induced electron donors. *Nature Chemical Biology* 5 (7):459–461.

Bokman, S.H. and W.W. Ward. 1981. Renaturation of *Aequorea* green fluorescent protein. *Biochemical and Biophysical Research Communications* 101:1372–1380.

Bomati, E.K., G. Manning, and D.D. Deheyn. 2009. Amphioxus encodes the largest known family of green fluorescent proteins, which have diversified into distinct functional classes—Art. no. 77. *BMC Evolutionary Biology* 9:77–77.

Bou-Abdallah, F., N.D. Chasteen, and M.P. Lesser. 2006. Quenching of superoxide radicals by green fluorescent protein. *Biochimica Et Biophysica Acta-General Subjects* 1760 (11):1690–1695.

Branchini, B.R., D.M. Ablamsky, M.H. Murtiashaw, L. Uzasci, H. Fraga, and T.L. Southworth. 2007. Thermostable red and green light-producing firefly luciferase mutants for bioluminescent reporter applications. *Analytical Biochemistry* 361 (2):253–262.

Branchini, B.R., A.R. Nemser, and M. Zimmer. 1998. A computational analysis of the unique protein-induced tight turn that results in posttranslational chromophore formation in green fluorescent protein. *Journal of the American Chemical Society* 120:1–6.

Brejc, K., T.K. Sixma, P.A. Kitts, S.R. Kain, R.Y. Tsien, M. Ormo, and S.J. Remington. 1997. Structural basis for dual excitation and photoisomerization of the *Aequorea victoria* green fluorescent protein. *Proceedings of the National Academy of Sciences of the United States of America* 94 (6):2306–2311.

Campbell, R.E., O. Tour, A.E. Palmer, P.A. Steinbach, G.S. Baird, D.A. Zacharias, and R.Y. Tsien. 2002. A monomeric red fluorescent protein. *Proceedings of the National Academy of Sciences of the United States of America* 99 (12):7877–7882.

Caysa, H., R. Jacob, N. Muther, B. Branchini, M. Messerle, and A. Soling. 2009. A redshifted codon-optimized firefly luciferase is a sensitive reporter for bioluminescence imaging. *Photochemical and Photobiological Sciences* 8 (1):52–56.

Chalfie, M., Y. Tu, G. Euskirchen, W.W. Ward, and D.C. Prasher. 1994. Green fluorescent protein as a marker for gene expression. *Science* 263:802–805.

Chang, B.S.W., J.A. Ugalde, and M.V. Matz. 2005. Applications of ancestral protein reconstruction in understanding protein function: GFP-like proteins. In *Molecular Evolution: Producing the Biochemical Data, Part B*, eds. E.A. Zimmer and E. Roalson, 1 edn., Waltham, MA: Academic Press.

Chirico, G., A. Diaspro, F. Cannone, M. Collini, S. Bologna, V. Pellegrini, and F. Beltram. 2005. Selective fluorescence recovery after bleaching of single E(2)GFP proteins induced by two-photon excitation. *Chemphyschem* 6:328–335.

Chudakov, D.M., M.V. Matz, S. Lukyanov, and K.A. Lukyanov. 2010. Fluorescent proteins and their applications in imaging living cells and tissues. *Physiological Reviews* 90 (3):1103–1163.

Cody, C.W., D.C. Prasher, W.M. Westler, F.G. Pendergast, and W.W. Ward. 1993. Chemical structure of the hexapeptide chromophore of the *Aequorea* green fluorescent protein. *Biochemistry* 32:1212–1218.

Crooks, G.E., G. Hon, J.M. Chandonia, and S.E. Brenner. 2004. WebLogo: A sequence logo generator. *Genome Research*. 14:1188–1190.

Cubitt, A.B., R. Heim, S.R. Adams, A.E. Boyd, L.A. Gross, and R.Y. Tsien. 1995. Understanding, improving and using green fluorescent proteins. *Trends in Biochemical Sciences* 20:448–455.

Day, R.N. and M.W. Davidson. 2009. The fluorescent protein palette: Tools for cellular imaging. *Chemical Society Reviews* 38 (10):2887–2921.

Dewet, J.R., K.V. Wood, D.R. Helinski, and M. Deluca. 1985. Cloning of firefly luciferase cDNA and the expression of active luciferase in *Escherichia coli. Proceedings of the National Academy of Sciences of the United States of America* 82 (23):7870–7873.

Dove, S. 2004. Scleractinian corals with photoprotective host pigments are hypersensitive to thermal bleaching. *Marine Ecology-Progress Series* 272:99–116.

Dove, S.G., O. Hoegh-Guldberg, and S. Ranganathan. 2001. Major colour patterns of reef-building corals are due to a family of GFP-like proteins. *Coral Reefs* 19 (3):197–204.

Fang, C., R.R. Frontiera, R. Tran, and R.A. Mathies. 2009. Mapping GFP structure evolution during proton transfer with femtosecond Raman spectroscopy. *Nature* 462 (7270):200–274.

Fischer, A.J. and J.C. Lagarias. 2004. Harnessing phytochrome's glowing potential. *Proceedings of the National Academy of Sciences of the United States of America* 101 (50):17334–17339.

Gross, L.A., G.S. Baird, R.C. Hoffman, K.K. Baldridge, and R.Y. Tsien. 2000. The structure of the chromophore within DsRed, a red fluorescent protein from coral. *Proceedings of the National Academy of Sciences of the United States of America* 97 (22):11990–11995.

Haddock, S.H.D., M.A. Moline, and J.F. Case. 2010. Bioluminescence in the sea. *Annual Review of Marine Science* 2:443–493.

History and perspective

Heim, R., A. Cubitt, and R.Y. Tsien. 1995. Improved green fluorescence. *Nature* 373:663–664.

Heim, R., D.C. Prasher, and R.Y. Tsien. 1994. Wavelength mutations and posttranslational autoxidation of green fluorescent protein. *Proceedings of the National Academy of Sciences of the United States of America* 91:12501–12504.

Jablonski, E. and M. DeLuca. 1976. Immobilization of bacterial luciferase and FMN reductase on glass rods. *Proceedings of the National Academy of Sciences of the United States of America* 73 (33):3848–3851.

Kennis, J.T.M. 2004. Uncovering the hidden ground state of green fluorescent protein. *Proceedings of the National Academy of Sciences of the United States of America* 101:17988–17993.

Kredel, S., F. Oswald, K. Nienhaus, K. Deuschle, C. Roecker, M. Wolff, R. Heilker, G. Ulrich Nienhaus, and J. Wiedenmann. 2009. mRuby, a bright monomeric red fluorescent protein for labeling of subcellular structures. *PLoS One* 4 (2):e4391.

Langhojer, F., F. Dimler, G. Jung, and T. Brixner. 2009. Ultrafast photoconversion of the green fluorescent protein studied by accumulative femtosecond spectroscopy. *Biophysical Journal* 96:2763–2770.

Litvinenko, K.L. and S.R. Meech. 2004. Observation of low frequency vibrational modes in a mutant of the green fluorescent protein. *Physical Chemistry Chemical Physics* 6 (9):2012–2014.

Litvinenko, K.L., N.M. Webber, and S.R. Meech. 2001. An ultrafast polarisation spectroscopy study of internal conversion and orientational relaxation of the chromophore of the green fluorescent protein. *Chemical Physics Letters* 346 (1–2):47–53.

Litvinenko, K.L., N.M. Webber, and S.R. Meech. 2002. Ultrafast excited state relaxation of the chromophore of the green fluorescent protein. *Bulletin of the Chemical Society of Japan* 75 (5):1065–1070.

Litvinenko, K.L., N.M. Webber, and S.R. Meech. 2003. Internal conversion in the chromophore of the green fluorescent protein: Temperature dependence and isoviscosity analysis. *Journal of Physical Chemistry A* 107 (15):2616–2623.

Maddalo, S.L. and M. Zimmer. 2006. The role of the protein matrix in green fluorescent protein fluorescence. *Photochemistry and Photobiology* 82 (2):367–372.

Matz, M.V., A.F. Fradkov, Y.A. Labas, A.P. Savitisky, A.G. Zaraisky, M. L. Markelov, and S.A. Lukyanov. 1999. Fluorescent proteins from nonbioluminescent *Anthozoa* species. *Nature Biotechnology* 17:969–973.

Matz, M.V., N.J. Marshall, and M. Vorobyev. 2006. Symposium-in-print: Green fluorescent protein and homologs. *Photochemistry and Photobiology* 82 (2):345–350.

Mazel, C.H., M.P. Lesser, M.Y. Gorbunov, T.M. Barry, J.H. Farrell, K.D. Wyman, and P.G. Falkowski. 2003. Green-fluorescent proteins in Caribbean corals. *Limnology and Oceanography* 48 (1):402–411.

McNaughton, B.R., J.J. Cronican, D.B. Thompson, and D.R. Liu. 2009. Mammalian cell penetration, siRNA transfection, and DNA transfection by supercharged proteins. *Proceedings of the National Academy of Sciences of the United States of America* 106 (15):6111–6116.

Megley, C.M., L.A. Dickson, S.L. Maddalo, G.J. Chandler, and M. Zimmer. 2009. Photophysics and dihedral freedom of the chromophore in yellow, blue, and green fluorescent protein. *Journal of Physical Chemistry B* 113 (1):302–308.

Mena, M.A., T.P. Treynor, S.L. Mayo, and P.S. Daugherty. 2006. Blue fluorescent proteins with enhanced brightness and photostability from a structurally targeted library. *Nature Biotechnology* 24 (12):1569–1571.

Mishin, A.S., F.V. Subach, I.V. Yampolsky, W. King, K.A. Lukyanov, and V.V. Verkhusha. 2008. The first mutant of the *Aequorea victoria* green fluorescent protein that forms a red chromophore. *Biochemistry* 47 (16):4666–4673.

Miyawaki, A. 2002. Green fluorescent protein-like proteins in reef Anthozoa animals. *Cell Structure and Function* 27 (5):343–347.

Mizuno, H., T. Kumar Mal, M. Waelchli, A. Kikuchi, T. Fukano, R. Ando, J. Jeyakanthan, J. Taka, Y. Shiro, M. Ikura, and A. Miyawaki. 2008. Light-dependent regulation of structural flexibility in a photochromic fluorescent protein. *Proceedings of the National Academy of Sciences of the United States of America* 105 (27): 9227–9232.

Nifosi, R., A. Ferrari, C. Arcangeli, V. Tozzini, V. Pellegrini, and F. Beltram. 2003. Photoreversible dark state in a tristable green fluorescent protein variant. *Journal of Physical Chemistry B* 107:1679–1684.

Niwa, H., S. Inouye, T. Hirano, T. Matsuno, S. Kojima, M. Kubota, M. Ohashi, and F.I. Tsuji. 1996. Chemical nature of the light emitter of the *Aequorea* green fluorescent protein. *Proceedings of the National Academy of Sciences of the United States of America* 93 (24):13617–13622.

Ong, W.J.H., S. Alvarez, I.E. Leroux, R.S. Shahid, A.A. Samma, P. Peshkepija, A.L. Morgan, S. Mulcahy, and M. Zimmer. 2011. Function and structure of GFP-like proteins in the protein data bank. *Molecular BioSystems* 7:984–992.

Ormoe, M., A.B. Cubitt, K. Kallio, L.A. Gross, R.Y. Tsien, and S.J. Remington. 1996. Crystal structure of the *Aequorea* victoria green fluorescent protein. *Science* 273:1392–1395.

Ow, D.W., K.V. Wood, M. Deluca, J.R. Dewet, D.R. Helinski, and S.H. Howell. 1986. Transient and stable expression of the firefly luciferase gene in plant-cells and transgenic plants. *Science* 234 (4778):856–859.

Palmer, C.V., C.K. Modi, and L.D. Mydlarz. 2009. Coral fluorescent proteins as antioxidants. *PLoS One* 4 (10):e7298.

Perozzo, M.A., K.B. Ward, R.B. Thompson, and W.W. Ward. 1988. X-ray-diffraction and time-resolved fluorescence analyses of *Aequorea* green fluorescent protein crystals. *Journal of Biological Chemistry* 263 (16):7713–7716.

Piatkevich, K.D., E.N. Efremenko, V.V. Verkhusha, and D.D. Varfolomeev. 2010. Red fluorescent proteins and their properties. *Russian Chemical Reviews* 79 (3):243–258.

Pletnev, S., N.G. Gurskaya, N.V. Pletneva, K.A. Lukyanov, D.M. Chudakov, V.I. Martynov, V.O. Popov, M.V. Kovalchuk, A. Wlodawer, Z. Dauter, and V. Pletnev. 2009. Structural basis for phototoxicity of the genetically encoded photosensitizer KillerRed. *Journal of Biological Chemistry* 284 (46):32028–32039.

Pletnev, S., F.V. Subach, Z. Dauter, A. Wlodawer, and V.V. Verkhusha. 2010. Understanding blue-to-red conversion in monomeric fluorescent timers and hydrolytic degradation of their chromophores. *Journal of the American Chemical Society* 132 (7):2243–2253.

Prasher, D.C., V.K. Eckenrode, W.W. Ward, F.G. Pendergast, and M.J. Cormier. 1992. Primary structure of the *Aequorea victoria* green fluorescent protein. *Gene* 111:229–233.

Reid, B.G. and G.C. Flynn. 1997. Chromophore formation in green fluorescent protein. *Biochemistry* 36:6786–6791.

Ridgway, E.B. and C.C. Ashley. 1967. Calcium transients in single muscle fibers. *Biochemical and Biophysical Research Communications* 29 (2):229–234.

Rosenow, M.A., H.A. Huffman, M.E. Phail, and R.M. Wachter. 2004. The crystal structure of the Y66L variant of green fluorescent protein supports a cyclization-oxidation-dehydration mechanism for chromophore maturation. *Biochemistry* 43 (15):4464–4472.

Roth, M.S., M.I. Latz, R. Goericke, and D.D. Deheyn. 2010. Green fluorescent protein regulation in the coral *Acropora yongei* during photoacclimation. *Journal of Experimental Biology* 213 (21):3644–3655.

Seward, H.E. and C.R. Bagshaw. 2009. The photochemistry of fluorescent proteins: Implications for their biological applications. *Chemical Society Reviews* 38 (10):2842–2851.

Shagin, D.A., E.V. Barsova, Y.G. Yanushevich, A.F. Fradkov, K.A. Lukyanov, Y.A. Labas, T.N. Semenova, J.A. Ugalde, A. Meyers, J.M. Nunez, E.A. Widder, S.A. Lukyanov, and M.V. Matz. 2004. GFP-like proteins as ubiquitous metazoan superfamily: Evolution of functional features and structural complexity. *Molecular Biology and Evolution* 21 (5):841–850.

Shaner, N.C., R.E. Campbell, P.A. Steinbach, B.N.G. Giepmans, A.E. Palmer, and R.Y. Tsien. 2004. Improved monomeric red, orange and yellow fluorescent proteins derived from *Discosoma* sp red fluorescent protein. *Nature Biotechnology* 22 (12):1567–1572.

Shcherbo, D., C.S. Murphy, G.V. Ermakova, E.A. Solovieva, T.V. Chepurnykh, A.S. Shcheglov, V.V. Verkhusha et al. 2009. Far-red fluorescent tags for protein imaging in living tissues. *Biochemical Journal* 418:567–574.

Shimomura, O. 1979. Structure of the chromophore of *Aequorea* green fluorescent protein. *FEBS Letters* 104:220–222.

Shimomura, O. 1995. A short story of aequorin. *Biological Bulletin* 189 (1):1–5.

Stiel, A.C., S. Trowitzsch, G. Weber, M. Andresen, C. Eggeling, S.W. Hell, S. Jakobs, and M.C. Wahl. 2007. 1.8 angstrom bright-state structure of the reversibly switchable fluorescent protein Dronpa guides the generation of fast switching variants. *Biochemical Journal* 402:35–42.

Strack, R.L., D.E. Strongin, L. Mets, B.S. Glick, and R.J. Keenan. 2010. Chromophore formation in DsRed occurs by a branched pathway. *Journal of the American Chemical Society* 132 (24):8496–8505.

Striker, G., V. Subramaniam, C.A.M. Seidel, and A. Volkmer. 1999. Photochromicity and fluorescence lifetimes of green fluorescent protein. *Journal of Physical Chemistry B* 103 (40):8612–8617.

Suto, K., H. Masuda, Y. Takenaka, F.I. Tsuji, and H. Mizuno. 2009. Structural basis for red-shifted emission of a GFP-like protein from the marine copepod *Chiridius poppei*. *Genes to Cells* 14 (6):727–737.

Wachter, R.M., M.A. Elsiger, K. Kallio, G.T. Hanson, and S.J. Remington. 1998. Structural basis of spectral shifts in the yellow emission variants of green fluorescent protein. *Structure* 6:1267–1277.

Wachter, R.M., B.A. King, R. Heim, K. Kallio, R.Y. Tsien, S.G. Boxer, and S.J. Remington. 1997. Crystal structure and photodynamic behavior of the blue emission variant Y66H/Y145F of green fluorescent protein. *Biochemistry* 36:9759–9765.

Wang, L., W.C. Jackson, P.A. Steinbach, and R.Y. Tsien. 2004. Evolution of new nonantibody proteins via iterative somatic hypermutation. *Proceedings of the National Academy of Sciences of the United States of America* 101 (48):16745–16749.

Warrant, E.J. and N.A. Locket. 2004. Vision in the deep sea. *Biological Reviews* 79 (3):671–712.

Webber, N.M., K.L. Litvinenko, and S.R. Meech. 2001. Radiationless relaxation in a synthetic analogue of the green fluorescent protein chromophore. *Journal of Physical Chemistry B* 105 (33):8036–8039.

Weber, W., V. Helms, J. McCammon, and P. Langhoff. 1999. Shedding light on the dark and weakly fluorescent states of green fluorescent proteins. *Proceedings of the National Academy of Sciences of the United States of America* 96:6177–6182.

Widder, E.A. 2010. Bioluminescence in the ocean: Origins of biological, chemical, and ecological diversity. *Science* 328 (5979):704–708.

Wiedenmann, J., S. Ivanchenko, F. Oswald, F. Schmitt, C. Rocker, A. Salih, K.D. Spindler, and G.U. Nienhaus. 2004. EosFP, a fluorescent marker protein with UV-inducible green-to-red fluorescence conversion. *Proceedings of the National Academy of Sciences of the United States of America* 101 (45):15905–15910.

Wu, L.X. and K. Burgess. 2008. Syntheses of highly fluorescent GFP-chromophore analogues. *Journal of the American Chemical Society* 130 (12):4089–4096.

Yang, F., L.G. Moss, and G.N. Phillips. 1996. The molecular structure of green fluorescent protein. *Nature Biotechnology* 14 (10):1246–1251.

Zimmer, M. 2008. The role of the tight-turn, broken hydrogen bonding, Glu22 and Arg96 in the post-transitional green fluorescent protein chromophore formation. *Chemical Physics* 348: 152–160.

Zimmer, M. 2009. Photophysics and dihedral freedom of the chromophore in yellow, blue and green fluorescent protein. *Journal of Physical Chemistry B* 113: 302.

Zimmer, M. 2011. Function and structure of GFP-like proteins in the protein data bank. *Molecular Biosystems* 7: 984.

Zimmer, M., B.R. Branchini, and J.O. Lusins. 1996. A computational analysis of the preorganization and the activation of the chromophore forming hexapeptide fragment in green fluorescent protein. In *Bioluminescence and Chemiluminescence: Proceedings of the Ninth International Symposium 1996*, eds. J.W. Hastings, L.J. Kricka, and P.E. Stanley. Chichester, U.K.: John Wiley.

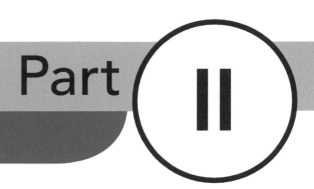

Part II

Photophysical properties of fluorescent proteins

Lessons learned from structural studies of fluorescent proteins

S. James Remington
University of Oregon

Contents

2.1 INTRODUCTION TO THE STRUCTURE AND MECHANISM OF GREEN FLUORESCENT PROTEIN

The spring and summer of 1996 were an intensely exciting time in our laboratory. The green fluorescent protein (GFP) revolution was already off to a bang, and we had good crystals of the GFP mutant $S^{65}T$. The crystals presented technical problems and the structure was slow to emerge, but during that period it was already clear that the molecular structure would be new. Furthermore, postdoctoral researcher Mats Ormö and I knew of no fewer than four other research groups that also had crystals. The race ended with the more-or-less simultaneous publication of GFP $S^{65}T$ (Ormo et al. 1996) and wild-type GFP (Yang et al. 1996), but in truth, those structures, and the many that rapidly followed, began the revolution in GFP engineering. In this chapter, I will focus on what we have learned from the structural analysis of fluorescent proteins (FPs) and how they can be reengineered to provide new and brighter colors, biosensors to report environmental conditions within cells, and to be optimized for new techniques in light microscopy.

2.1.1 OVERALL STRUCTURE OF GREEN FLUORESCENT PROTEIN

As befits a protein that revolutionized cell biology, the molecular structure of GFP is strikingly unique and seems perfectly suited to its multiple tasks. Indeed, the icon of the GFP revolution is a ribbon diagram representing the fold (Figure 2.1a). The so-called β-can or β-barrel fold of GFP (Yang et al. 1996) has become the most easily recognized protein structure known.

The iconic backbone fold of GFP is most simply described as a 238-amino acid, 11-stranded beta barrel surrounding a central alpha-helix, with short, distorted helical segments capping the barrel ends. The all-atom representation is a nearly perfect cylinder about 25 Å wide and 40 Å tall. Closer inspection of the GFP fold (Figure 2.1b) shows that the structure can be conceptually divided into sequential segments, for example, an N-terminal portion (roughly residues 1–80) consisting of three beta strands plus the central helix and a larger C-terminal portion (roughly residues 81–238) that consists of eight beta strands arranged in a "Greek key" motif.

This formal compartmentalization helps to understand the initially surprising observation that GFP, despite its monolithic appearance, can be expressed as two separate segments that combine within cells to yield a functional FP (Ghosh et al. 2000; Kent et al. 2008). Termed "bimolecular fluorescence complementation" (BIFC), the promising technique has led to numerous applications, for example, in the so-called two-hybrid scheme, the segments are fused to two other proteins and the resulting genes expressed in parallel. The appearance of visible fluorescence indicates that the two fusion partners interact *in vivo*. As implied by the schematic diagram of Figure 2.1b, which shows that the N- and C-terminus are close together in the wild-type protein, the amino acid sequence of GFP can also be circularly permuted, resulting in new N- and C-termini at various places in the polypeptide backbone, without loss of function (Baird et al. 1999; Topell et al. 1999).

Protein folding is absolutely essential for both chromophore formation and for subsequent fluorescence; see reviews (Tsien 1998; Zimmer 2002; Remington 2006). As is now well known, three amino acids within the central helix (serine[65], tyrosine[66], and glycine[67]) rearrange covalently in the presence of molecular oxygen to form the chromophore (see Section 2.3). The chemistry takes place late in protein folding (Reid and Flynn 1997) and is required for fluorescence. Denatured GFP is colorless but fluorescence is regained upon renaturation (Bokman and Ward 1981). The GFP chromophore is not fluorescent when isolated, either as the naked chromophore (McCapra et al. 1988) or as the isolated hexapeptide (Cody et al. 1993). Nevertheless, when model compounds are frozen in ethanol glass, they are strongly fluorescent (Niwa et al. 1996).

These observations make it clear that in addition to directing the chemistry of chromophore formation, a major function of the β-can fold is to tightly constrain the chromophore, so that excitation energy is dissipated as light, rather than by thermal or other processes. This implies that the folding of GFP and subsequent chromophore formation imparts a great deal of mechanical stability to the protein, which is consistent with the early experimental observations that the protein is remarkably inert toward proteases (Bokman and Ward 1981) and denatures only under extreme conditions, such as in 6 M urea at 65°C (Ward and Bokman 1982).

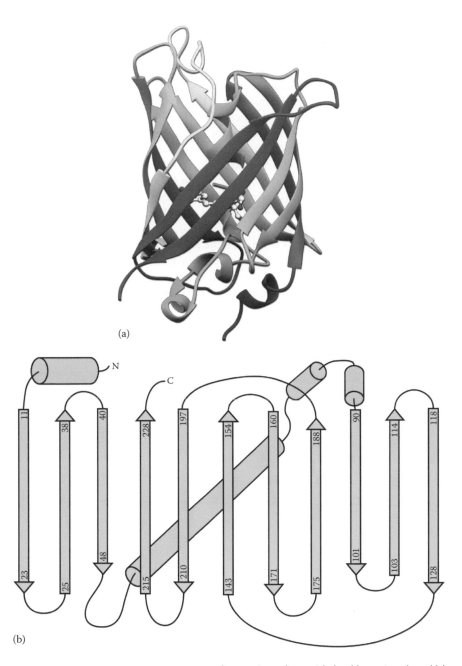

(a)

(b)

Figure 2.1 (a) Schematic backbone cartoon of GFP. The polypeptide backbone is colored blue to red, from the N-terminus to the C-terminus. The chromophore is shown as a ball-and-stick figure with carbon atoms white, nitrogen atoms blue, and oxygen red. (b) Flat sheet cartoon showing the relative organization of the secondary structure elements in GFP. Residue numbers mark the approximate beginning and ending of beta sheet strands. Helices are represented by cylinders but are more distorted than suggested by the drawing.

2.1.2 GFP CHROMOPHORE AND ENVIRONMENT

The *p*-hydroxybenzilideneimidazolidinone chromophore of GFP (Figures 2.2 and 2.3) consists of two rings linked by a double bond and is closely related to other dyes of biological origin, such as cinnamic acid. For example, the chromophore of the bacterial blue light photosensor Photoactive Yellow Protein (PYP) contains a covalently bound cinnamic acid moiety (Borgstahl et al. 1995). The chemical structure of the chromophore and the absorbance spectrum of PYP are very similar to those of GFP. In both cases, the chromophore is expected to be planar in solution but can exist in two conformations, *cis* and *trans*, which are related by a rotation of 180° about the double bond. As expected, the chromophore within GFP is found by crystallographic analyses to be approximately planar, in the lowest energy *cis* configuration (Ormo et al. 1996; Yang et al. 1996). However in some other FPs, the *trans* isomer is found to predominate (e.g., eqFP611 (Petersen et al. 2003), see Section 2.3 of this chapter). The isomeric state therefore does not determine fluorescence efficiency, but as discussed later in this chapter, chromophore planarity is a key factor in efficient fluorescence.

In GFP the *cis* chromophore configuration and overall planarity are enforced by the surrounding protein matrix using a combination of hydrogen bonds and

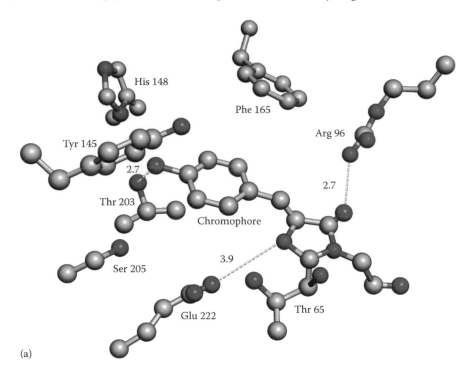

(a)

Figure 2.2 (a) Ball-and-stick representation of the GFP chromophore and its immediate environment within the protein. Carbon atoms are white spheres, nitrogen blue, and oxygen red. Not all side chains are shown. A few key hydrogen bonds, such as those between the chromophore and the catalytic residues Glu 222 and Arg 96, are shown as dashed lines and have the indicated lengths in Å. Finally, a key hydrogen bond from Thr 203 to the chromophore hydroxyl is shown. This latter hydrogen bond is important for stabilizing the anionic form of the chromophore over the neutral, protonated form.

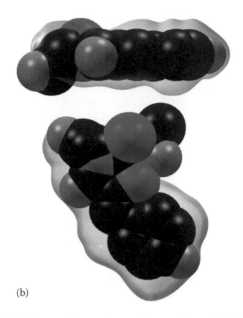

(b)

Figure 2.2 (continued) (b) Edge and face views of the protein cavity in GFP S⁶⁵T are shown as a transparent surface. The surface was calculated using the program package MSROLL (Connolly 1993) after removal of the chromophore atoms and rendered using RASTER3D and PyMOL. Superimposed within the cavity is the van der Waals surface of the chromophore, represented using atomic spheres with the appropriate van der Waals radii. Nitrogen atoms are blue, oxygens are red, and carbon is black.

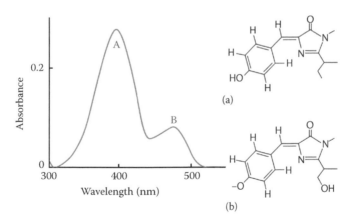

Figure 2.3 The absorbance spectrum of wild-type GFP. Band A is at about 395 nm, while band B is at about 475 nm. (a) and (b) are schematic representations of the chromophore species corresponding to the two peaks. The B chromophore form is negatively charged.

hydrophobic side chains, such as Phe165 and Tyr145, which in turn determine the shape of the enclosing cavity (see Figure 2.2a and b). The cavity, as determined directly from the van der Waals surface of the time-averaged crystal structure after removal of the atoms constituting the chromophore, fits the chromophore like a hand fits a glove (Figure 2.2b). A similarly high level of complementarity between the chromophore and encapsulating cavity is observed for most other FPs as well, with the notable exception of red fluorescent eqFP611 (see later section). It is clear from the crystal structures that cavity shape is an important determinant of the chromophore configuration.

Shortly after the gene for GFP was cloned (Prasher et al. 1992), GFP mutants at the Tyr 66 position (e.g., Phe 66, His 66, Trp 66) provided evidence that a close cavity fit is important for high fluorescence efficiency, as all have substantially reduced brightness (Heim et al. 1994). In such mutants, brightness can be improved by appropriate mutagenesis of the immediate chromophore environment, which presumably improves the complementarity between the chromophore and the surrounding cavity. This result is consistent with the notion that a loose cavity permits energy dissipation via vibrational relaxation or internal conversion (Kummer et al. 1998; Megley et al. 2009). More dramatic results were obtained with red FPs, where the *cis* versus *trans* isomeric state of the chromophore (see later sections) can be switched at will by varying the size and hydrophobicity of groups located on the edge of the cavity (Chudakov et al. 2003b; Nienhaus et al. 2008). In many of these mutants, such positional switching effectively eliminates fluorescence in one of the isomeric states.

In addition to the groups that determine cavity shape and hence chromophore configuration and rigidity, two potentially charged side chains are found in all FPs. In GFP, these are Glu 222 and Arg 96 (Figure 2.2). They are located on opposite edges of the chromophore, such that the polar side chains are close to the site where the chemical rearrangements that form the chromophore take place. Both groups are catalytic, that is, they have been determined to be vitally important but not absolutely essential for chromophore formation (see Section 2.3). In Figure 2.2a, some of those interactions are shown, with key hydrogen bonds drawn as dashed lines.

The GFP chromophore can exist in two charge states, anionic or neutral, which explains the spectroscopic observation of two absorbance and excitation bands (Figure 2.3). The neutral protonated or A form of GFP absorbs in the near UV at about 395 nm, while the anionic or B form absorbs at about 475 nm. Curiously, in wild-type GFP, both bands are present at equilibrium in an A/B intensity ratio of about 36:1. This ratio is not strongly sensitive to the pH of the environment (Ward et al. 1982) implying that in wild-type GFP, the chromophore is somehow protected from environmental changes. This effect is not fully understood. However, any number of point mutations within the cavity (and some outside the cavity) can affect the population ratio, with two of the most important examples being GFP $S^{65}T$ (Heim et al. 1994) (later improved as enhanced green fluorescent protein [EGFP]), showing only B band excitation, and the photoactivatible-GPF (PA-GFP) $T^{203}H$ (Patterson and Lippincott-Schwartz 2002; Henderson et al. 2009), showing only A band excitation.

A clear explanation for this change in excitation behavior is provided in each case by the relevant crystal structure. For $S^{65}T$, rearrangements of the hydrogen bond configuration associated with Glu^{222} lead to suppression of the negative charge on its side chain, in turn allowing the chromophore to become ionized (Ormo et al. 1996). For PA-GFP, the exchange of His for Thr^{203} reduces hydrogen bonding opportunities for the chromophore hydroxyl group so that it preferentially remains neutral (Henderson et al. 2009) with Glu^{222} being negatively charged. Upon irradiation with UV light, Glu^{222} decarboxylates and becomes electrically neutral, allowing the chromophore to ionize (Henderson et al. 2009).

2.1.3 MECHANISM OF FLUORESCENCE EMISSION AND EXCITED-STATE PROTON TRANSFER

It is generally accepted that for most FPs, emission from the anionic chromophore is rather simply described as a transition from the first singlet excited state to ground (e.g. a transition between the lowest unoccupied and highest occupied molecular orbital, see for example the recent study [Hasegawa et al. 2010]).

However the situation in GFP and some of the other FPs is quite a bit more complicated as the chromophore populates two different protonation states. Indeed, the earliest biophysical studies of wild-type GFP (Ward et al. 1982) presented the excitation spectrum as an unsolved puzzle. As mentioned previously, the absorbance and the excitation spectrum are coincident and exhibit a major band (A) at about 395 nm and a minor band (B) at about 475 nm (Figure 2.3). The intensity ratio A/B is typically about 3:1; however, variations in protein concentration, salt, and pH influence the ratio somewhat. It has now been firmly established that the A band corresponds to the protonated phenol form of the chromophore, while the B band arises from the phenolate form. Because the anionic form of the chromophore has extinction coefficient about twice as large as the protonated form, the actual population ratio of the A and B states is about 6:1. Comparison with model compounds suggests that the A state should emit in the blue. However, in GFP, only green emission is observed under steady-state conditions, regardless of which band is excited.

The puzzle of dual excitation was solved in 1996 by the Boxer group at Stanford (Chattoraj et al. 1996). They used ultrafast fluorescence upconversion spectroscopy to show that upon excitation by 397 nm light, blue emission at 460 nm is observed, but it decays with biphasic time constants of about 3 and 12 ps. The decay of blue emission is matched by the rise in 510 nm green fluorescence. Both rates are slowed by a factor of 5–6 upon substitution of exchangeable hydrogens with deuterium, suggesting that the A species is converted into an excited anionic intermediate I* by proton transfer. The I* state then decays to a B′ anionic state by emission of a green photon. B′ differs from the steady-state B form of the chromophore, probably because the B′ configuration of the protein takes some time to reach thermal equilibrium. Chattoraj et al. (1996) also demonstrated the photoconversion of GFP, where prolonged exposure to blue or ultraviolet light enhances the B excitation band at the expense of the A band. The dark recovery of the A band is slow and incomplete, suggesting that light excitation drives some permanent impairment of the protein structure.

The mechanism of A band fluorescence in GFP, although incompletely understood, is generally accepted and is the first biologically relevant example of excited-state proton transfer (ESPT) to be discovered. In solution, the chromophore pK_a is about 8.0, but similar to tyrosine, the pK_a drops to less than 1.0 upon excitation (Bent and Hayon 1975). The GFP chromophore is thus considered to be a strong photoacid (Tolbert and Solntsev 2002). The mechanism has since been elaborated by several other groups, who demonstrated a photocycle with at least two excited-state intermediates and multiple ground states (see, e.g., Lossau et al. 1996; Kennis et al. 2004; Agmon 2005; Stoner-Ma et al. 2005; van Thor et al. 2008).

2.1.4 STRUCTURAL BASIS FOR EXCITED-STATE PROTON TRANSFER

Shortly after determining the structure of GFP S⁶⁵T, we became interested in ESPT and collaborated with Titia Sixma's group at the Netherland Cancer Institute to compare the S⁶⁵T structure with their crystal structure of wild-type GFP. The result (Brejc et al. 1997) was a proposal in atomic detail for a proton transfer pathway from the

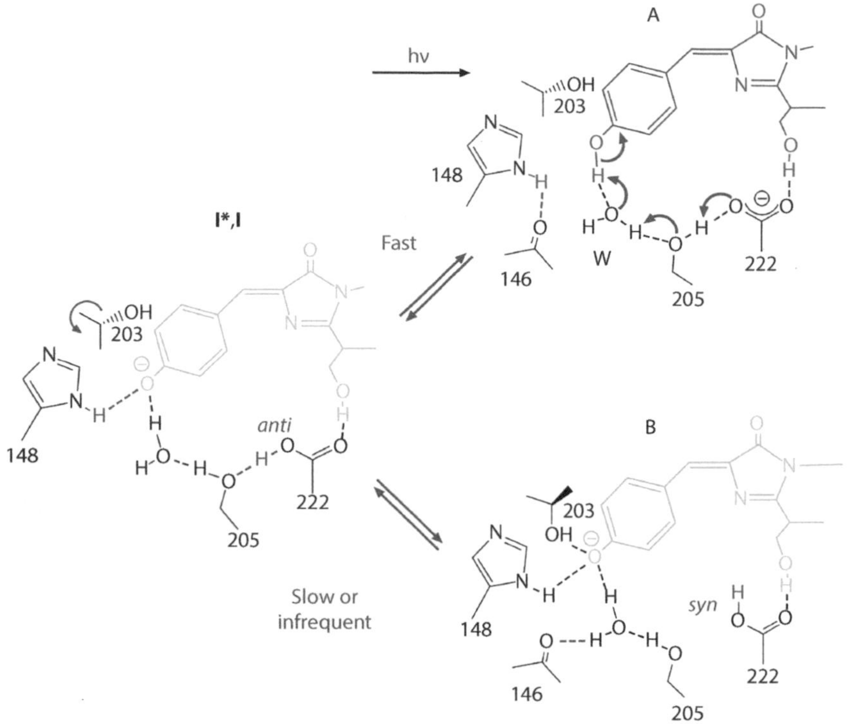

Figure 2.4 The reaction scheme proposed for ESPT and reversible photoconversion in wild-type GFP. The A form chromophore absorbs a long-wave UV photon, which dramatically lowers the chromophore pK_a. The hydroxyl proton is then effectively transferred via a water molecule and Ser 205 to Glu 222, forming the excited-state species I*. After emission of a green photon, state I can rearrange to a second ground state B, which is stabilized by reorientation of Thr 203.

chromophore OH, through a water molecule and the hydroxyl of Ser[205] to the acceptor Glu[222]. For details, the reader is referred to Figure 2.4, which also summarizes a proposal for the slow A → B photoconversion.

Subsequently, Stoner-Ma et al. (2005) used ultrafast vibrational spectroscopy to confirm that Glu[222] is the proton acceptor. Remarkably, the rate of protonation at Glu[222] matches the decay of blue fluorescence, suggesting that the proton transfer process is concerted (i.e., all three protons in Figure 2.4A move simultaneously). Alternatively, proton transfer could be stepwise, but the individual, intermediate steps may take place too quickly for these techniques to resolve. Studies of the proton transfer pathways in GFP are currently being conducted by us (Shu et al. 2007) and by others (Stoner-Ma et al. 2008; Fang et al. 2009), using a combination of mutagenesis, high-resolution crystallography, and ultrafast spectroscopic techniques.

Thus, in addition to its invaluable properties as a visible label, GFP constitutes an outstanding model system to study proton transfer processes, which are among the most fundamental chemical reactions in biology. The entire reaction is isolated from bulk solvent, and the atomic positions of the participating groups are known at extremely high resolution. Furthermore, these groups may be changed at will using mutagenesis. As will be seen in Section 2.4, ESPT is very useful for the construction of ratiometric biosensors.

2.1.5 PRACTICAL APPLICATIONS OF EXCITED-STATE PROTON TRANSFER: FLUORESCENT PROTEIN BIOSENSORS

ESPT is extremely valuable for development of ratiometric FP biosensors. Some years ago, we and others began to explore the possibility that the internal charge state equilibrium of the chromophore could be used to create useful biosensors. The most obvious application would be a pH indicator consisting of an FP that responds via a spectroscopic change in emission or excitation to changes in pH. Miesenbock and colleagues (Miesenbock et al. 1998) described pHluorins, a series of GFP mutants that respond to changes in the external pH by changes in protonation state of the chromophore. Similar pH indicators can be constructed from other FPs, because the chromophore pK_a ranges from 4 to 11, depending on the environment (Miesenbock et al. 1998; Hanson et al. 2002; Shu et al. 2006, 2007) and thus in principle covers the entire physiological range.

Importantly, these pH biosensors are ratiometric. The excitation and emission intensity depends on the concentration of the chromophore, but because it is a two-state process, measurements at two wavelengths followed by simple division removes or reduces the concentration dependence from the measurement, as well as the dependences on changes in sample thickness, illumination intensity, etc. With the pHluorins one can use a dual excitation scheme, for example monitoring fluorescence intensity at 510 nm as a function of excitation at typically either 390 nm [I^{510}_{390}] or 475 nm [I^{510}_{475}]) and the isosbestic point I^{510}_{425}. The emission ratio $I^{510}_{390}/I^{510}_{425}$ (or $I^{510}_{475}/I^{510}_{425}$) yields the ambient pH upon comparison with a standard calibration curve obtained by titration of the indicator *in vitro*. The principle of ratiometry is discussed extensively for a class of calcium indicators (Grynkiewicz et al. 1985). Furthermore, the approach can be generalized such that any two wavelengths can be chosen for ratiometric measurements. One must only choose wavelengths such that the observed changes in the absorption or emission spectrum oppose one another at those wavelengths (Grynkiewicz et al. 1985; Lohman and Remington 2008).

Hanson et al. (2002) later described dual-emission pH sensors called deGFPs. These biosensors switch from blue emission (~460 nm) at low pH to green, 510 nm emission at high pH, allowing them to also be analyzed by ratiometric imaging. With deGFPs, one can excite at 280 nm (which takes advantage of Förster energy transfer from internal tryptophan and tyrosine residues to the chromophore) and measure the emission ratio I^{460}/I^{510} to determine the pH.

The mechanism of deGFPs stands in contrast to the pHluorin example, which is simply explained by direct titration of the chromophore. The crystal structures of the deGFPs at high and low pH revealed dramatic structural changes in the protein. In turn, these changes modify the chromophore environment such that ESPT is possible at high pH, but is blocked at low pH. Consequently at low pH, blue emission from the chromophore predominates (Hanson et al. 2002; McAnaney et al. 2002, 2005) whereas at high pH, green emission predominates. The pH-sensitive biosensors have not seen many practical applications, possibly because most researchers are not fully aware of their potential.

Two more general and important classes of biosensors that can take advantage of ESPT include redox-sensitive GFPs, which report the local thiol–disulfide equilibrium (Ostergaard et al. 2001; Hanson et al. 2004) and calcium indicators; for a recent review, see Hasan et al. (2004). The most useful of these are ratiometric indicators that respond to changes in the environment. The ratiometric redox-sensitive GFP indicators were first described by Hanson et al. (2004). To create these indicators, surface-exposed

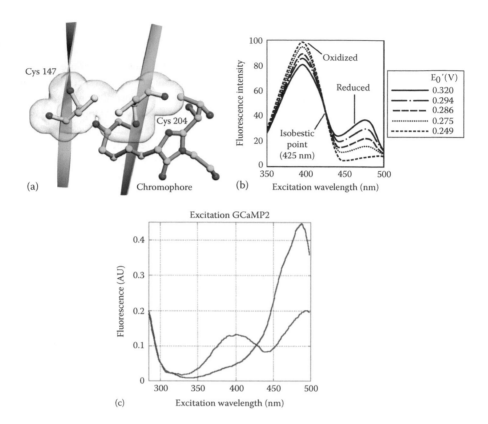

(a)   Chromophore   (b)   Excitation wavelength (nm)

(c)   Excitation wavelength (nm)

Figure 2.5 (a) A model of the oxidized state of redox biosensor roGFP2, showing a portion of the final electron density map superimposed. The intent is to show that the disulfide bridge is close to the hydroxyl group of the chromophore (pictured) and thus can exert control on the solvent accessibility and hydrophobicity of the chromophore environment. The contour level of the electron density map is 1 standard deviation of the electron density map. (b) The excitation spectrum of the redox biosensor roGFP1 as a function of the thiol–disulfide chemical potential of the milieu, as defined by a dithiothreitol redox buffer system. roGFP1 is a ratiometric biosensor with an isosbestic wavelength of 425 nm. (c) Spectral changes associated with binding of Ca^{2+} to the calcium biosensor GCaMP2. Shown is the relative fluorescence emission at about 510 nm as a function of excitation wavelength. The calcium-free form is associated with the red curve, while the calcium-bound form is associated with the blue curve. Structural rearrangements that take place upon calcium binding favor the negatively charge chromophore form over the neutral form. Similar to roGFP1, GCaMP2 is a ratiometric biosensor with an isosbestic wavelength of 425 nm.

cysteines were incorporated into neighboring strands of the GFP barrel, adjacent to the chromophore, where disulfide bonds C^{147}–C^{204} (see Figure 2.5a) and C^{149}–C^{202} could form. The disulfide formation increases the hydrophobicity of the environment and reduces solvent accessibility of the chromophore, in turn changing the pK_a, which results in a readily detectible change in the excitation spectrum. Thus, the roGFPs (reduction–oxidation-sensitive GFPs) created using this strategy rapidly equilibrate with the local thiol–disulfide pool, and ratiometric changes in the excitation spectrum indicate the ambient redox potential (Figure 2.5b). Titrations against a redox buffer demonstrate that the transition is accurately modeled as a two-state process and, depending on other modifications (Lohman and Remington 2008), redox indicators have been produced with midpoint potentials ranging from –240 to –290 mV. These are all rather negative for a disulfide, indicating an energetically stable S–S linkage.

Structural studies of roGFPs (Hanson et al. 2004; Lohman and Remington 2008) demonstrated the reversible formation of the disulfide and revealed changes in the chromophore environment that can account for the observed changes in the average protonation state of the chromophore. The more reducing indicators, such as roGFP1 and roGFP2, are well matched to the extremely reducing environments of the mitochondria and cytoplasm of mammalian cells (midpoint potentials roughly –360 and –320 mV). For detailed reviews of genetically encoded redox-sensitive indicators, their applications, and some of the results, see Cannon and Remington (2007) and Meyers and Dick (2010).

As a final note for this section, the gCaMP2 Ca^{2+} indicator is an ESPT-based ratiometric reporter of calcium concentration. Originally, gCaMP was created as a fusion protein consisting of a circularly permuted enhanced GFP (cpEGFP) attached to the calcium-binding protein calmodulin (CaM) and the CaM-binding peptide M13 from myosin light chain kinase (Nakai et al. 2001). Directed evolution of the starting construct led to gCaMP2, which has dramatically improved brightness and stability (Tallini et al. 2006). The crystal structures of Ca-bound and Ca-free forms of gCaMP2 revealed dramatic conformational changes (Akerboom et al. 2009), but the spectral response is remarkably simple. It is a two-state transition, with the GFP chromophore switching between the A and B protonation states in response to changes in $[Ca^{2+}]$. Thus, $[Ca^{2+}]$ can be directly determined in real time from the excitation/emission ratio $I^{510}_{390}/I^{510}_{425}$ or $I^{510}_{475}/I^{510}_{425}$ as described previously for pHluorins.

2.2 FLUORESCENT PROTEIN RAINBOW

Mutagenesis of the GFP cDNA led to the creation of useful blue, cyan (Heim et al. 1994), and yellow (Ormo et al. 1996) variants, but useful red mutants of GFP have never been discovered. Red fluorescence is preferred over other visible colors due to increased penetration of tissue and low cellular autofluorescence background, but the discovery of red FPs was some time in coming. Investigations of various bioluminescent organisms have so far revealed only other green FPs, for example, from the sea pansy *Renilla* (Ward et al. 1980).

Contrary to popular belief, the discovery of an entire rainbow of FPs from coral reef organisms (Matz et al. 1999) was not made on or even near a warm, sunny beach. The idea instead developed at a party in a Moscow apartment featuring a saltwater aquarium containing fluorescent anemones (M. V. Matz, personal communication). It is rumored that alcohol contributed to the speculation regarding the nature of the molecules responsible for the fluorescence, but no matter, as the result is now historic (Matz et al. 1999).

Considerable excitement followed the discovery (Matz et al. 1999) and the early commercial distribution of DsRed, but this was rapidly tempered by several inconvenient facts. DsRed and most of its reef-derived cousins are obligate tetramers (Baird et al. 2000; see Figure 2.6) that do not dissociate even when diluted to single molecules. This can be disastrous for protein labeling purposes, because anything that is successfully tagged with wild type DsRed is generally also aggregated into higher-order

Photophysical properties of fluorescent proteins

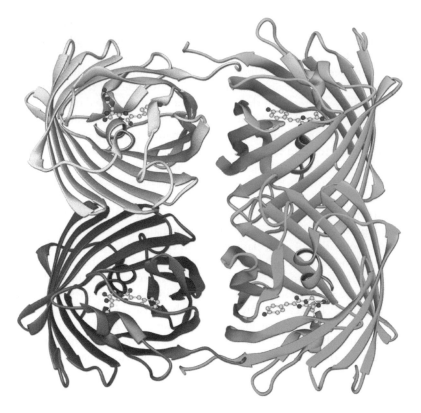

Figure 2.6 Top view of a backbone schematic of the DsRed tetramer (PDB ID 1G7K). The A, B, C, and D chains are shown in counterclockwise order, starting in the upper left-hand corner. The chromophore arrangement is shown in ball-and-stick representation. The plane of the illustration is approximately the same as the mean plane of the chromophore arrangement.

complexes. Furthermore, the red FPs are slow to mature and frequently contain a large percentage of dead-end green fluorophores (Baird et al. 2000), which complicates dual-labeling experiments involving GFP. Finally, the red-emitting proteins seem to be more toxic to certain hosts, possibly because of reactive oxygen or free radical species that are produced during illumination (for a review, see Dixit and Cyr 2003). A stunning and useful example of this side effect is provided by KillerRed, which can kill a host cell upon brief illumination (Bulina et al. 2006).

2.2.1 COLOR CLASSES AND MOLECULAR STRUCTURES

At last count, no less than seven evolutionary classes are required to account for the diversity of the FP family (Alieva et al. 2008; Wachter et al. 2010). These include five color classes (cyan, green, yellow, and two red classes); a class of colorless proteins found in the extracellular protein matrix of animals, including humans, that lack a chromophore entirely (Hopf et al. 2001); and a class of nonfluorescent chromoproteins (Lukyanov et al. 2000) that are generally deep purple or blue in appearance. More classes probably remain to be discovered, but so far, the evidence suggests that the GFP fold arose only once during the course of evolution.

Photophysical properties of fluorescent proteins

2.2.1.1 Fluorescent protein folds and assemblies

All members of the family discovered to date have detectable amino acid sequence similarity to GFP. Thus it is not surprising that between the classes, the basic backbone folds are very similar, with only minor differences in loops and C-terminal extensions. As new data become available, reconstructions of the evolutionary pathways are being revised but consistently point to a common ancestor that diverged several times. The evidence now suggests that red emitting FPs developed along two different pathways: the Kaede pathway and the DsRed pathway (Alieva et al. 2008). It is apparent, however, that there is diversity in the tertiary structures of these proteins. Most of the proteins characterized thus far assemble as tetramers, as exemplified by DsRed (Wall et al. 2000; Yarbrough et al. 2001) but dimers have also been encountered. Note that although Kaede and DsRed have apparently diverged with regard to the mechanism of chromophore formation (presumably from a green ancestor; see review by Wachter et al. 2010), their tetrameric assemblies are essentially identical.

There are two striking features of the DsRed tetramer. First, it is an extremely close-packed entity, appearing as a squat, square prism with remarkably flat sides and a small elliptical hole directly through the center. The hole is lined with polar residues and salt bridges that localize a number of water molecules. The tetramer is a dimer of dimers, with the A and C chains making close contacts via their surfaces and extended carboxy termini, while the A and B chains interact via large buried hydrophobic patches. The AB interface consists of hydrophobic interactions between small side chains, although a few hydrogen bonds and salt bridges are present. On the other hand, interactions in the AC (and BD) interface consists largely of salt bridges and hydrogen bonds, many of which are mediated by buried water molecules, as well a surprisingly large number of interactions involving aromatic residues. DsRed and several other FPs also have an extended carboxy terminus, which in GFP is flexible (only residues 2–229 are visible of the 238 in the mature protein [Ormo et al. 1996]). In DsRed, the carboxy terminus of the A monomer embraces the C monomer and vice versa, forming a "clasp" about a local twofold axis. The clasp arrangement is clearly evident in Figure 2.6, top and bottom center.

A second striking feature of the DsRed tetramer is the chromophore arrangement. The four chromophores are antiparallel and form an antenna-like rectangular array approximately 27 by 34 Å from center to center (see Figure 2.6), suggesting very efficient energy transfer between chromophores. Evidence from the electron density map (Yarbrough et al. 2001) and other studies (Gross et al. 2000) indicate that in DsRed, approximately 50% of the chromophores are green instead of red. Thus, in hybrid tetramers, nonradiative energy transfer from green-emitting to red-emitting centers should be essentially 100% efficient, which complicates spectroscopic analyses.

The function of the reef FPs remains unknown, but considering that they are ubiquitous, it is safe to assume that they provide an evolutionary advantage. Therefore, it is important to consider that such energy transfer, as facilitated by the oligomeric state of the protein, and/or incomplete maturation may have an important biological function.

2.2.2 REEF-DERIVED FLUORESCENT PROTEIN CHROMOPHORES AND ENVIRONMENT

Shortly after the discovery of DsRed, it was established (Gross et al. 2000) that the red chromophore differed chemically from that found in GFP and that an additional oxidative modification takes place during maturation. In DsRed, the backbone is oxidized so that

Photophysical properties of fluorescent proteins

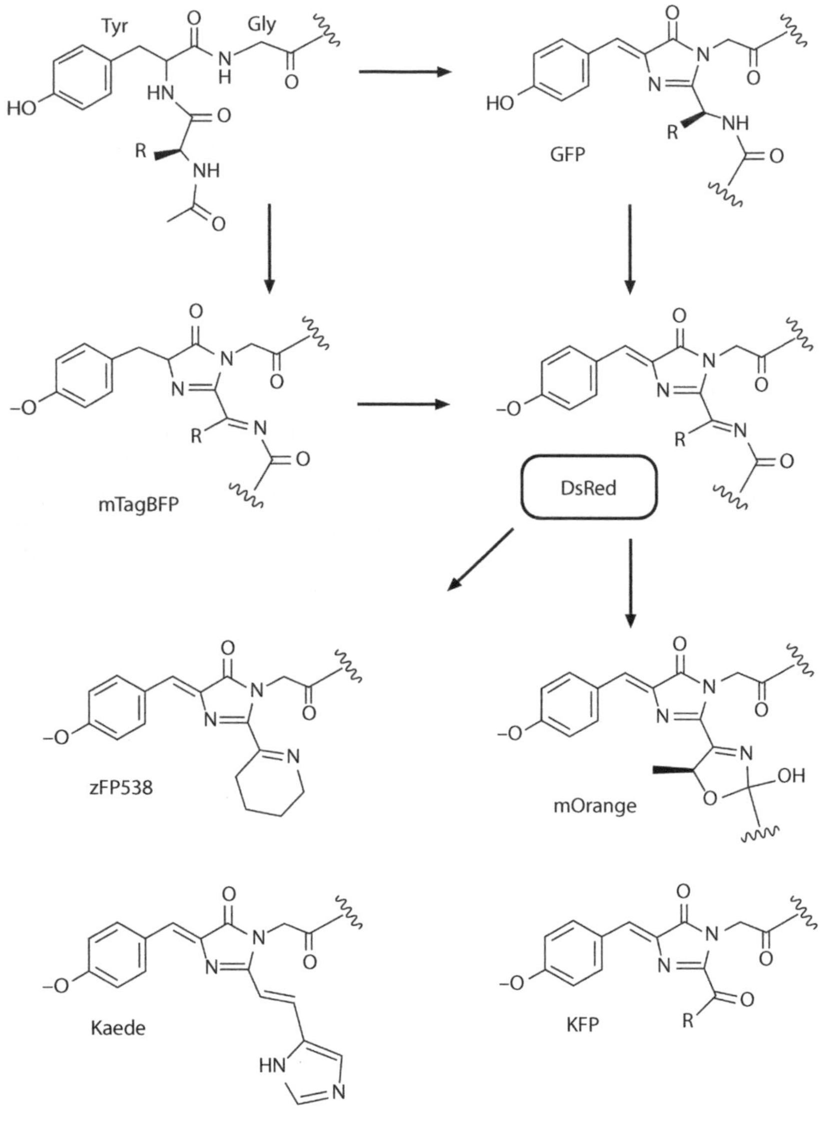

Figure 2.7 Schematic representations of the various chromophores found in fluorescent proteins and nonfluorescent chromoproteins. The arrows among the top six diagrams are intended to represent multistep chemical pathways that may be branched, for example, the chromophore shown for mTagBFP (note the reduced bond forming the tyrosine α-β linkage) may be a precursor for the DsRed chromophore, which in turn may be a precursor for the zFP538 and mOrange chromophores. KFP is the "kindling fluorescent protein," also called asFP595.

chromophore conjugation is physically extended by one double bond compared to GFP (Figure 2.7). Comparison with the spectra of model compounds shows that this physical increase in size adequately accounts for the increased excitation and emission wavelengths (He et al. 2002). An easily detected hallmark of backbone oxidization is the appearance of multiple bands (usually three) on a denaturing SDS/PAGE gel, as harsh conditions such as

Photophysical properties of fluorescent proteins

heat and pH extremes lead to partial hydrolysis of the protein backbone adjacent to the chromophore (Gross et al. 2000). More recent structural and spectroscopic studies have expanded the repertoire of chromophores to seven (summarized in Figure 2.7), but it is very likely that others remain to be discovered.

The protein cavity of the reef-derived FPs is generally more polar than that of GFP, with additional acidic and basic groups (shown in detail for DsRed in Figure 2.8). These groups form a complex network of salt links and hydrogen bonds, so it is difficult to assign charge states to individual groups. The high polarity of this cavity is evidently an artifact of evolutionary history, as most of the other polar groups may be replaced with hydrophobic groups without major consequences, provided that the protein folds correctly. Furthermore, many such modifications lead to red shifts in the emission. An excellent example of this principle is provided by the family of mFruits, monomeric red FPs which were derived by extensive mutagenesis from DsRed (see later section, Shaner et al. 2004).

In the red FPs (and most other reef-derived FPs as well), an additional positively charged group, either Arg or Lys, is found adjacent to the chromophore and seems to be required for the development of red fluorescence. In DsRed, Lys 70 occupies this key position, and replacement with anything other than Arg results in either a colorless protein or formation of a green chromophore (Baird et al. 2000). In a few other proteins, including some of the nonfluorescent chromoproteins, the additional positive charge is found to originate

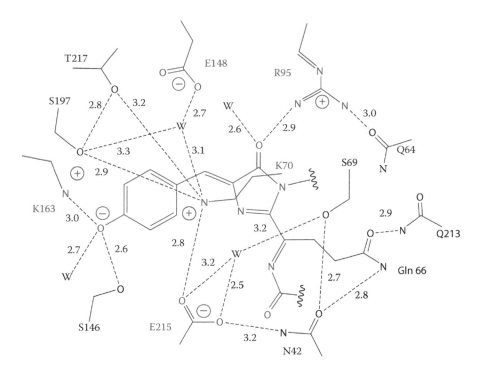

Figure 2.8 Schematic diagram showing details of the DsRed chromophore and environment. Potential hydrogen bonds and salt links are shown as dashed lines, with their approximate lengths indicated in Å. Note the unusual quadrupole arrangement of charges contributed by Lys 70, Glu 148, Glu 215, and Lys 163, Lys 70, and Glu 215 that are required for the development of red fluorescence, while Glu 148 and Lys 163 are not required.

Photophysical properties of fluorescent proteins

elsewhere in the chain. For example, in the chromoprotein pocilloporin (described in Section 2.2.4), this function is provided by Arg197 (Prescott et al. 2003).

All examples of the cyan and green families of FPs examined to date have chromophores chemically identical to that found in GFP, for example, asFP499, amFP486, KikG, and dsFP483 (Henderson and Remington 2005; Tsutsui et al. 2005; Nienhaus et al. 2006; Malo et al. 2008), which suggests that the protein environment is responsible for the differences in emission and excitation between the cyan and green classes. The placement of charged residues and polar groups is believed to play a key role distinguishing the two color classes. Surprisingly, structural analysis of these proteins revealed that compared to DsRed, all recognized catalytic groups are conserved. This includes the additional catalytic positive charge at position 70 of DsRed, so something in the folding process must determine whether the green or the red pathway(s) for chromophore formation is prioritized.

2.2.3 ENGINEERED MONOMERIC FLUORESCENT PROTEINS

As mentioned previously (in the introduction to Section 2.2), the tendency of reef-derived FPs to form tight tetramers is a serious problem for *in vivo* labeling applications. Furthermore, chromophore formation is linked to protein folding, so early attempts at mutagenesis to inhibit tetramer formation resulted in colorless protein or, more often, a white precipitate. Campbell et al. (2002) undertook the heroic efforts required to solve this problem and, beginning with DsRed, derived the first monomeric red FP, mRFP1. This required several rounds of directed evolution to first create a weakly fluorescent dimer, followed by cycles of semirandom mutagenesis and colony selection to improve the brightness. The dimer was then broken into monomers by mutation at the subunit interface, and again, brightness had to be improved by subsequent cycles of mutagenesis. The result was 33 mutations incorporated into the parent gene to yield the first monomeric RFP, mRFP1. mRFP1 has been extremely popular but is rather dim, with a quantum yield (QY) of about 0.25 (about 1/3 as bright as DsRed or GFP). mRFP1 also has a strong absorbance peak at 500 nm with no corresponding emission peak, suggesting that an alternative pathway can lead to a formation of a nonfluorescent arrangement of a GFP-like chromophore (Campbell et al. 2002).

Roger Tsien and coworkers extended the mRFP1 effort using an innovative combination of directed evolution, somatic hypermutation, and flow cytometry to select for color and brightness variants in individual cells (Shaner et al. 2004; Wang et al. 2004). This effort resulted in an invaluable collection of bright, photostable monomeric proteins called "mFruits" ($\lambda_{max}^{em} = 550 - 650$ nm). Atomic resolution structures have been determined for three members of the family (Shu et al. 2006), which allowed us to rationalize the observed differences in emission and excitation wavelengths. One particularly interesting result of that study is the popular mOrange that was discovered to have a new type of chromophore (shown in Figure 2.7). The threonine Oγ of the chromogenic tripeptide cyclizes with a presumed acylimine in the backbone to generate a novel three-ring chromophore. A similar chromophore that develops from a Cys[65]-Tyr[66]-Gly[67] tripeptide and forms a 2-hydroxy-3-thiazoline ring was later discovered in the naturally occurring orange-emitting FP mKO (Kikuchi et al. 2008).

Other groups, including those led by Campbell, Wiedenmann, and Miyawaki, have utilized the same general approaches to produce bright, monomeric FPs with valuable properties. For example, mTFP1 (teal, $\lambda_{max}^{em} = 492$ nm; see Ai et al. 2006) is optimized for use in FRET systems with a yellow or orange FP, while tandem dimers of EosFP offer high brightness and the capability of green to red photoconversion (Nienhaus et al. 2006), albeit with larger molecular mass. Finally, mKeima (Kogure et al. 2006) offers an extremely large Stokes shift, with peak excitation at 440 nm producing red fluorescence at 620 nm, making it useful for dual-color labeling. Interestingly, it was recently demonstrated that mKeima utilizes ESPT to achieve this large Stokes shift, so that excitation of the neutral chromophore results in transfer of a proton to an acceptor prior to light emission, but the ESPT pathway is completely different from that of GFP (Henderson et al. 2009; Violot et al. 2009).

On a final note for this section, most engineered, monomeric FPs described to date are significantly less bright and less photostable than the naturally evolved FPs (for a recent review, see Piatkevich and Verkhusha [2010]). To achieve a given purpose, many mutations are typically introduced into the cavity vicinity, so it seems very likely that the cavity shape is less than optimal for maximum QY. An obvious solution is to conduct a directed evolution screen that targets a significant fraction of the amino acids lining the chromophore cavity, but for most research groups, such a brute force approach has generally been considered to be prohibitive in terms of personnel, time, and cost.

2.2.4 NONFLUORESCENT AND REVERSIBLY PHOTOSWITCHABLE FLUORESCENT PROTEINS

The first example of a nonfluorescent GFP homolog to be discovered seems to be pocilloporin (Dove et al. 1995), although at the time of publication, the authors had not yet recognized that pocilloporin is closely related to GFP. These proteins strongly absorb green or yellow light and thus appear purple or blue. They are potent pigments and are often found as decorations on coral reef denizens, such as on the tentacle tips of *Anemonia sulcata*. The latter observation led to the isolation of several other members of the group, such as asFP595 (Lukyanov et al. 2000).

The crystal structures of pocilloporin, asFP595, and other nonfluorescent chromoproteins have been determined (Prescott et al. 2003; Quillin et al. 2005; Wilmann et al. 2005). From these studies, it is clear that the lack of fluorescence is due to a highly distorted chromophore configuration, which in turn is enforced by the protein cavity (see Figure 2.9). Such structural deformations are well known to reduce fluorescence QY, but the effectiveness of this strategy is remarkable. A close analog of the asFP595 chromophore has been synthesized (Yampolsky et al. 2005), and for the naked chromophore in solution, the QY is 10 times higher than the QY of the naturally occurring protein. This observation, combined with the abundance and diversity of nonfluorescent pigments, argues for a strong evolutionary pressure that favors efficient coloration, but not visible fluorescence in that context. One wonders why this should be so.

Mutants of asFP595 are reversibly photoswitchable, in that exposure to intense green light can temporarily induce a weak orange fluorescent state that decays with time constants of seconds to minutes depending on the mutation (Chudakov et al. 2003a,b). Presumably, light energy can induce rearrangements of the protein cavity that permit the chromophore to adopt a more planar configuration with a higher QY, which then thermally relaxes to the ground state. The process is reversible, as it is possible to switch the dark state back to the light state, usually by illumination with blue or ultraviolet light. The cycle may be repeated hundreds of times, and it has been suggested that photoswitchable FPs could be used for information storage at the single molecule level (Ando et al. 2004). Erasable labels have also been proposed as tools to study motion of molecules within cells

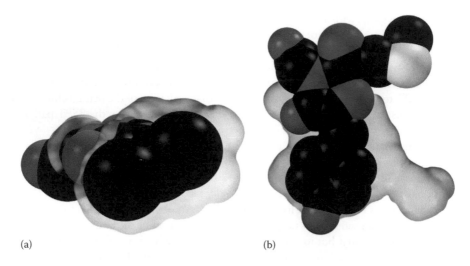

(a)   (b)

Figure 2.9 (a) The internal cavity of Rtms5 (Protein Data Bank ID 1MOU), a nonfluorescent chromoprotein, with superimposed van der Waals shape of the chromophore (end view). The figure was calculated exactly as described in Figure 2.2b and is intended to show that the pronounced twist of the chromophore is enforced by a close fit of the protein cavity. (b) The internal cavity of eqFP611 (Protein Data Bank ID 1UIS), a red fluorescent protein with *trans* chromophore configuration. The diagram illustrates that the chromophore cavity is in principle capable of accommodating either the *cis* or the *trans* isomer, supporting experimental results obtained by mutagenesis. (From Nienhaus, K. et al., *J. Am. Chem. Soc.*, 130, 12578, 2008.)

(Ando et al. 2004; Chudakov et al. 2004; Hofmann et al. 2005; Nienhaus et al. 2006). Finally, photoswitchable FPs have been shown to be useful for superresolution light microscopy (Hofmann et al. 2005; Betzig et al. 2006).

While these new techniques are extremely exciting and promising, it is fair to say that no major discoveries in cell biology have yet been made using photoswitchable proteins (the superresolution techniques are providing novel insights into cellular structures). Perhaps it is just too early to judge, and time will tell. On the other hand, one notes that GFP and its variants were discovered to be photoswitchable some years earlier (see, e.g., Dickson et al. [1997]), but the effect was at the time considered to be problematic, rather than potentially useful.

2.2.5 PHOTOSWITCHING MECHANISMS

Without the benefit of crystallographic studies, Chudakov et al. (2003b) carried out mutagenesis of photoswitchable asFP595 based on a homology model of the protein. These workers introduced small changes into the chromophore cavity and studied the photoswitching behavior of the FP. The results implicated *cis–trans* isomerization as the structural basis for the light and dark states, but of course the situation is not that simple.

Later studies of asFP595 by Andreson et al. (2005) provided the first crystallographic studies of a photoswitchable protein. These studies suggested that *cis–trans* chromophore isomerization initiates the cavity rearrangement, but the exact nature of the fluorescent form of the chromophore, and a clear consensus on the mechanism of photoswitching in asFP595 has yet to emerge. It is difficult to draw conclusions from the available data, as the effect is weak (i.e., the QY of <0.01 for the fluorescent form implies that only a small fraction of the molecules become fluorescent) and structural data reveal a superposition of structures in the putative activated state (Henderson and Remington 2006).

On the other hand, studies of reversibly photoswitchable green FPs such as Dronpa and mTFP0.7 have been more fruitful regarding the mechanism of photoswitching. The photoswitching behavior of these proteins is the opposite of that observed for asFP595, in that the "resting" state is brightly fluorescent. With continued excitation, the proteins quickly convert to a dark state, but bright fluorescence is rapidly restored upon exposure to blue or ultraviolet light. The structures of green fluorescent Dronpa (Wilmann et al. 2006; Andresen et al. 2007; Stiel et al. 2007) and mTFP0.7 (Henderson et al. 2007) have been determined for both the light and dark states with strikingly similar results, which provide an explanation for the reversible photoswitching behavior. In the fluorescent state, the chromophore is *cis* and planar, while in the dark state, the chromophore is *trans* and highly twisted. Furthermore, in the dark state, the tyrosine moiety of the chromophore relocates to a rather hydrophobic environment and becomes protonated. In the protonated state, absorption of a 380 nm photon is expected to ionize the tyrosine hydroxyl group and cause it to be repelled from the hydrophobic cavity, as observed experimentally.

2.3 CHEMISTRY AND EVOLUTION OF CHROMOPHORE FORMATION

The key feature of the FPs, distinguishing them from all other colored molecules of biological origin, is that chromophore formation is an autocatalytic process requiring only the presence of molecular oxygen. In short, visible fluorescence is encoded by a single gene that can be expressed in any organism. It was this discovery that led to the GFP revolution and the award of the 2008 Nobel Prize for Chemistry to Osamu Shimomura, Roger Tsien, and Marty Chalfie.

In order to engineer new colors and to make existing FPs more useful in biotechnology (e.g., to improve brightness and photostability), it is vitally important to understand the maturation process. Significant advances in our understanding have been made, but many mechanistic details of the process remain to be determined. Space does not permit an exhaustive summary in this chapter, so for details beyond those provided, the reader is referred to recent comprehensive reviews (Craggs 2009; Wachter et al. 2010).

As with most proteins, important clues about chromophore formation can be derived from conserved elements in the amino acid sequence and hundreds of gene sequences are now available for FP homologs. Analysis of a 105-member collection of sequences that was generously placed in the public domain by Alieva et al. (Alieva, Konzen et al. 2008) reveals that, with just one or two poorly understood exceptions, seven amino acids are absolutely conserved. In GFP numbering, these are Gly[20], Phe[27], Gly[40], Gly[67], Tyr[66], Arg[96] and Glu[222] (Note: at least two sequences have been discovered in which Gln is substituted for Glu[222]). Tyr[66] and Gly[67] form the tripeptide that becomes the chromophore; Gly[67] is the nucleophile in the initial backbone cyclization reaction and Arg[96] and Glu[222] (see Figure 2a for spatial relationships) are catalytic, as mutation of either severely compromises maturation. Allowing two additional very conservative substitutions Phe/Tyr and Ile/Val, then one can include Phe[100], Phe[130], and Ile[136]. Together with Gly[20] and Phe[27], these latter residues form a hydrophobic core at one end of the barrel, which is probably a key for proper chain folding. Figure 2.10 provides a

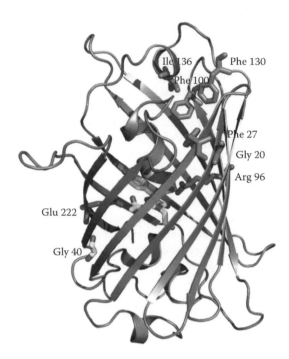

Figure 2.10 Schematic diagram of the GFP backbone, showing the location of side chains that are conserved in a representative sample of 105 amino acid sequences. The blue side chains on upper right are proposed to form a hydrophobic core that is required for proper folding of the polypeptide backbone, while most of the others are involved in the autocatalytic reactions leading to chromophore formation.

schematic diagram of the GFP backbone, with conserved side chains represented by stick bonds. The side chains of the proposed conserved hydrophobic core are shown in blue.

It now seems certain that in all FPs, the first step in chromophore formation is backbone cyclization to form the five-membered ring. The most recent evidence suggests that oxidation to produce a double bond is the next step (in GFP, the oxidation of the tyrosine α-β linkage), with dehydration of the five-membered ring occurring last (Wachter et al. 2010). The situation is quite a bit more complicated with the reef-derived FPs as at least one additional oxidation step is required, possibly in concert with the hydrolysis, cyclization and elimination reactions, producing a remarkable variety of chromophores. The chemical structures of the known chromophores are summarized in Figure 2.7.

The roles of Glu[222] and Arg[96] have been investigated by the Getzoff and Wachter research groups (Sniegowski et al. 2005; Wood et al. 2005). Arg96 can be replaced by lysine without much effect and since both are positively charged, it seems that its role is an electrostatic catalyst. An important advance was provided by structural studies of the severely compromised GFP mutant R96A (Barondeau et al. 2003), which takes months to mature. Crystals were obtained at several stages in the process. Those at an early stage of the process revealed a proposed precyclization intermediate. The results suggested that the backbone fold acts to preorganize the tripeptide to favor ring closure, but I note that in the mammalian GFP homolog nidogen (Hopf et al. 2001), ring closure fails to take place, despite the essential identity of the backbone folds. The precise position of the positively charged group is not important, as substitution of Gln[183] with Arg in the R[96]A background restores fluorescence (Wood et al. 2005). Arg[96] is thought to stabilize the enolate of Gly[67] in preparation for the backbone cyclization reaction, a result supported by crystallographic studies of dithionite-reduced GFP (Barondeau et al. 2006).

The absolute conservation of tyrosine at position 66 is puzzling, since it can be replaced by almost any side chain. Indeed, substitution of aromatic residues leads to useful color variants such as a blue (histidine and phenylalanine) or cyan (tryptophan) FPs (Tsien 1998). Alternatively, when nonaromatic side chains such as alanine, serine, glycine, or leucine are substituted for Tyr[66], a great variety of other interesting reactions are observed. While none of these proteins are fluorescent, Y[66]L slowly becomes yellow (Rosenow et al. 2005). The crystal structure of Y[66]L revealed the formation of an unprecedented leucine-histidine cross-link within the chromophore cavity. This is an important result, because the aliphatic leucine side chain becomes oxidized in the process, demonstrating that the overall maturation reaction does not require and is not driven by the aromatic character of residue 66.

In addition to the steps described previously, the red and yellow reef-derived FPs require an additional oxidation step. As discussed earlier (Section 2.2.2), in DsRed the protein backbone is oxidized to form an acylimine, but very little is known about the protein features that are responsible for initiating and controlling the reaction. Also, as pointed out earlier, most green and cyan FPs contain essentially the same complement of potentially catalytic side chains as do the red and yellow proteins. The reaction to form the red chromophore is often slow and incomplete, suggesting that it is difficult to organize the components productively.

Until recently, it was thought that formation of the red chromophore proceeded through a neutral GFP-like intermediate, as a blue-emitting intermediate has been detected (Verkhusha et al. 2004). This ordering implies that the tyrosine α-β linkage becomes oxidized before the polypeptide backbone becomes oxidized. However, the recent structural characterization of the blue FP mTagBFP (Subach et al. 2010) may require these ideas to be revised. Evidently, in mTagBFP the backbone is oxidized, forming the acylimine linkage, but the tyrosine Cα-Cβ linkage remains in the reduced state (see Figure 2.7). This opens the possibility that the mTagBFP chromophore is the blue intermediate that has been detected in the normal course of formation of the red chromophore.

This proposal is supported by a recent paper by Strack et al. (2010) describing a clever set of experiments involving isotopic substitution and reaction rate studies. These authors present an outline for a branched pathway of red chromophore formation, in which the two oxidation steps compete to produce the red chromophore (in addition to off-pathway dead-end products). Many details remain to be clarified, such as identification of the precise roles played by the protein in the individual steps. For example, compared to GFP, an additional positively charged side chain appears to be necessary (but is not sufficient) for the development of red fluorescence. As pointed out earlier, in red and yellow fluorescent proteins Lys or Arg is almost always found at position 5 of the chromogenic pentapeptide -X-Y-G-X-K/R.

2.4 CONCLUSIONS

The genes for fluorescent proteins encode visible fluorescence, which requires that the resulting polypeptide chain fold efficiently and properly, in the process catalyzing the formation of a unique cofactor and then, when folded, physically stabilize that cofactor

Photophysical properties of fluorescent proteins

within a rigid framework, so that light is absorbed and re-emitted rather than being transformed into heat.

The wide distribution of these proteins, combined with the wide range of emission and absorption wavelengths found in Nature suggests that they evolved to provide numerous different and useful functions. In particular the generation of different colors requires different configurations within the chromophore cavity and/or changes in the chemistry of chromophore formation, which seems at odds with the requirements for efficient protein folding, catalysis and subsequent stability of the fold.

Prior to the discovery of GFP, one might have thought it impossible for a single protein to fulfill all of these requirements. Yet we have learned from structure/function studies that fluorescent proteins are remarkably amenable to genetic manipulation and that rational modifications are indeed possible. In fact, as demonstrated by the remarkable variety of biosensors that have been conceived and implemented, entirely new and very useful functions can be generated. It is not often easy to accomplish the goal that one has envisioned, but persistence, coupled with directed evolution ("let Nature show the way"), has usually resulted in success. Fluorescent protein research has transformed many areas in biological research, and I don't expect that transformative process to come to an end any time soon.

REFERENCES

Agmon, N. (2005). Proton pathways in green fluorescence protein. *Biophys. J.* **88**: 2452–2461.

Ai, H. W., J. N. Henderson et al. (2006). Directed evolution of a monomeric, bright, and photostable version of *Clavularia* cyan fluorescent protein: Structural characterization and applications in fluorescence imaging. *Biochem. J.* **400**: 531–540.

Akerboom, J., J. D. Rivera et al. (2009). Crystal structures of the GCaMP calcium sensor reveal the mechanism of fluorescence signal change and aid rational design. *J. Biol. Chem.* **284**: 6455–6464.

Alieva, N. O., K. A. Konzen et al. (2008). Diversity and evolution of coral fluorescent proteins. *PLoS One* **3**: e2680.

Ando, R., H. Mizuno et al. (2004). Regulated fast nucleoplasmic shuttling observed by reversible protein highlighting. *Science* **306**: 1370–1373.

Andresen, M., A. C. Stiel et al. (2007). Structural basis for reversible photoswitching in Dronpa. *Proc. Natl Acad. Sci. USA* **104**: 13005–13009.

Baird, G. S., D. A. Zacharias et al. (1999). Circular permutation and receptor insertion within green fluorescent proteins. *Proc. Natl Acad. Sci. USA* **96**: 11241–11246.

Baird, G. S., D. A. Zacharias et al. (2000). Biochemistry, mutagenesis and oligomerization of DsRed, a red fluorescent protein from coral. *Proc. Natl Acad. Sci. USA* **97**: 11984–11989.

Barondeau, D. P., C. D. Putnam et al. (2003). Mechanism and energetics of green fluorescent protein chromophore synthesis revealed by trapped intermediate structures. *Proc. Natl Acad. Sci. USA* **100**: 12111–12116.

Barondeau, D. P., J. A. Tainer et al. (2006). Structural evidence for an enolate intermediate in GFP fluorophore biosynthesis. *J. Am. Chem. Soc.* **128**: 3166–3168.

Bent, D. V. and E. Hayon (1975). Excited state chemistry of aromatic amino acids and related peptides. I. Tyrosine. *J. Am. Chem. Soc.* **97**: 2599–2606.

Betzig, E., G. H. Patterson et al. (2006). Imaging intracellular fluorescent proteins at nanometer resolution. *Science* **313**: 1642–1645.

Bokman, S. H. and W. W. Ward (1981). Renaturation of *Aequorea* green-fluorescent protein. *Biochem. Biophys. Res. Commun.* **101**: 1372–1380.

Brejc, K., T. K. Sixma, P. A. Kitts, S. R. Kain, R. Y. Tsien, M. Ormo, and S. J. Remington (1997). Structural basis for dual excitation and photoisomerization of the *Aequorea victoria* green fluorescent protein. *Proc. Natl Acad. Sci. USA* **94**: 2306–2311.

Bulina, M. E., D. M. Chudakov et al. (2006). A genetically encoded photosensitizer. *Nat. Biotechnol.* **24**: 95–99.

Campbell, R. E., O. Tour et al. (2002). A monomeric red fluorescent protein. *Proc. Natl Acad. Sci. USA* **99**: 7877–7882.

Cannon, M. B. and S. J. Remington (2007). Redox-sensitive green fluorescent protein: Probes for dynamic intracellular redox responses. In J. T. Hancock, ed., *Methods in Molecular Biology: Redox Mediated Signal Transduction*, vol. 476. Humana Press Inc., New York, pp. 51–65.

Chattoraj, M., B. A. King et al. (1996). Ultra-fast excited state dynamics in green fluorescent protein: Multiple states and proton transfer. *Proc. Natl Acad. Sci. USA* **93**: 8362–8367.

Chudakov, D. M., V. V. Belousov et al. (2003a). Kindling fluorescent proteins for precise in vivo photolabeling. *Nat. Biotechnol.* **21**: 191–194.

Chudakov, D. M., A. V. Foefanov et al. (2003b). Chromophore environment provides clue to kindling fluorescent protein riddle. *J. Biol. Chem.* **278**: 7215–7219.

Chudakov, D. M., V. V. Verkhusha et al. (2004). Photoswitchable cyan fluorescent protein for protein tracking. *Nat. Biotechnol.* **22**: 1435–1439.

Cody, C. W., D. C. Prasher, W. M. Westler, F. G. Prendergast, and W. W. Ward (1993). Chemical structure of the hexapeptide chromophore of the *Aequorea* green-fluorescent protein. *Biochemistry* **32**: 1212–1218.

Connolly, M. L. (1993). The molecular surface package. *J. Mol. Graph.* **11**: 139–141.

Craggs, T. D. (2009). Green fluorescent protein: Structure, folding and chromophore maturation. *Chem. Soc. Rev.* **38**: 2865–2875.

Dickson, R. M., A. B. Cubitt et al. (1997). On/off blinking and switching behaviour of single molecules of green fluorescent protein. *Nature* **388**: 356–358.

Dixit, R. and R. Cyr (2003). Cell damage and reactive oxygen species production induced by fluorescence microscopy: Effect on mitosis and guidelines for non-invasive fluorescence microscopy. *Plant J.* **36**: 280–290.

Dove, S., M. Takabayashi et al. (1995). Isolation and characterization of the pink and blue pigments of pocilloporid and acroporid corals. *Biol. Bull.* **189**(3): 288–297.

Ghosh, I., A. D. Hamilton et al. (2000). Antiparallel leucine zipper-directed protein reassembly: Application to the green fluorescent protein. *J. Am. Chem. Soc.* **122**: 5658–5659.

Gross, L. A., G. S. Baird et al. (2000). The structure of the chromophore within DsRed, a red fluorescent protein from coral. *Proc. Natl Acad. Sci. USA* **87**: 11990–11995.

Grynkiewicz, G., P. Martin et al. (1985). A new generation of Ca2+ indicators with greatly improved fluorescence properties. *J. Biol. Chem.* **260**: 3440–3450.

Hanson, G. T., R. Aggeler et al. (2004). Investigating mitochondrial redox potential with redox-sensitive green fluorescent protein indicators. *J. Biol. Chem.* **279**: 13044–13053.

Hanson, G. T., T. B. McAnaney et al. (2002). Green fluorescent protein variants as ratiometric dual emission pH sensors. 1. Structural characterization and preliminary application. *Biochemistry* **41**: 15477–15488.

Hasan, M. T., R. W. Friedrich et al. (2004). Functional fluorescent Ca2+ indicator proteins in transgenic mice under TET control. *PLoS Biol.* **2**: e163.

Hasegawa, J., T. Ise et al. (2010). Excited states of fluorescent proteins, mKO and DsRed: Chromophore-protein electrostatic interaction behind the color variations. *J. Phys. Chem. B* **114**: 2971–2979.

He, X., A. F. Bell et al. (2002). Synthesis and spectroscopic studies of model red fluorescent protein chromophores. *Org. Lett.* **9**: 1523–1526.

Heim, R., D. C. Prasher et al. (1994). Wavelength mutations and posttranslational autoxidation of green fluorescent protein. *Proc. Natl Acad. Sci. USA* **91**: 12501–12504.

Henderson, J. N., H. W. Ai et al. (2007). Structural basis for reversible photobleaching in a green fluorescent protein homologue. *Proc. Natl Acad. Sci. USA* **104**: 6672–6677.

Henderson, J. N., R. Gepshtein et al. (2009). Structure and mechanism of the photoactivatable green fluorescent protein. *J. Am. Chem. Soc.* **131**: 4176–4177.

Henderson, J. N., M. F. Osborne et al. (2009). Excited state proton transfer in the red fluorescent protein mKeima. *J. Am. Chem. Soc.* **131**: 13212–13213.

Henderson, J. N. and S. J. Remington (2005). Crystal structures and mutational analysis of amFP486, a cyan fluorescent protein from *Anemonia majano*. *Proc. Natl Acad. Sci. USA* **102**: 12712–12717.

Henderson, J. N. and S. J. Remington (2006). The kindling fluorescent protein: A transient photoswitchable marker. *Physiology* **21**: 162–170.

Hofmann, M., C. Eggeling et al. (2005). Breaking the diffraction barrier in fluorescence microscopy at low light intensities by using reversibly photoswitchable proteins. *Proc. Natl Acad. Sci. USA* **102**: 17565–17569.

Hopf, M., W. Gohring et al. (2001). Crystal structure and mutational analysis of a perlecan-binding fragment of nidogen-1. *Nat. Struct. Biol.* **8**: 573–574.

Kennis, J. T., D. S. Larsen et al. (2004). Uncovering the hidden ground state of green fluorescent protein. *Proc. Natl Acad. Sci. USA* **101**: 17988–17993.

Kent, K. P., W. Childs et al. (2008). Deconstructing green fluorescent protein. *J. Am. Chem. Soc.* **130**: 9664–9665.

Kikuchi, A., E. Fukumura et al. (2008). Structural characterization of a thiazoline-containing chromophore in an orange fluorescent protein, monomeric Kusabira Orange. *Biochemistry* **47**: 11573–11580.

Kogure, T., S. Karasawa et al. (2006). A fluorescent variant of a protein from the stony coral Montipora facilitates dual-color single-laser fluorescence cross-correlation spectroscopy. *Nat. Biotechnol.* **24**: 577–581.

Kummer, A. D., C. Kompa et al. (1998). Dramatic reduction in fluorescence quantum yield in mutants of green fluorescent protein due to fast internal conversion. *Chem. Phys.* **237**: 183–193.

Lohman, J. R. and S. J. Remington (2008). Development of a family of redox-sensitive green fluorescent protein indicators for use in relatively oxidizing subcellular environments. *Biochemistry* **47**: 8678–8688.

Lossau, H., A. Kummer et al. (1996). Time-resolved spectroscopy of wild-type and mutant green fluorescent proteins reveals excited state deprotonation consistent with fluorophore-protein interactions. *Chem. Phys.* **213**: 1–16.

Lukyanov, K. A., A. F. Fradkov et al. (2000). Natural animal coloration can be determined by a nonfluorescent green fluorescent protein homolog. *J. Biol. Chem.* **275**: 25879–25882.

Malo, G. D., M. Wang et al. (2008). Crystal structure and Raman studies of dsFP483, a cyan fluorescent protein from *Discosoma striata*. *J. Mol. Biol.* **378**: 871–886.

Matz, M. V., F. F. Arkady et al. (1999). Fluorescent proteins from nonbioluminescent *Anthozoa* species. *Nat. Biotechnol.* **17**: 969–973.

McAnaney, T. B., E. S. Park et al. (2002). Green fluorescent protein variants as ratiometric dual emission pH sensors. 2. Excited-state dynamics. *Biochemistry* **41**: 15489–15495.

McAnaney, T. B., X. Shi et al. (2005). Green fluorescent protein variants as ratiometric dual emission pH sensors: 3. Temperature dependence of proton transfer. *Biochemistry* **44**: 8701–8711.

McCapra, F., Z. Razavi et al. (1988). The fluorescence of the chromophore of the green fluorescent protein of Aequorea and Renilla. *J. Chem. Soc. Chem. Commun.* **12**: 790–791.

Megley, C. M., L. A. Dickson et al. (2009). Photophysics and dihedral freedom of the chromophore in yellow, blue, and green fluorescent protein. *J. Phys. Chem. B* **113**: 302–308.

Meyers, A. and T. P. Dick (2010). Genetically encoded redox probes. *Antioxid. Redox Sign.* **13**: 1–30.

Miesenbock, G., D. A. De Angelis et al. (1998). Visualizing secretion and synaptic transmission with pH-sensitive green fluorescent proteins. *Nature* **394**: 192–195.

Nakai, J., M. Ohkura et al. (2001). A high signal-to-noise Ca(2+) probe composed of a single green fluorescent protein. *Nat. Biotechnol.* **19**: 137–141.

Nienhaus, G. U., K. Nienhaus et al. (2006). Photoconvertible fluorescent protein EosFP: Biophysical properties and cell biology applications. *Photochem. Photobiol.* **82**: 351–358.

Nienhaus, K., H. Nar et al. (2008). *Trans-cis* isomerization is responsible for the red-shifted fluorescence in variants of the red fluorescent protein eqFP611. *J. Am. Chem. Soc.* **130**: 12578–12579.

Nienhaus, K., F. Renzi et al. (2006). Chromophore-protein interactions in the anthozoan green fluorescent protein asFP499. *Biophys. J.* **91**: 4210–4220.

Niwa, G., S. Inouye et al. (1996). Chemical nature of the light emitter of the *Aequorea* green fluorescent protein. *Biochemistry* **93**: 13617–13622.

Ormo, M., A. B. Cubitt et al. (1996). Crystal structure of the *Aequorea victoria* green fluorescent protein. *Science* **273**: 1392–1395.

Ostergaard, H., A. Henriksen et al. (2001). Shedding light on disulfide bond formation: Engineering a redox switch in green fluorescent protein. *EMBO J.* **20**: 5853–5862.

Patterson, G. H. and J. Lippincott-Schwartz (2002). A photoactivatable GFP for selective photolabeling of proteins and cells. *Science* **297**: 1873–1877.

Petersen, J., P. G. Wilmann et al. (2003). The 2.0A crystal structure of eqFP611, a far-red fluorescent protein from the sea anemone *Entacmaea quadricolor*. *J. Biol. Chem.* **278**: 44626–44631.

Piatkevich, K. D. and V. V. Verkhusha (2010). Advances in engineering of fluorescent proteins and photoactivatable proteins with red emission. *Curr. Opin. Chem. Biol.* **14**: 23–29.

Prescott, M., M. Ling et al. (2003). The 2.2 A crystal structure of a pocilloporin pigment reveals a nonplanar chromophore conformation. *Structure* **11**: 275–284.

Quillin, M. L., D. M. Anstrom et al. (2005). The kindling fluorescent protein from *Anemonia sulcata*: Dark state structure at 1.38 angstroms resolution. *Biochemistry* **44**: 5774–5787.

Reid, B. G. and G. C. Flynn (1997). Chromophore formation in green fluorescent protein. *Biochemistry* **36**: 6786–6791.

Remington, S. J. (2006). Fluorescent proteins: Maturation, photochemistry and photophysics. *Curr. Opin. Struct. Biol.* **16**: 714–721.

Rosenow, M. A., H. N. Patel et al. (2005). Oxidative chemistry in the GFP active site leads to covalent cross-linking of a modified leucine side chain with a histidine imidazole: Implications for the mechanism of chromophore formation. *Biochemistry* **44**: 8303–8311.

Shaner, N. C., R. E. Campbell et al. (2004). Improved monomeric red, orange and yellow fluorescent proteins derived from *Discosoma* sp. red fluorescent protein. *Nat. Biotechnol.* **22**: 1567–1572.

Shu, X., K. Kallio et al. (2007). Ultrafast excited-state dynamics in the green fluorescent protein variant S65T/H148D 1. Mutagenesis and structural studies. *Biochemistry* **46**: 12005.

Shu, X., P. Leiderman et al. (2007). An alternative excited-state proton transfer pathway in green fluorescent protein variant S205V. *Protein Sci.* **16**: 2703–2710.

Shu, X., N. C. Shaner et al. (2006). Structural basis for spectral variations in mFruits: Monomeric orange and red fluorescent proteins. *Biochemistry* **45**: 9639–9647.

Sniegowski, J. A., J. W. Lappe et al. (2005). Base catalysis of chromophore formation in Arg96 and Glu222 variants of green fluorescent protein. *J. Biol. Chem.* **280**: 26248–24255.

Sniegowski, J. A., M. E. Phail et al. (2005). Maturation efficiency, trypsin sensitivity, and optical properties of Arg96, Glu222, and Gly67 variants of green fluorescent protein. *Biochem. Biophys. Res. Commun.* **332**: 657–663.

Stiel, A. C., S. Trowitzsch, G. Weber, M. Andresen, C. Eggeling, S. W. Hell, S. Jakobs, and M. C. Wahl. (2007). 1.8 A bright-state structure of a reversibly photoswitchable fluorescent protein Dronpa guides the generation of fast switching mutants. *Biochem. J.* **402**: 35–42.

Stoner-Ma, D., A. A. Jaye et al. (2005). Observation of excited-state proton transfer in green fluorescent protein using ultrafast vibrational spectroscopy. *J. Am. Chem. Soc.* **127**: 2864–2865.

Stoner-Ma, D., A. A. Jaye et al. (2008). An alternate proton acceptor for excited-state proton transfer in green fluorescent protein: Rewiring GFP. *J. Am. Chem. Soc.* **130**: 1227–1235.

Strack, R. L., D. E. Strongin et al. (2010). Chromophore formation in DsRed occurs by a branched pathway. *J. Am. Chem. Soc.* **132**: 8496–8505.

Subach, O. M., V. N. Malashkevich et al. (2010). Structural characterization of acylimine-containing blue and red chromophores in mTagBFP and TagRFP fluorescent proteins. *Chem. Biol.* **17**: 333–341.

Tallini, Y. N., M. Ohkura et al. (2006). Imaging cellular signals in the heart in vivo: Cardiac expression of the high-signal Ca2+ indicator GCaMP2. *Proc. Natl Acad. Sci. USA* **103**: 4753–4758.

Tolbert, L. M. and K. M. Solntsev (2002). Excited-state proton transfer: From constrained systems to super photoacids to superfast proton transfer. *Acc. Chem. Res.* **35**: 19–27.

Topell, S., J. Hennecke et al. (1999). Circularly permuted variants of the green fluorescent protein. *FEBS Lett.* **457**: 283–289.

Tsien, R. Y. (1998). The green fluorescent protein. *Annu. Rev. Biochem.* **67**: 509–544.

Tsutsui, H., S. Karasawa et al. (2005). Semi-rational engineering of a coral fluorescent protein into an efficient highlight. *EMBO Rep.* **6**: 233–238.

van Thor, J. J., K. L. Ronayne et al. (2008). Balance between ultrafast parallel reactions in the green fluorescent protein has a structural origin. *Biophys. J.* **95**: 1902–1912.

Verkhusha, V. V., D. M. Chudakov et al. (2004). Common pathway for the red chromophore formation in fluorescent proteins and chromoproteins. *Chem. Biol.* **11**: 845–854.

Violot, S., P. Carpentier et al. (2009). Reverse pH-dependence of chromophore protonation explains the large stokes shift of the red fluorescent protein mKeima. *J. Am. Chem. Soc.* **131**: 10356–10357.

Wachter, R. M., J. L. Watkins et al. (2010). Mechanistic diversity of red fluorescence acquisition by GFP-like proteins. *Biochemistry* **49**: 7417–7427.

Wall, M. A., M. Socolich et al. (2000). The structural basis for red fluorescence in the tetrameric GFP homolog DsRed. *Nat. Struct. Biol.* **7**: 1133–1138.

Wang, L., W. C. Jackson et al. (2004). Evolution of new nonantibody proteins via iterative somatic hypermutation. *Proc. Natl Acad. Sci. USA* **101**: 16745–16749.

Ward, W. W. and S. H. Bokman (1982). Reversible denaturation of *Aequorea* green-fluorescent protein: Physical separation and characterization of the renatured protein. *Biochemistry* **21**: 4535–4540.

Ward, W. W., C. W. Cody, R. C. Hart, and M. J. Cormier (1980). Spectrophotometric identity of the energy transfer chromophores in *Renilla* and *Aequorea* green proteins. *Photochem. Photobiol.* **31**: 611–615.

Ward, W. W., H. J. Prentice et al. (1982). Spectral perturbations of the *Aequorea* green-fluorescent protein. *Photochem. Photobiol.* **35**: 803–808.

Wilmann, P. G., J. Petersen et al. (2005). A polypeptide fragmentation within the chromophore revealed in the 2.1 A crystal structure of a nonfluorescent chromoprotein from *Anemonia sulcata*. *J. Biol. Chem.* **280**: 2401–2404.

Wilmann, P. G., K. Turcic et al. (2006). The 1.7 A crystal structure of Dronpa: A photoswitchable green fluorescent protein. *J. Mol. Biol.* **364**: 213–224.

Wood, T. I., D. P. Barondeau et al. (2005). Defining the role of arginine 96 in green fluorescent protein fluorophore biosynthesis. *Biochemistry* **44**: 16211–16220.

Yampolsky, I. V., S. J. Remington et al. (2005). Synthesis and properties of the chromophore of asFP595 chromoprotein from *Anemonia sulcata*. *Biochemistry* **44**: 5788–5793.

Yang, F., L. G. Moss et al. (1996). The molecular structure of green fluorescent protein. *Nat. Biotechnol.* **14**: 1246–1251.

Yarbrough, D., R. M. Wachter et al. (2001). Refined crystal structure of DsRed, a red fluorescent protein from coral, at 2.0 A resolution. *Proc. Natl Acad. Sci. USA* **98**: 462–467.

Zimmer, M. (2002). Green fluorescent protein (GFP): Applications, structure and related photophysical behavior. *Chem. Rev.* **102**: 759–781.

Optimization of fluorescent proteins

Mark A. Rizzo
University of Maryland School of Medicine

Contents

3.1 INTRODUCTION

The great utility of genetically encoded fluorescent proteins (FPs) can be directly traced to their ease of use and their ability to specifically label cellular targets using standard molecular biology cloning techniques. New investigation techniques can now explore cellular dynamics in ways that have been, to a large degree, inaccessible by other means. Although the potential of this approach often seems limitless, in practice, the information that can be gathered from any fluorescence experiment is constrained, first, by the instrumentation used for detection and, second, by the fluorescence properties of the probes.

Similar to any new technology, the first FPs were noteworthy more for their potential than their utility. Even the original wild-type *Aequorea* green FP (wtGFP) (Chalfie al., 1994) has seldom been used because of its inherent limitations: the natural sequence does not express well in mammalian cells, it is optimally excited with ultraviolet light, it exhibits unusual photoswitching characteristics, and it is not very bright. Three generations of development were required to produce the highly successful enhanced green fluorescent protein (EGFP) variant. Not only did these rounds of optimization improve brightness, they also improved expression in mammalian cells (Yang et al., 1996), folding at 37°C (Cormack et al., 1996), and, importantly, enhanced the red-shifted component of its excitation spectra to create a probe that was fluorescein-like in its spectral characteristics (Heim and Tsien, 1996). The result was an FP that could be imaged using the green FITC filter set found on nearly every modern fluorescence microscope. This factor facilitated widespread adoption of EGFP as a tool for fluorescence imaging. Optimization of photophysical and biological properties is thus valuable for improving the overall *usefulness* of FPs in the context of specific applications in addition to improving their intrinsic fluorescence properties. This chapter will highlight some of the ways FPs can be improved and the strategies used to create optimized variants.

3.2 PROBE FLUORESCENCE

3.2.1 WHAT MAKES A PROBE USEFUL?

The central premise of FP development is to make a better reagent. But what makes one FP preferable over another? Certainly, there are many ways to choose an FP for a given application (Snapp, 2009), and other chapters in this volume provide guidance for advanced applications, such as superresolution imaging (see Chapter 12) or whole animal imaging (see Chapter 13). In a simple sense, obtaining high-quality data is a goal for any experiment. With fluorescence, a high-quality data set is one for which the fluorescence signal far exceeds the detection noise. Thus, better dyes or FPs generate more photons. Indeed, all fluorescence experiments are fundamentally limited in what they can accomplish by the number of photons that can be collected. More photons yield images with higher signal-to-noise, enable faster data collection, and provide more accurate quantification. Examination of FP brightness is a good place to start.

3.2.2 MEASURING MOLECULAR BRIGHTNESS

Fluorescence involves absorption of a photon of light and subsequent re-emission of a lower energy photon. Absorbance (A) can be quantified experimentally, and a useful constant for comparative purposes can be calculated using Beer's law:

$$\varepsilon = \frac{A}{cl} \tag{3.1}$$

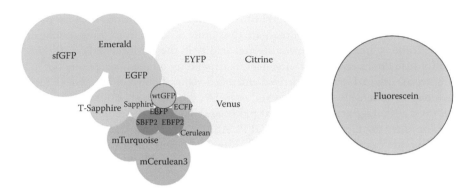

Figure 3.1 Molecular brightness of common FPs. The molecular brightness of popular FPs derived from *Aequorea* green fluorescent protein (GFP) is represented schematically by the relative circle size (left). Although optimization of FPs has improved molecular brightness, even the brightest available FPs are substantially dimmer than commonly used organic fluorophores such as fluorescein.

The extinction coefficient (ε) describes how much light a fluorophore solution of concentration c can absorb over the length (l) of a suitably sized cuvette. Absorption relocates electrons within the chromophore to a higher energy excited state for a brief amount of time, typically on the order of nanoseconds. Relaxation to the more stable ground state can either occur nonradiatively, via vibrational relaxation, or through release of a photon. The fraction of the total number of absorption events that actually produce an emission photon is termed the quantum yield (QY). Molecular brightness can then be simply expressed as the product of the extinction coefficient and the quantum yield. This expression serves as a fairly convenient parameter for comparing the brightness of different FPs. For example, the second-generation EBFP2 has an improved molecular brightness number of 18 compared to 4.5 for the previous generation EBFP (Ai et al., 2007).

Molecular brightness among FPs can vary substantially even among derivatives of the same parent. Figure 3.1 displays the relative molecular brightness of commonly used variants derived from *Aequorea* wtGFP. Green and yellow FPs tend to be much brighter than blue and cyan, although the most recent generations of cyan FPs (Goedhart et al., 2010; Markwardt et al., 2011) are comparable to EGFP. Nonetheless, even the best yellow FPs (Griesbeck et al., 2001; Nagai et al., 2002) have not yet achieved the brightness of fluorescein (Shaner et al., 2005), indicating that there is still considerable room for further improvement among the current set of FPs to bring them closer to the brightness of commonly used organic fluorophores.

3.2.2.1 Limitations of molecular brightness measurements

The molecular brightness of a new FP can be impressive (e.g., superfolder or sfGFP has a molecular brightness of 54 [Pédelacq et al., 2005] compared to 34 for EGFP [Patterson et al., 1997, Figure 3.1]). Nonetheless, this simple expression is limited in its ability to predict experimental performance, especially expression and maturation in mammalian cells. One source of error arises from the experimental conditions used for quantifying FP extinction coefficient and quantum yields. Typically, molecular brightness parameters are calculated using purified proteins at room temperature in

Photophysical properties of fluorescent proteins

a buffer of alkaline pH (commonly pH 8). These standard conditions used for the determination of molecular brightness are thus quite different from those in many live cell applications. Increased temperature and acidity in the experimental conditions can diminish fluorescence to varying degrees depending on the fluorophore. For example, the original *Aequorea*-derived enhanced yellow fluorescent protein (EYFP) is much more acid sensitive than green and cyan FPs (Shaner et al., 2005), and thus, its brightness advantage at an intracellular pH is less than what the molecular brightness numbers imply (51 for EYFP, 34 for EGFP). Another mitigating factor is that extinction coefficients are calculated using the "total" protein concentration. Since unfolded protein does not form a mature chromophore, measured extinction coefficients tend to favor proteins that fold well in bacterial preparations. The measured extinction coefficient for sfGFP (Pédelacq et al., 2005) is about 50% greater than EGFP (Patterson et al., 1997) despite the fact that no changes were made to the chromophore structure. Although improved folding can enhance the effective brightness of FPs expressed in cells as well as in vitro (Kremers et al., 2006), the reported brightness of EGFP (Hillesheim et al., 2006) and sfGFP (Cotlet et al., 2006) at the single-molecule level are similar. Furthermore, the dependence of the extinction coefficient on protein folding is a significant source of error. As such, calculated extinction coefficients can vary greatly between laboratories. Values for enhanced cyan fluorescent protein (ECFP), for example, vary by as much as 25% (Rizzo et al., 2004).

Molecular brightness calculations also have a limited ability to predict the relative effectiveness of FP variants across different color categories. To some degree, it is obvious that substituting a brighter FP of a different color does not guarantee better results in the absence of suitably optimized detection filters. What is less obvious, however, is that each optical element in a measurement device, from the excitation source to the detector, has its own spectral profile, such that the playing field between colors is uneven. Confocal imaging with GFPs is inherently more successful than with cyan and blue FPs because the optics and excitation sources are better matched for GFPs than their blue-shifted cousins. The mismatch between available detection options and FP spectra is perhaps most evident at the fringes of the optical range (particularly at wavelengths lower than 450 nm or higher than 600 nm), as the optics and detection devices used for fluorescence are commonly optimized for green light. Thus, bright GFPs will generally outperform bright blue or red FPs, even if they have equivalent molecular brightness values. Molecular brightness comparisons are thus most useful for identifying better FPs within a particular color category.

3.2.3 PHOTOSTABILITY

A good fluorophore not only is bright but also retains that brightness over the duration of the measurement. Once excited, a number of events can occur to the high-energy state electrons of a fluorophore besides emission of a photon, including photochemical reactions that permanently alter the molecular composition of the dye. If this light-induced reaction produces a structure that no longer fluoresces, it is said to be photobleached. Thus, FPs that are resistant to photobleaching are valuable, particularly for time-resolved measurements. As such, photostability is a parameter that can be selectively optimized. Screening strategies focused on photostability have proven successful for developing better red and orange FP (Shaner et al., 2008). Furthermore, a targeted strategy has also been successful for improving the photostability of cyan FPs

(Markwardt et al., 2011). Improving the photostability of FPs is a key ingredient for producing the best performers in any particular color class.

3.2.4 FLUORESCENCE LIFETIME

The properties of FP fluorescence lifetimes have become increasingly important, owing to advances in instrumentation for measuring fluorescence lifetimes in living cells and the utility of such measures for FRET analysis (see Chapter 11). Probes with longer lifetimes are desirable in general, because longer lifetimes can be more accurately measured. Further, probes with a simple, single-component lifetime decay curve provide data that is simpler to analyze and interpret than FPs that display more complex fluorescence lifetimes (Rizzo et al., 2004). One way to generate FPs with longer fluorescence lifetimes is to optimize the quantum yield since quantum yield (Φ) and fluorescence lifetime (τ) are related:

$$\Phi = \Phi_* k_e^0 \tau \tag{3.2}$$

where

Φ_* represents efficiency of formation for the excited state

k_e^0 is the rate constant for fluorescence emission (Turro, 1991)

Alternatively, fluorescence lifetime microscopy can be incorporated into the developmental and screening strategy (Goedhart et al., 2010), although both approaches achieve similar results (Markwardt et al., 2011).

3.3 PROTEINS AS PROBES

While the standard measurement methodology of probe fluorescence has been generally useful for generating optimized FPs, these measurements have been far less useful for predicting performance in actual experiments. Furthermore, the experimental performance of an individual FP can vary greatly between applications. One of the more mystifying examples is the red-emitting mCherry FP. In vivo, its fluorescence can vary by 30-fold depending on the tissue it is expressed in (Chen et al., 2011). Why do we see such variation in performance? To answer that, we need to look a little more closely into what exactly an FP is and, perhaps just as importantly, what an FP is not.

3.3.1 GFP VERSUS FLUORESCEIN

Fluorescein is a green fluorescent dye commonly used in fixed cell staining that is readily visualized in fluorescence microscopy using a standard FITC filter set. Under a microscope, cells labeled with a fluorescein-conjugated antibody appear very similar to cells expressing an EGFP-fusion protein. Given that both dyes have a bright green fluorescence that can be captured under identical imaging conditions, it is very easy to compare the two. Nonetheless, it is important to consider that these two dyes are fundamentally different and the difference between the two is helpful in understanding ways to improve and optimize FPs.

Fluorescein is a synthetic dye of known composition and structure that varies little regardless of the application, be it labeling an antibody or coloring the Chicago River green on St. Patrick's Day. The advantage of a known quantity is that its behavior is generally both consistent and predictable from one application to another. FPs, on the other hand, are genetically expressed by the cell being under experimentation. Further, they are often fused to a second, targeting protein, requiring that the FP fold and mature while attached to another protein that must also then properly fold and mature. This introduces a considerable level of complications and uncertainty.

3.3.2 IMPORTANCE OF PROTEIN STABILITY

Denatured or unfolded FPs are completely nonfluorescent, even with a properly formed chromophore (Ward and Bokman, 1982), such that the ability of a particular FP to fold properly is critical to its performance in a given application. This relationship between folding and performance is perhaps most evident with the development of FPs from new species. Unfortunately, many of the FPs derived from other organisms have proven to be less robust than their *Aequorea*-derived counterparts, despite fluorescence properties that can be superior from a molecular brightness perspective. Figure 3.2 shows a simple example of such a problem. In this experiment, mCerulean cyan FP was fused in tandem to a yellow FP derived from either *Aequorea* (mVenus) or *Phialidium* (phiYFPm). In the absence of the fusion, purified FPs for all three variants produced bright fluorescence. In tandem, however, both mCerulean and mVenus retained bright fluorescence (Figure 3.2a). Fluorescence of phiYFPm, on the other hand, was barely detectable when expressed as a tandem fusion with mCerulean (Figure 3.2b), despite having similar fluorescence properties as mVenus. Although the precise mechanism underlying the disappearance of phiYFPm fluorescence is not fully understood, be it unfolded protein, an improperly formed chromophore, or direct inhibition of fluorescence, the fusion of an additional sequence to phiYFPm profoundly altered its fluorescence. Thus, limitations in protein folding can directly affect the usefulness of an FP.

3.3.3 RELATIONSHIP BETWEEN FP STRUCTURE AND FLUORESCENCE

The fact that denatured FPs are no longer fluorescent suggests the importance of the tertiary structure in FP fluorescence. In fact, protein structure is an active participant in the fluorescence process and gives rise to some surprising characteristics that makes the fluorescence properties of FPs distinct from their organic counterparts. A preliminary examination of the relationship between FP structure and fluorescence is shown in Figure 3.3. The sequences of 32 FPs derived from the *Aequorea* wtGFP parent were aligned, and the extent of amino acid conservation over the course of development is represented by sequence logo and color coding (Figure 3.3a). Upon inspection of the primary sequence, mutations that have been useful for optimizing FPs are primarily restricted to just a few locations. Not surprisingly, the triplet of amino acids comprising the chromophore (residues 64–66) has received extensive attention. Furthermore, substitutions in the β-barrel amino acids are generally confined to ~6 or so stretches of amino acids within the sequence. These regions are confined to 1/3 of the β-strands in the tertiary structure that line the cavity surrounding the aromatic moiety of residue 66 (either Y, W, or H) (Figure 3.3b), suggesting that the local chromophore environment greatly influences FP fluorescence.

To understand the extent that amino acid moieties in the β-barrel can modulate fluorescence, it is important to consider some fundamental differences between FPs and their organic counterparts. The chromophore of FPs is generally derived from only

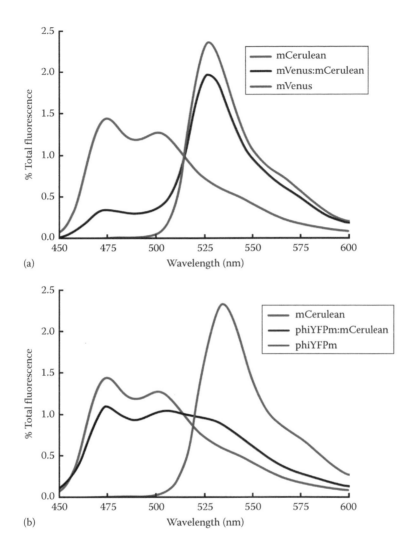

Figure 3.2 FP fluorescence in a fusion partner. The fluorescence emission spectra of a cyan FP (blue line), a yellow FP (red line), and a fusion protein containing the cyan–yellow pair in tandem (black line) are shown for mCerulean and mVenus (a) and mCerulean and phiYFPm (b). Discernible spectral components from the individual FPs are present in the mVenus–mCerulean protein, whereas the yellow FP component is largely missing from the phiYFPm–mCerulean fusion, despite fluorescence properties that are very similar to mVenus (molecular brightness of phiYFPm = 48, mVenus = 50). The ability to maintain its fluorescence properties when expressed in a fusion protein varies greatly among the available FPs.

3 of the ~240 amino acids, and generally consists of two conjugated ring structures: the first supplied by the aromatic moiety of an incorporated amino acid and the second arising from heterocyclization of the neighboring amino acids on either side of the aromatic amino acid. Figure 3.4 shows a top view of the EGFP chromophore compared to that of fluorescein, with the chromophore portions represented using space-filling molecular models. Of note, the fluorescein ring structure is compact

Photophysical properties of fluorescent proteins

Photophysical properties of fluorescent proteins

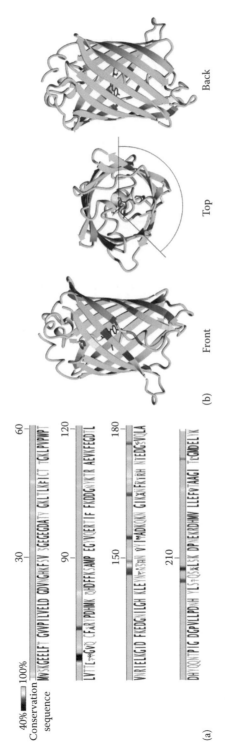

(a)

40% ▆▆▆ 100%

Conservation
sequence

MVSKGEELFT GVVPILVELD GDVNGHKFSV SGEGEGDATY GKLTLKFICT TGKLPVPWPT

LVTTL··GVQ CFARYPDHMK QHDFFKSAMP EGYVQERTIF FKDDGNYKTR AEVKFEGDTL

VNRIELKGID FKEDGNLIGH KLEYNYNSHN VYIMADKQKN GIKVNFKIRH NIEDGSVQLA

DHYQQNTPIG DGPVLLPDNH YLSYQSALSK DPNEKRDHMV LLEFVTAAGI TGMDELYK

(b)

Front   Top   Back

Figure 3.3 Location of changes made to Aequorea FPs during optimization. The sequences of 32 FPs derived from the Aequorea wtGFP were aligned. (a) Conservation of each residue is indicated by color, and the sequence logo indicates the relative consensus of a particular amino acid by size. (b) The structure of the green FP (1EMA.pdb) was colored using the conservation LUT shown in A and is shown from arbitrary view points. Sequence substitutions made to the Aequorea FP during optimization are primarily constrained to structure elements in the chromophore or in the barrel structure directly facing the aromatic moieties in the chromophore, as indicated by the annotation in the "top" view.

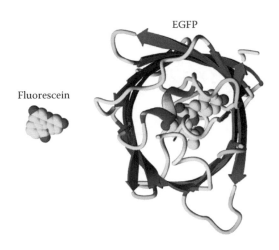

Figure 3.4 The structure of FPs compared to organic fluorophores. The structures of fluorescein and EGFP (2Y0G.pdb) are represented. Cartoon representations are shown for the nonfluorescent parts of GFP. Fluorescein consists of a highly conjugated multiple-ring structure that is planar and inflexible, resulting in highly efficient fluorescence. In FPs, the chromophore consists of two-ring structures conjugated by a flexible linker. The 3D structure of the surrounding barrel is required to position these ring structures in a conformation that produces fluorescence.

and rigid owing to a third conjugated ring linking the structures analogous to those in the EGFP chromophore. The rigid nature of the structure stabilizes the geometry of molecular orbitals that participate in fluorescence, producing bright, reliable fluorescence. For FP chromophores, the orbitals that participate in fluorescence also span conjugated ring structures, but unlike their organic counterparts, they lack the central, stabilizing ring. Thus, the linkage between rings is flexible and subject to influence by the surrounding protein tertiary structure. Figure 3.5 shows an alignment of the chromophores of related yellow FPs (EYFP, Citrine, and Venus) produced using the MUSTANG algorithm (Konagurthu et al. 2006). In general, the chromophores of FPs have slightly nonplanar geometries, owing to the heavy influence of surrounding barrel cavity (Ong et al., 2011). The Citrine yellow FP has a more planar geometry than its counterparts, and a corresponding higher quantum yield. Nonetheless, this advantage is fragile and can be negated by a single point mutation to position 206 in the β-barrel (Rizzo et al., 2006).

In theory, knowledge of the precise conformation of the ring structure can be used to optimize FP fluorescence by targeted mutagenesis. Nonetheless, analyses on chromophore conformations based on crystallography are limited by the fact that the structure exclusively reflects the ground state configuration. The excited state geometry can be quite different (Zimmer, 2002), and it is thought that the conformational dynamics that occur in that higher energy state heavily influence the quantum yield (Megley et al., 2009). As computational power and the understanding of various FP photophysical properties increase, calculation of these states should prove useful for structure-guided optimization of FPs with low quantum yields.

Photophysical properties of fluorescent proteins

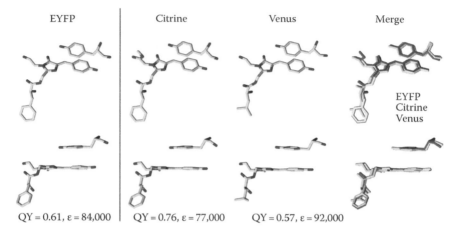

EYFP Citrine Venus Merge

EYFP
Citrine
Venus

QY = 0.61, ε = 84,000 | QY = 0.76, ε = 77,000 QY = 0.57, ε = 92,000

Figure 3.5 Chromophore conformations of related yellow FPs in x-ray structures. Chromophore and the π-stacking Tyr203 structures from EYFP (1YFP.pdb) (Wachter et al., 1998), Citrine (1HUY.pdb) (Griesbeck et al., 2001; Bevis and Glick, 2002), and Venus (1MYW.pdb-) (Rekas et al., 2002) were aligned using MUSTANG (Konagurthu et al., 2006). The differences in chromophore positioning are subtle, yet the differences in quantum yield (QY) are quite large, owing to the intricate relationship between the chromophores with the surrounding cavity.

3.3.4 PHOTOSTABILITY OF FPs

The fundamental differences in composition between organic dyes and FPs also give rise to differences in photostability and photobleaching behavior. The photostability of fluorophores such as fluorescein is generally dependent on the rate of entry into the triplet excited state. Unlike the primary fluorescence state that relaxes on the timescale of nanoseconds (Bailey and Rollefson, 1953), relaxation from a triplet state is slower by several orders of magnitude (Widengren et al., 1995) and can extend into the millisecond range. The long lifetime of the triplet state makes it prone to chemical reactions, for example, with oxygen species, which can result in bleaching (Kasche and Lindqvist, 1964). Furthermore, the long lifetime can act as a sink and lead to accumulation of fluorophores in the triplet state (Song et al., 1996), hence providing linearity to the photobleaching reaction as well (Patterson and Piston, 2000). Thus, for organic dyes like fluorescein, photobleaching can be controlled and corrected for in a straightforward manner during data collection and can also be suppressed by reagents that can quench triplet states or reduce oxygen radicals. The bleaching behavior of FPs, on the other hand, is not so straightforward.

3.3.4.1 Triplet-free bleaching

FP triplet states tend to have much shorter lifetimes, on the order of microseconds (Visser and Hink, 1999; Widengren et al., 1999; Heikal et al., 2001); thus, FP photobleaching is likely independent from a fluorescein-like triplet state mechanism. This is supported by the ineffectiveness of triplet state quenchers and suppression of molecular oxygen in enhancing FP photostability (Swaminathan et al., 1997). Although the β-barrel provides steric protection to such reagents, it is thought that the internal quenching from side chain residues may prevent triplet state accumulation (Widengren et al., 1999).

An alternative bleaching mechanism for fluorescein under two-photon illumination is instructive for considering the bleaching of FPs. Similar to the FPs, bleaching of fluorescein under two-photon illumination is resistant to both triplet-state quenchers and antioxidants. Importantly, the excitation power dependence exceeds three (Patterson and Piston, 2000), suggesting that the sequential interaction of the excited state with additional photons triggers the bleaching reaction. Bleaching of EGFP under two-photon excitation conditions also proceeds through a similar mechanism (Chen et al., 2002; Kalies et al., 2011).

Nonetheless, the bleaching mechanism under single photon conditions has proven more difficult to study. Experimental evidence does not support a classical triplet-state mechanism or reaction with oxygen free radicals. Furthermore, characterization of bleaching rates is notoriously difficult. Large discrepancies in reported photobleaching rates tend to be the norm, and the data are frequently contradictory. For example, the Cerulean cyan FP was initially described as being more photostable than the first generation ECFP by about 30% (Rizzo et al., 2004), whereas later studies have reported almost exactly the opposite, with Cerulean found to be ~40% less photostable than the ECFP (Shaner et al., 2005). Taken together with the insensitivity to triplet quenchers and oxygen, the high variance in the bleaching is reminiscent of a sequential mechanism, although the photobleaching of EGFP has been shown to display linear power dependence (Chen et al., 2002), leaving chemical reaction with the barrel side chains as the most likely mechanism. Unfortunately, the high level of variance among studies makes it difficult to draw definitive conclusions from individual reports, particularly in light of additional complications provided by the presence of a phenomenon known as reversible photoswitching.

3.3.4.2 Reversible photoswitching

The development of fluorescence recovery after photobleaching (FRAP) methods to study diffusion and transport of molecules in cells led to an interesting discovery: The speed of fluorescence recovery after bleaching sometimes exceeded what is theoretically possible by diffusion. This so-called "reversible photobleaching" in fluorescein conjugates involves repopulation of the ground state through oxygen-assisted triplet state relaxation (Periasamy et al., 1996). The kinetics of this process are indeed slow enough to be measurable by FRAP and yet faster than diffusive recovery to account for the high-speed FRAP component.

FPs also display a similar fluorescence behavior, known as reversible photoswitching (Figure 3.6). Upon illumination, FP fluorescence rapidly decreases, but can subsequently return following a brief dark interval. Almost all of the FPs examined can photoswitch under the right conditions (Shaner et al., 2008), but like photobleaching, the behavior can be highly variable.

One mechanism behind reversible photoswitching is likely related to the twisting of the chromophore to a nonplanar conformation during excitation (Henderson et al., 2007; Olsen et al., 2010). Crystal structures of a teal fluorescent protein suggested that photoswitching is associated with a light-induced isomerization from a planar *cis* configuration to a nonplanar *trans* configuration (Henderson et al., 2007). Nonetheless, it is unclear if this is a universal mechanism, as a number of photochemical dark states have been identified that arise from the complex interplay between protonation of the chromophore and side chains, chromophore twisting, and alternative zwitterionic

Photophysical properties of fluorescent proteins

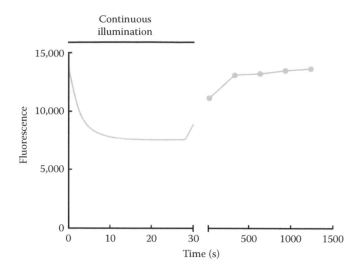

Figure 3.6 Reversible photoswitching. The fluorescence decay of a cyan FP-labeled bead was observed under a continuous illumination period. The steep initial decay of fluorescence reverses when the illumination is changed to intermittent periods. This transient decay of fluorescence is known as "reversible photoswitching," and the rate of recovery differs among FPs.

energy states (Zimmer, 2002). The internal pH of the β-barrel is also likely to influence photoswitching. Protonation (or deprotonation) of the chromophore can change the underlying resonance structure and can directly affect which portion of the chromophore may twist or alternatively promote entry into zwitterionic "dark" states (Weber et al., 1999). Furthermore, FP photoswitching behaviors are quite diverse, suggesting that more than one mechanism may be in play. Some FPs, such as TagRFP and mVenus, more prominently display photoswitching behavior under high intensity illumination conditions (Shaner et al., 2008). Others, such as mCerulean and mEGFP, show more photoswitching behavior at lower intensities rather than high ones. Of note, mCerulean3 displays greatly reduced photoswitching compared to mCerulean2, which is a fairly active photoswitcher (Markwardt et al., 2011). The difference between the two is merely the subtraction of a methyl group in the chromophore, demonstrating that FP changes that are structurally subtle can have profound consequences on the fluorescence properties.

The consequences of photoswitching on quantitative applications can be quite evident. For applications that directly employ photobleaching, such as FRAP (Chapter 8), single-molecule superresolution, or calculation of FRET efficiency by acceptor photobleaching, reversible photoswitching can directly affect the measurement accuracy. The precision of steady-state measurements can also be affected to a surprisingly large degree. Substitution of an FP with reduced photoswitching character can improve the standard deviation of fluorescence measurements by twofold and also improve fluorescence collection over integration times ranging from 100 ms to 1 s (Markwardt et al., 2011). Thus, a rapid decrease in fluorescence associated with photoswitching phenomena can affect brightness and quantification even in applications that utilize conventional steady-state collection conditions.

Photophysical properties of fluorescent proteins

3.3.5 MATURATION OF FLUORESCENCE

Chromophore formation results from spontaneous cyclization and oxidation reactions, and our understanding of the underlying chemical mechanism is constantly evolving (Wachter, 2007). From the perspective of applied FP biology, "better" folding proteins are generally the most useful, as they are more likely to independently fold when attached to a fusion and permit detection at earlier time points. In theory, a fast maturing and folding FP is less likely to perturb not only the function of a fusion protein partner, but also cellular function if experimentation on the system can be achieved with fewer FP molecules. Thus, there is a great interest in generating rapidly maturing and structurally stable FPs. Venus yellow FP (Nagai et al., 2002) and sfGFP (Pédelacq et al., 2005) are two notable examples of proteins that have been optimized for folding, particularly at mammalian temperatures, and this has indeed improved their performance in terms of cellular brightness. Venus expression can be visualized a few hours post transfection in some systems, as opposed to the more typical ~24–48 h time frame.

The focus on maturation kinetics takes on increased importance for red FPs. Red fluorescence requires increased double-bond conjugation within the chromophore, and, thus, can require additional chemical reactions to form the final chromophore (Bevis and Glick, 2002). If the maturation process is slow enough, green intermediate chromophores can often be observed, which are disadvantageous for multicolor experiments (Bevis and Glick, 2002). Fortunately, the latest generations of monomeric red FPs, such as mCherry, mature very rapidly (Shaner et al., 2004) such that the intermediate states are not observed.

3.3.6 UNWANTED INTERACTIONS

The wild-type parent proteins that variant FPs are derived from are generally oligomeric (dimeric or tetrameric) to some degree. The first useful red FP, from *Dicosoma*, is an obligate tetramer (Wall et al., 2000) and many of the non-*Aequorea*-derived FPs are naturally dimers (Yang et al., 1996). For *Aequorea* FPs, the tendency to dimerize is quite weak, with a k_d of approximately 0.1 mM (Zacharias et al., 2002). Even with such low affinity, dimerization can be observed inside living cells in cases where the mobility of FPs is limited. Notably, restriction of an FP localization to a surface that constrains diffusion to two dimensions, such as the plasma membrane, can create circumstances that facilitate dimerization of FPs (Zacharias et al., 2002). Mutation of position Ala206 to lysine or another charged amino acid is generally sufficient to inhibit dimerization in these cases.

In addition to protein:protein interactions, FPs can also interact with their environment in ways that affect their fluorescence. Notably, an endoplasmic reticulum-targeted EGFP can form disulfide linkages that prevent proper folding and reduce cellular fluorescence (Aronson et al., 2011). Interestingly, sfGFP performs particularly well in the endoplasmic reticulum (ER) and does not form inappropriate disulfide linkages. Some FPs also tend to be especially sensitive to acidic conditions. Yellow FPs derived from *Aequorea*, for example, have a pKa around six. Early variants also suffered from a peculiar increase in acid sensitivity in the presence of chloride (Griesbeck et al., 2001). The citrine mutation, Q69M, sterically inhibits binding to chloride ions and improves acid sensitivity, although yellow FPs in general are still more sensitive to acid compared to red and cyan fluorescent proteins (Shaner et al., 2005).

Photophysical properties of fluorescent proteins

3.4 METHODS FOR OPTIMIZING FPs

The fact that FPs are genetically encoded makes them easy to manipulate using molecular cloning techniques as compared to synthetic fluorophores. Mutations can be introduced using a wide variety of strategies, and new FPs can be expressed in cultured bacterial or mammalian cells for screening. This section will describe some of the common methods and strategies used to optimize FPs.

3.4.1 RANDOM MUTAGENESIS

The simplest approach for optimizing FP fluorescence is to introduce random mutations over the entire cDNA encoding an FP. Frequently regarded as "evolutionary," this strategy involves random mutation of an FP sequence and subsequent screening for increasingly refined and improved variants. The methods for introducing mutations are varied, ranging from degenerate polymerase chain reaction (PCR) (Griesbeck et al., 2001) and specialized PCR protocols (Sawano and Miyawaki, 2000) to DNA shuffling (Crameri et al., 1996). Typically, mutant proteins are encoded in a plasmid designed for inducible expression in bacteria. The leakiness of these expression vectors, although low, is generally sufficient for screening. Mutagenesis can also be performed in mammalian cells. As an example, FPs with far-red fluorescence were derived from the *Anthozoa* red FPs by harnessing the somatic hypermutation process in B lymphocytes (Wang et al., 2004).

Effective utilization of the molecular evolution approach requires effective screening. Fluorescence cell sorting (Cormack et al., 1996; Wang et al., 2004) can be quite effective for many applications. Some of the more inventive screens have employed solar simulators to develop photostable FPs (Shaner et al., 2008) and elaborate devices to examine fluorescent lifetimes (Goedhart et al., 2010). Fusion constructs have also been employed in screening. The sfGFP was developed by fusing mutagenized FP cDNAs to a poorly folding ferritin protein prior to screening (Pédelacq et al., 2005), the idea being that the brightest variants would be the best independent folders.

A drawback of such innovative approaches is that the random nature of the directed evolution process can sometimes produce unexpected results. Notably, a method was developed to optimize FRET between cyan and yellow FPs by mutagenizing both pairs and screening for improved FRET-ratio contrast (Nguyen and Daugherty, 2005). While the screen resulted in an FP combination with an impressive 20-fold dynamic range between FRET and non-FRET conditions, further investigation revealed that the contrast change was only partially related to FRET (Ohashi et al., 2007). Finally, screening protocols do not necessarily require advanced detection technologies to be effective. One of the earliest DsRed optimizations found a fast maturing DsRed variant among 100,000 bacterial colonies employing a slide projector for illumination and manual screening by observation with rose-tinted goggles (Bevis and Glick, 2002).

3.4.2 TARGETED APPROACH

FP optimization can also be achieved by targeting specific amino acids for mutation. Commonly, substitutions that have proven valuable in one FP are introduced into a close relative. A mutation at position 206 (Zacharias et al., 2002), for example, is generally introduced into modern *Aequorea* derivatives to reduce dimerization. Occasionally, substitutions can improve results across species as was shown for the "superfolder" mutations, which not only improved *Aequorea* color variants, but the *Anthozoa*-derived dsRed FP as well (Pédelacq et al., 2005). Conservation among FP variants is often great enough that substitutions that improve cyan and yellow FPs (Kremers et al., 2006) can also improve blue and green FPs (Kremers et al., 2007), although this is not universal. In particular, substitutions to amino acids adjacent to the chromophore can dramatically affect fluorescence, and reversion to the wild-type amino acids was favorable for the Venus yellow FP (Leu64) (Nagai et al., 2002). Reversion to Ser65 within the chromophore has also been shown to improve fluorescence of blue (Kremers et al., 2007) and cyan *Aequorea* FP derivatives (Goedhart et al., 2010; Markwardt et al., 2011). Perhaps, among the most ambitious examples of the targeted approach arose from Steven Kain's group at Clontech, who introduced over 190 silent mutations to humanize the jellyfish codons in an optimized green FP (Yang et al., 1996). This effort improved fluorescence ~4-fold in transfected cells and ~18-fold over the native jellyfish protein.

3.4.3 COMBINATION APPROACH

Both the large-scale and the targeted mutagenesis approaches can have their drawbacks. Mutagenesis over the entire full-length sequence can be too broad and inefficient, while the targeted approach can be too narrow. Even so, expanding the targeted approach to include broadly defined structures can also be a successful approach for optimizing FPs. One of the earliest attempts at FP optimization targeted 10 amino acids on each side of the chromophore using a specially designed degenerate PCR strategy (Cormack et al., 1996). This screen not only resulted in the EGFP mutations, but was the first to identify S72A as a useful mutation. The latter mutation is found in approximately half of the optimized *Aequorea* FP derivatives.

   The wealth of x-ray crystallographic data for FPs has also enabled optimization approaches based on structure. The identification of two distinct conformations of Tyr145 and His148 in crystal structures for ECFP (Bae et al., 2003) led to the optimization strategy that resulted in the development of Cerulean (Rizzo et al., 2004). Similarly, the structure of Cerulean underpinned the mutagenesis strategy that produced mCerulean3 (Markwardt et al., 2011). The spacing between β strands 7 and 8 in the Cerulean structure is greater than 3 Å in certain portions. This exceeds the distances associated with efficient hydrogen bonding, which are close to 2 Å. The mutagenesis strategy was thus directed on optimizing the ~8 amino acids found in regions with aberrant spacing. Six optimal substitutions were found that improved the brightness by approximately 30%. Interestingly, structural optimization by this method was synergistic with reversion of Thr65 to Ser in the chromophore to produce mCerulean3, which has even greater brightness and photostability.

3.5 FUTURE DIRECTIONS

With new classes of FPs continuously being discovered, the role of optimization will be of continued importance, as the increased use of novel FPs tends to reveal new limitations. For example, the general tendency of FPs to misfold in the ER (Aronson et al., 2011) and the error introduced into quantitative experiments by reversible photoswitching (Markwardt et al., 2011) have only recently become appreciated. Further advancements in molecular modeling and computing should also make structure-based refinements more accessible to investigators looking to fine tune the properties of FPs for specific applications. Finally, as imaging technology becomes further refined and breaks free from traditional color categories through the use of white light lasers and filterless imaging strategies, optimization of nontraditional colors such as teal and orange should take on greater importance. Together, these trends should improve the FP color palette to accommodate experiments of ever-increasing complexity and help pave the way for the next generation of discoveries.

REFERENCES

Ai, H.W., N. Shaner, Z. Cheng, R. Tsien, and R. Campbell. 2007. Exploration of new chromophore structures leads to the identification of improved blue fluorescent proteins. *Biochemistry* 46:5904–5910.

Aronson, D.E., L.M. Costantini, and E.L. Snapp. 2011. Superfolder GFP is fluorescent in oxidizing environments when targeted via the Sec translocon. *Traffic* 12:543–548.

Bae, J.H., M. Rubini, G. Jung, G. Wiegand, M.H. Seifert et al. 2003. Expansion of the genetic code enables design of a novel "gold" class of green fluorescent proteins. *J Mol Biol* 328:1071–1081.

Bailey, E.A. and G.K. Rollefson. 1953. The determination of the fluorescence lifetimes of dissolved substances by a phase shift method. *J Chem Phys* 21:1315–1323.

Bevis, B.J. and B.S. Glick. 2002. Rapidly maturing variants of the *Discosoma* red fluorescent protein (DsRed). *Nat Biotechnol* 20:83–87.

Chalfie, M., Y. Tu, G. Euskirchen, W.W. Ward, and D.C. Prasher. 1994. Green fluorescent protein as a marker for gene expression. *Science* 263:802–805.

Chen, S.X., A.B. Osipovich, A. Ustione, L.A. Potter, S. Hipkens et al. 2011. Quantification of factors influencing fluorescent protein expression using RMCE to generate an allelic series in the ROSA26 locus in mice. *Dis Model Mech* 4:537–547.

Chen, T.-S., S.-Q. Zeng, Q.-M. Luo, Z.-H. Zhang, and W. Zhou. 2002. High-order photobleaching of green fluorescent protein inside live cells in two-photon excitation microscopy. *Biochem Biophys Res Commun* 291:1272–1275.

Cormack, B.P., R.H. Valdivia, and S. Falkow. 1996. FACS-optimized mutants of the green fluorescent protein (GFP). *Gene* 173:33–38.

Cotlet, M., P.M. Goodwin, G.S. Waldo, and J.H. Werner. 2006. A comparison of the fluorescence dynamics of single molecules of a green fluorescent protein: One- versus two-photon excitation. *ChemPhysChem* 7:250–260.

Crameri, A., E.A. Whitehorn, E. Tate, and W.P. Stemmer. 1996. Improved green fluorescent protein by molecular evolution using DNA shuffling. *Nat Biotechnol* 14:315–319.

Goedhart, J., L. Van Weeren, M. Hink, N. Vischer, K. Jalink et al. 2010. Bright cyan fluorescent protein variants identified by fluorescence lifetime screening. *Nat Methods* 7:137–139.

Griesbeck, O., G.S. Baird, R.E. Campbell, D.A. Zacharias, and R.Y. Tsien. 2001. Reducing the environmental sensitivity of yellow fluorescent protein. Mechanism and applications. *J Biol Chem* 276:29188–29194.

Heikal, A.A., S.T. Hess, and W.W. Webb. 2001. Multiphoton molecular spectroscopy and excited-state dynamics of enhanced green fluorescent protein (EGFP): Acid-base specificity. *Chem Phys* 274:37–55.

Heim, R. and R.Y. Tsien. 1996. Engineering green fluorescent protein for improved brightness, longer wavelengths and fluorescence resonance energy transfer. *Curr Biol* 6:178–182.

Henderson, J.N., H.W. Ai, R.E. Campbell, and S.J. Remington. 2007. Structural basis for reversible photobleaching of a green fluorescent protein homologue. *Proc Natl Acad Sci USA* 104:6672–6677.

Hillesheim, L.N., Y. Chen, and J.D. Müller. 2006. Dual-color photon counting histogram analysis of mRFP1 and EGFP in living cells. *Biophys J* 91:4273–4284.

Kalies, S., K. Kuetemeyer, and A. Heisterkamp. 2011. Mechanisms of high-order photobleaching and its relationship to intracellular ablation. *Biomed Opt Express* 2:805–816.

Kasche, V. and L. Lindqvist. 1964. Reactions between the triplet state of fluorescein and oxygen. *J Phys Chem* 68:817–823.

Konagurthu, A.S., J.C. Whisstock, P.J. Stuckey, and A.M. Lesk. 2006. MUSTANG: A multiple structural alignment algorithm. *Proteins* 64:559–574.

Kremers, G., J. Goedhart, E. Van Munster, and T.W. Gadella. 2006. Cyan and yellow super fluorescent proteins with improved brightness, protein folding, and FRET Förster radius. *Biochemistry* 45:6570–6580.

Kremers, G.J., J. Goedhart, D.J. van den Heuvel, H.C. Gerritsen, and T.W. Gadella. 2007. Improved green and blue fluorescent proteins for expression in bacteria and mammalian cells. *Biochemistry* 46:3775–3783.

Markwardt, M.L., G.-J. Kremers, C.A. Kraft, K. Ray, P.J.C. Cranfill et al. 2011. An improved cerulean fluorescent protein with enhanced brightness and reduced reversible photoswitching. *PLoS One* 6:e17896.

Megley, C.M., L.A. Dickson, S.L. Maddalo, G.J. Chandler, and M. Zimmer. 2009. Photophysics and dihedral freedom of the chromophore in yellow, blue, and green fluorescent protein. *J Phys Chem* B 113:302–308.

Nagai, T., K. Ibata, E.S. Park, M. Kubota, K. Mikoshiba et al. 2002. A variant of yellow fluorescent protein with fast and efficient maturation for cell-biological applications. *Nat Biotechnol* 20:87–90.

Nguyen, A.W. and P.S. Daugherty. 2005. Evolutionary optimization of fluorescent proteins for intracellular FRET. *Nat Biotechnol* 23:355–360.

Ohashi, T., S.D. Galiacy, G. Briscoe, and H.P. Erickson. 2007. An experimental study of GFP-based FRET, with application to intrinsically unstructured proteins. *Protein Sci* 16:1429–1438.

Olsen, S., K. Lamothe, and T.J. Martínez. 2010. Protonic gating of excited-state twisting and charge localization in GFP chromophores: A mechanistic hypothesis for reversible photoswitching. *J Am Chem Soc* 132:1192–1193.

Ong, W.J.H., S. Alvarez, I.E. Leroux, R.S. Shahid, A.A. Samma et al. 2011. Function and structure of GFP-like proteins in the protein data bank. *Mol BioSyst* 7:984.

Ormö, M., A.B. Cubitt, K. Kallio, L.A. Gross, R.Y. Tsien et al. 1996. Crystal structure of the *Aequorea victoria* green fluorescent protein. *Science* 273:1392–1395.

Patterson, G.H., S.M. Knobel, W.D. Sharif, S.R. Kain, and D.W. Piston. 1997. Use of the green fluorescent protein and its mutants in quantitative fluorescence microscopy. *Biophys J* 73:2782–2790.

Patterson, G.H. and D.W. Piston. 2000. Photobleaching in two-photon excitation microscopy. *Biophys J* 78:2159–2162.

Pédelacq, J.-D., S. Cabantous, T. Tran, T.C. Terwilliger, and G.S. Waldo. 2005. Engineering and characterization of a superfolder green fluorescent protein. *Nat Biotechnol* 24:79–88.

Periasamy, N., S. Bicknese, and A.S. Verkman. 1996. Reversible photobleaching of fluorescein conjugates in air-saturated viscous solutions: Singlet and triplet state quenching by tryptophan. *Photochem Photobiol* 63:265–271.

Rekas, A., J.R. Alattia, T. Nagai, A. Miyawaki, and M. Ikura. 2002. Crystal structure of venus, a yellow fluorescent protein with improved maturation and reduced environmental sensitivity. *J Biol Chem* 277:50573–50578.

Rizzo, M.A., G. Springer, K. Segawa, W.R. Zipfel, and D.W. Piston. 2006. Optimization of pairings and detection conditions for measurement of FRET between cyan and yellow fluorescent proteins. *Microsc Microanal* 12:238–254.

Rizzo, M.A., G.H. Springer, B. Granada, and D.W. Piston. 2004. An improved cyan fluorescent protein variant useful for FRET. *Nat Biotechnol* 22:445–449.

Royant, A. and M. Noirclerc-Savoye. 2011. Stabilizing role of glutamic acid 222 in the structure of enhanced green fluorescent protein. *J Struct Biol* 174:385–390.

Sawano, A. and A. Miyawaki. 2000. Directed evolution of green fluorescent protein by a new versatile PCR strategy for site-directed and semi-random mutagenesis. *Nucleic Acids Res* 28:E78.

Shaner, N.C., R.E. Campbell, P.A. Steinbach, B.N. Giepmans, A.E. Palmer et al. 2004. Improved monomeric red, orange and yellow fluorescent proteins derived from *Discosoma* sp. red fluorescent protein. *Nat Biotechnol* 22:1567–1572.

Shaner, N.C., M. Lin, M. McKeown, P.A. Steinbach, K. Hazelwood et al. 2008. Improving the photostability of bright monomeric orange and red fluorescent proteins. *Nat Methods* 5:545–551.

Shaner, N.C., P.A. Steinbach, and R.Y. Tsien. 2005. A guide to choosing fluorescent proteins. *Nat Methods* 2:905–909.

Snapp, E.L. 2009. Fluorescent proteins: A cell biologist's user guide. *Trends Cell Biol* 19:649–655.

Song, L., C.A. Varma, J.W. Verhoeven, and H.J. Tanke. 1996. Influence of the triplet excited state on the photobleaching kinetics of fluorescein in microscopy. *Biophys J* 70:2959–2968.

Swaminathan, R., C.P. Hoang, and A.S. Verkman. 1997. Photobleaching recovery and anisotropy decay of green fluorescent protein GFP-S65T in solution and cells: Cytoplasmic viscosity probed by green fluorescent protein translational and rotational diffusion. *Biophys J* 72:1900–1907.

Turro, N.J. 1991. *Modern Molecular Photochemistry.* University Science Books, Sausalito, CA, pp. 76–152.

Visser, A.J.W.G. and M.A. Hink. 1999. New perspectives of fluorescence correlation spectroscopy. *J Fluoresc* 9:81–87.

Wachter, R.M. 2007. Chromogenic cross-link formation in green fluorescent protein. *Acc Chem Res* 40:120–127.

Wachter, R.M., M.A. Elsliger, K. Kallio, G.T. Hanson, and S.J. Remington. 1998. Structural basis of spectral shifts in the yellow-emission variants of green fluorescent protein. *Structure* 6:1267–1277.

Wall, M.A., M. Socolich, and R. Ranganathan. 2000. The structural basis for red fluorescence in the tetrameric GFP homolog DsRed. *Nat Struct Biol* 7:1133–1138.

Wang, L., W.C. Jackson, P.A. Steinbach, and R.Y. Tsien. 2004. Evolution of new nonantibody proteins via iterative somatic hypermutation. *Proc Natl Acad Sci USA* 101:16745–16749.

Ward, W.W. and S.H. Bokman. 1982. Reversible denaturation of *Aequorea* green-fluorescent protein: Physical separation and characterization of the renatured protein. *Biochemistry* 21:4535–4540.

Weber, W., V. Helms, J.A. McCammon, and P.W. Langhoff. 1999. Shedding light on the dark and weakly fluorescent states of green fluorescent proteins. *Proc Natl Acad Sci USA* 96:6177–6182.

Widengren, J., Ü. Mets, and R. Rigler. 1995. Fluorescence correlation spectroscopy of triplet states in solution: A theoretical and experimental study. *J Phys Chem* 99:13368–13379.

Widengren, J., Ü. Mets, and R. Rigler. 1999. Photodynamic properties of green fluorescent proteins investigated by fluorescence correlation spectroscopy. *Chem Phys* 250:171–186.

Yang, F., L.G. Moss, and G.N.J. Phillips. 1996. The molecular structure of green fluorescent protein. *Nat Biotechnol* 14:1246–1251.

Yang, T.T., L. Cheng, and S.R. Kain. 1996. Optimized codon usage and chromophore mutations provide enhanced sensitivity with the green fluorescent protein. *Nucleic Acids Res* 24:4592–4593.

Zacharias, D.A., J.D. Violin, A.C. Newton, and R.Y. Tsien. 2002. Partitioning of lipid-modified monomeric GFPs into membrane microdomains of live cells. *Science* 296:913–916.

Zimmer, M. 2002. Green fluorescent protein (GFP): Applications, structure, and related photophysical behavior. *Chem Rev* 102:759–782.

Development of new colors from coral fluorescent proteins

Nathan C. Shaner
The Scintillon Institute

Contents

4.1 ART AND SCIENCE OF FLUORESCENT PROTEIN COLOR VARIANTS

Though many would be reluctant to admit it, much of the effort put forth to engineer the current wide range of fluorescent protein (FP) colors has been motivated nearly as much by the visual appeal of these proteins as by their utility as research tools. The aesthetic allure of FPs is undeniable—every cell biologist recalls the sense of awe inspired by peering through a microscope objective for the first time at living cells with their inner workings illuminated. The organisms that naturally produce these proteins are themselves exquisitely beautiful, and scientists have created innumerable works of art in their engineering and application to cell biology. It is no wonder, then, that FPs quickly captured the imagination and enthusiasm of the scientific community like few other technologies have.

Development of new colors from coral fluorescent proteins

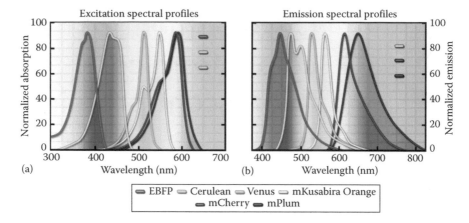

Figure 4.1 Wavelength diagram comparing the (a) excitation and (b) emission spectra of three avGFP-derived FPs with those of three coral-derived FPs. Note that most avGFP-derived FPs tend to emit in the blue-green portion of the visible light spectrum, while early coral-derived FPs were almost exclusively red-emitting.

FPs provide unique opportunities for protein engineering. Their easy expression in high-throughput systems such as *Escherichia coli* and suitability to *in situ* optical characterization facilitate screening of large libraries. Highly conserved structural features and the direct link between sequence and an optically measurable phenotype make directed evolution of useful variants relatively straightforward. Of the properties that may be manipulated by structure-guided engineering, wavelength presents itself as one of the most obvious as well as one of the most tractable. Soon after its discovery and characterization, *Aequorea victoria* green fluorescent protein (avGFP) was engineered to produce variants spanning from blue to yellow-green (Heim et al. 1994; Cubitt et al. 1995; Heim et al. 1995; Tsien 1998; Griesbeck et al. 2001). Work on optimizing avGFP-derived proteins has not completely stopped, and improvements are still being made, while many of the early avGFP variants still serve as useful tools in many laboratories. Today, researchers enjoy a diverse palette of FPs that spans virtually the entire visual spectrum, extending even to the near-infrared. The excitation and emission spectra of selected FPs spanning the visible spectrum are given in Figure 4.1. The full extent of color diversification of FPs was achieved through many years of focused work by several laboratories using FPs cloned from a large number of different organisms.

4.1.1 BEYOND *AEQUOREA* GFP VARIANTS

In retrospect, the discovery of the first FP in *A. victoria* in particular was extremely lucky. The Swiss Army knife of cell biology, avGFP can be expressed and detected successfully in virtually any organism. It is relatively small and forms only weak dimers, and so it can be attached to most proteins without measurably altering their function. Beyond simple fluorescent labeling, avGFP variants can be adapted relatively easily to optically report a wide variety of biochemical functions in living cells. Without these unusually user-friendly properties, it is likely that FP technology would never have attracted enough attention to gain wide acceptance. Unlike avGFP, FPs from most other organisms are rather unfit for use in heterologous organisms without substantial amounts of modification from their wild-type forms. Perhaps most critically, most FPs found in nature are obligate tetramers, making them unsuitable for use as fusion tags. In addition to its nearly monomeric nature, avGFP shows generally low toxicity and low tendency to aggregate compared with most other FPs that have been subsequently discovered. However, despite these advantages, avGFP has the notable limitation of not being easily evolvable into wavelength variants beyond yellow-green. The first red fluorescent mutant of avGFP was reported more than 14 years after the wild-type protein was cloned and is not sufficiently bright to be used as a reporter (Mishin et al. 2008). As the broad utility of FPs for live-cell imaging became apparent, the need soon grew for a wider range of wavelengths for multilabel imaging.

Because of avGFP's function as a bioluminescence energy acceptor in *Aequorea*, it seemed a logical hypothesis that FPs were unique to luminescent animals and possibly very rare in nature. However, the discovery of FP genes in nonluminescent anthozoans (corals) soon illustrated this expectation to be false. Indeed, FPs are a widely distributed and highly diverse family of proteins. Among the novel coral FPs first discovered was drFP583, which became commonly known as DsRed, the first "red" FP to be cloned (Matz et al. 1999). DsRed held the promise of expanding multilabel imaging with GFP variants but was soon found to have its own limitations, including slow maturation, substantial amounts of residual green fluorescence due to incomplete chromophore maturation, and obligate tetramerization, preventing its use as a fusion tag (Baird et al. 2000). After x-ray crystal structures of DsRed shed light on its oligomeric arrangement and chromophore environment (Gross et al. 2000; Yarbrough et al. 2001), several groups began work on evolving more useful variants that addressed many of these limitations. The quaternary structure of DsRed, as well as its chromophore structure and local environment, is illustrated in Figure 4.2.

4.1.2 DETERMINANTS OF CHROMOPHORE OPTICAL PROPERTIES

Several factors influence the optical properties of mature FPs. All known FPs share the same 11-strand beta-barrel fold with a central alpha-helical segment that contains the chromophore. The chromophore is formed via autocatalytic modification of the peptide backbone at an invariant XYG motif, leading to an extended pi-conjugated system with excited-state energy levels lying within the visible light spectrum. All known FP chromophores share the same imidazolinone core, differing only in its substituents (see Pakhomov and Martynov 2008, an excellent review of chromophore diversity). In large part, the extent of conjugation in the mature chromophore determines the excitation and emission wavelengths, with greater degrees of conjugation (a larger network of alternating single and double bonds) leading to longer wavelengths. This is shown by Figure 4.3, which illustrates different chromophore motifs of FPs spanning the visible spectrum, with greater degrees of conjugation readily apparent in more red-shifted varieties. The largest variations in wavelength thus come from variations in chromophore structure, and substitutions to amino acids that form part of the chromophore often influence the optical properties of the resulting mutant. Interacting side chains in the interior of the beta-barrel also have the potential to influence chromophore chemistry. For instance, DsRed contains a chromophore that has further modified its peptide backbone to produce an acylimine moiety, which extends the chromophore conjugation by two additional double bonds compared with the avGFP chromophore (Figure 4.3), but mutants of DsRed at some positions adjacent to the chromophore prevent the final chemical modification and thus lead to formation of a chromophore identical to that of avGFP (Baird et al. 2000).

Development of new colors from coral fluorescent proteins

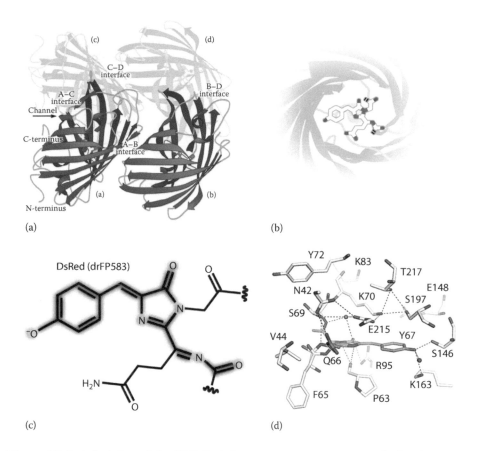

Figure 4.2 Detailed view of the FP DsRed, from quartenary structure to the local chromophore environment. (a) The quartenary organization of DsRed, note that DsRed is an obligate tetramer and that original efforts toward engineering dimeric and monomeric RFPs concentrated on disrupting the illustrated interfaces. (b) A "bird's-eye" view of the DsRed chromophore inside of the beta-barrel. (c) The molecular structure of the tripeptide-derived mature DsRed chromophore. (d) The local chemical environment surrounding the DsRed chromophore, with water molecules illustrated by red spheres and polar interactions denoted by the dashed grey lines. (From Strongin, D.E., Bevis, B., Khuong, N., Downing, M.E., Strack, R.L., Sundaram, K., Glick, B.S., and Keenan, R.J., Structural rearrangements near the chromophore influence the maturation speed and brightness of DsRed variants, *Protein Eng. Des. Sel.*, 20(11), 525–534, 2007 by permission of Oxford University Press.)

Other, more subtle features within the beta-barrel of FPs fine-tune the spectrum and help determine other optical properties such as quantum yield (QY, the probability of fluorescence emission per excitation) and extinction coefficient (EC, the efficiency of absorbing incident photons). Interactions with side chains proximal to the chromophore thus have a large influence over the final optical properties. For instance, it is possible to produce red shifts by stacking aromatic side chains with the chromophore through interactions between the pi orbitals in the chromophore and the aromatic side chain, as was done to produce "yellow" variants of avGFP (Ormo et al. 1996). The planarity

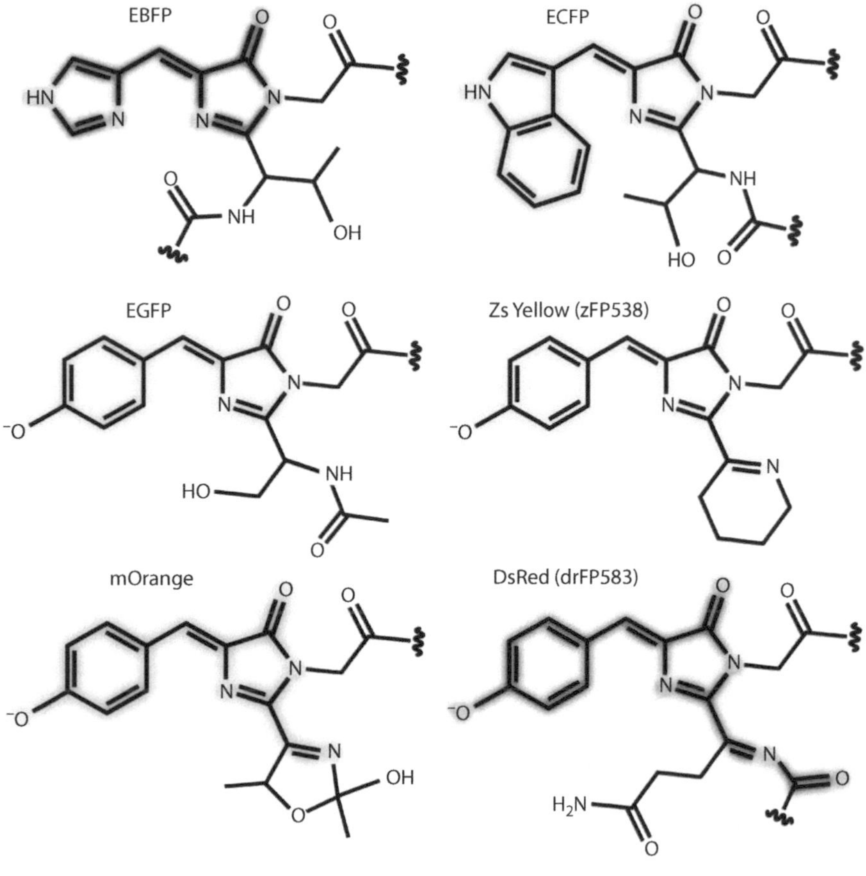

Figure 4.3 Example chromophore structures of different FPs with emission spectra spanning from blue to red, specifically EBFP, ECFP, EGFP, ZsYellow (zFP538), mOrange, and DsRed (drFP583). FPs are ordered by increasing wavelength, note that the first three FPs are avGFP-derived, while the last three are coral-derived. Structures are highlighted to illustrate the conjugated portions of each chromophore.

and rigidity of the chromophore are also largely determined by neighboring side chains and have a strong influence over vibrational modes available for excited-state energy dissipation, influencing the fluorescence QY of the chromophore. Electrostatic interactions between the chromophore and neighboring side chains can have numerous subtle effects on wavelength but can also control the charge state of the chromophore phenolate, which has a large effect on its Stokes shift, the distance between excitation and emission wavelength. Sapphire mutants of avGFP have a neutral (protonated) ground-state chromophore, which is excited by ~400 nm light but which emits green light at a similar wavelength to standard GFPs (Tsien 1998). Similar mutants have been produced from red FPs, producing variants such as Keima (derived from a *Montipora* sp. chromoprotein), which also has a long Stokes shift but emits in the red region of the visible spectrum (Kogure et al. 2006).

4.2 mRFP1 AND ITS OPTIMIZED RED VARIANTS

The first major improvement to DsRed came with the development of fast-maturing variants with mutations near the chromophore (Bevis and Glick 2002). Subsequently, one of these variants, DsRed.T1, was used as the starting material to design dimeric and monomeric mutants. Engineering efforts used the crystal structure to select mutations expected to disrupt oligomerization of DsRed. This was done by introducing positively charged amino acids at positions central to the two dimer interfaces present in the DsRed tetramer, with the hope that electrostatic repulsion would make dimerization at these interfaces less favorable. This approach was highly successful in preventing oligomerization but required many rounds of mutagenesis and selection to improve maturation and brightness of the resulting mutants. This led first to a bright dimer (eventually named "dimer2"), which became the starting material to produce a monomer, mRFP1 (Campbell et al. 2002). The first monomeric red FP, mRFP1, was quickly adopted by many researchers as a fusion tag that could be co-imaged with avGFP variants. While mRFP1 was a welcome advance, it was far from optimal in its original form. Among the shortcomings of mRFP1 were incomplete red chromophore maturation, leading to small but detectable green fluorescence, unpredictable brightness when used as a fusion partner due to folding defects, and poor photostability compared with other commonly used FPs such as EGFP. The primary aim of further efforts to evolve mRFP1 variants were thus focused on improving these properties (Shaner et al. 2004). Refer to Figure 4.4 for a timeline of the development of many of the most popular DsRed-derived FPs that will be discussed.

One of the first issues with mRFP1 to be addressed was its incomplete maturation. DsRed-derived chromophores mature via a branched pathway that yields some proportion of GFP-type chromophore. In tetrameric DsRed, these green fluorescent chromophores are efficient resonance energy transfer donors to red chromophores and so do not contribute a large portion of green emission. In mRFP1, these GFP-type chromophores are nearly nonfluorescent but represent a large portion of the protein population. So, while not a major source of bleedthrough into the GFP fluorescence channel during imaging, the large proportion of nonfluorescent mRFP1 protein effectively lowers the EC of the protein, limiting its total brightness. Mutagenesis experiments on wild-type DsRed suggested that the first chromophore position substitution Q66M would lead to a larger proportion of red chromophore formation (Baird 2001). When introduced into mRFP1, the Q66M mutation dramatically improved chromophore maturation with minimal impact to its optical properties. Several other libraries of mRFP1-Q66M with randomization of amino acids with chromophore-interacting side chains were then screened, leading to identification of the M163Q mutation, which conferred nearly complete red chromophore maturation (Shaner et al. 2004).

While successful when used with many proteins, mRFP1's performance as a fusion tag was not as reliable as many had come to expect from experience with avGFP variants. Many anecdotal reports indicated that mRFP1 did not always produce a bright fluorescent signal consistent with its measured optical properties when fused to a target protein. It was additionally observed that mRFP1 and later mutants folded poorly when expressed without an N-terminal 6xHis tag in *E. coli*. As an attempt to remedy this problem, a rather naïve approach was taken—appending the first and last seven amino

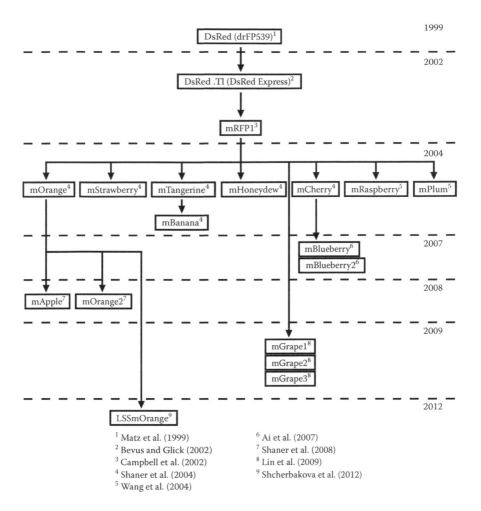

Figure 4.4 Flowchart detailing the development of DsRed-derived FP variants. One of the key developments was the development of mRFP1, which required 33 amino acid substitutions compared to the original DsRed. From mRFP1 many of the most popular and high performance YFPs, OFPs, and RFPs used today were developed, notably the "mFruits" (including mCherry).

acids from avGFP to the N and C terminus of mRFP1 and its derivatives. Amazingly, this approach was highly successful, restoring the ability of mRFP1 derivatives to fold regardless of their fusion state. This technique has since been applied to monomeric FPs from a number of different anthozoans and appears to be a nearly universal solution to fusion-dependent folding issues in FPs. In the case of mRFP1 derivatives, further enhancement was achieved with a four amino acid spacer (screened from a randomized library) inserted after the initial seven amino acids (Shaner et al. 2004).

The third major shortcoming of mRFP1 was its lack of sufficient photostability for long-term imaging. FPs vary widely in their bleaching rate upon illumination with a high-intensity light source. Photobleaching in FPs remains a poorly understood phenomenon but is thought to result from a combination of chemical inactivation

Photophysical properties of fluorescent proteins

(oxidation, etc.) of the chromophore, *cis* to *trans* isomerization of the double bond adjacent to the chromophore phenolate, and other covalent modifications (such as glutamate decarboxylation) to side chains that form part of the chromophore environment. Such complexity makes prediction of photostability-conferring mutations extremely difficult. Luckily, in the case of mRFP1, one of the mutations selected for improved chromophore maturation, M163Q, also improved its photostability more than 10-fold. After further rounds of random mutagenesis to improve protein yield from *E. coli* expression, work was complete on the second-generation mRFP1 variant, mCherry, which was adopted very widely as a reliable monomeric red fusion tag (Shaner et al. 2004).

4.3 STRUCTURE-GUIDED EVOLUTION OF WAVELENGTH VARIANTS OF mRFP1: THE "mFRUITS"

When work began on the mRFP1 variants that would come to be known as the "mFruits," it was unclear whether monomerizing other coral FPs would be a practical approach to expanding the available FP color palette. Since mRFP1 already showed considerable promise as a fusion partner, it was thus reasoned that this protein might also provide useful starting material for developing a wider range of color variants. Lessons learned from avGFP engineering provided useful starting points for exploration of wavelength variants of mRFP1. Although sequence conservation is minimal between anthozoan FPs and avGFP, x-ray crystal structures have revealed a highly conserved 3D structure with several invariant features among all FPs. Because the crystal structure of mRFP1 was never solved, the DsRed crystal structures served as a proxy for the design of variants with altered optical properties.

4.3.1 SUBSTITUTIONS TO THE CHROMOPHORE TYROSINE RESIDUE

Substitutions to the central chromophore-forming tyrosine present in all naturally occurring FPs are perhaps the most obvious way to dramatically alter the peak wavelengths of the chromophore. In general, substitutions with other aromatic amino acids yield variants with shorter excitation and emission wavelengths than their corresponding tyrosine-containing versions. Substitutions with nonaromatic amino acids may also be possible in some cases, but these variants have not been found to produce FPs with practically useful properties. In avGFP, aromatic substitutions give rise to cyan (Y66W), blue (Y66H), and ultramarine (Y66F) variants (Tsien 1998), with cyan variants being particularly useful for many imaging applications. Thus, it was reasoned that making similar aromatic substitutions to the mRFP1 chromophore would produce wavelength shifts similar to those seen in GFP, but with a red shift due to the presence of the acylimine component of the mature chromophore.

The first substitution attempted was Y67W, producing an mRFP1 variant that is homologous to the cyan variants of avGFP. As expected, the initial mutant matured poorly and was not particularly bright. Further rounds of directed evolution to rescue brightness yielded some improvements but never produced a variant with sufficient brightness for demanding imaging applications. Nevertheless, the final variant, mHoneydew, is an interesting example of an RFP-derived homolog of CFP. As with CFPs, mHoneydew has double-peaked excitation and emission spectra due to the tryptophan in its chromophore (Shaner et al. 2004). It occupies a unique place in the visual spectrum between yellow-green and true yellow, an area still lacking any bright variants. Compared with other mRFP1 derivatives, it has a relatively low EC and QY, and its photostability is rather poor.

Initial attempts to introduce other aromatic substitutions in the mRFP1 chromophore led to nearly nonfluorescent results, which were not pursued further. However, once the second-generation variant mCherry had been produced and progress had been made on identifying other mutations that improved folding efficiency, it became possible to develop variants containing the Y67F mutation. While avGFP Y66F variants have not been very popular among researchers due to their impractically short excitation wavelengths, mCherry-derived homologs have peaks in a wavelength range between those of standard BFPs and CFPs, making them potentially more useful for multicolor imaging. As with mHoneydew, initial Y67F mutants were not particularly bright, and several rounds of directed evolution with selection for brightness were required to produce the final variant. The brightest resulting variant, mBlueberry2, has excitation and emission peaks slightly longer than EBFP variants and is substantially brighter (Ai et al. 2007). Because it is derived from phenylalanine rather than histidine, this variant's peaks are also somewhat sharper than those of either BFPs or CFPs. mBlueberry2 suffers from rather low photostability, limiting its usefulness for many applications. However, it is possible that further efforts to improve this lineage of proteins with selection for high photostability could yield a superior blue FP. Other routes could produce useful coral-derived FPs with excitation and emission at the short wavelength end of the visual spectrum. Efforts to produce Y67H mutants from the mRFP1 lineage have thus far only produced nonfluorescent or very weakly fluorescent variants (Ai et al. 2007). However, it seems likely that other coral-derived FPs may be capable of tolerating this substitution.

4.3.2 SUBSTITUTIONS AT THE FIRST CHROMOPHORE POSITION

While the central tyrosine of the chromophore tripeptide is invariant among naturally occurring FPs, a far more diverse collection of amino acids can be found in the preceding position. A less obvious target than the central aromatic residue of the chromophore, substitutions to this first amino acid in the chromophore-forming tripeptide also have the potential to influence its optical properties. In avGFP variants, substitutions at this position can shift the balance between neutral and anionic chromophore, as in the S65T substitution that has formed the basis for the most commonly used EGFP variants (Tsien 1998). In DsRed-derived monomers, substitutions at this position not only influence chromophore maturation (as in Q66M) but can also lead to alternative posttranslational modifications to the chromophore, giving rise to large changes in excitation and emission wavelength.

During screening an mRFP1-based library in which position 66 had been randomized, it was noted that serine, threonine, and cysteine at this position produced proteins with strongly blue-shifted excitation and emission peaks as well as higher QY than the mRFP1 parent. Further rounds of directed evolution to improve brightness and maturation led to the development of the variants mTangerine (Q66C) and mStrawberry (Q66T), both of which emit in the orange-red region of the visual spectrum (Shaner et al. 2004). While also somewhat brighter than mRFP1 and mCherry, these variants suffer from relatively poor photostability. However, it was also

Photophysical properties of fluorescent proteins

noted that for early clones in the mStrawberry lineage, a much higher QY species with even further blue-shifted excitation and emission peaks appeared at high pH. Additional rounds of mutagenesis and selection focusing on lowering the pK_a of the transition to this bright orange species led to the development of mOrange, an unexpected color variant from the mRFP1 lineage. Interestingly, crystal structures revealed that mOrange possesses a novel chromophore containing an oxazole moiety formed by cyclization of the threonine side chain with the acylimine component of the immature chromophore, which reduces the extent of conjugation by one double bond and accounts for the shorter wavelength excitation and emission peaks of this protein. Unfortunately, as with many other mRFP1 variants, mOrange suffers from poor photostability (a problem later addressed by further directed evolution).

4.3.3 VARIANTS WITH MULTIPLE CHROMOPHORE-INTERACTING SUBSTITUTIONS

Along with the variants with substitutions directly in the chromophore, several variants of mRFP1 were engineered with multiple chromophore-interacting mutations that led to additional changes in optical properties. While the position of side chains within the beta-barrel is relatively trivial to determine, the influence of mutations, especially when targeted to multiple positions simultaneously, is much more difficult to predict. Experience with avGFP variants provided some clues about how the chromophore environment could be modified to influence optical properties, but the altered geometry and more complex chemistry of the DsRed-type chromophore added a great deal of uncertainty.

Following the example of avGFP-derived yellow variants such as enhanced YFP (EYFP) (Ormo et al. 1996), position 197, which has the potential to stack with the chromophore, was targeted to produce mutants with altered properties. The I197Y mutation, homologous to T203Y in EYFP, produces slightly red-shifted excitation and emission when introduced to mRFP1 variants, as expected (Lin et al. 2009). Additional rounds of selection for brightness of I197Y mutants produced mGrape1, the first in a series of far-red mRFP1 derivatives. Further selection from randomly mutagenized libraries for mutants with additional red-shift produced mGrape2, which has additional substitutions near the chromophore that probably affect packing of other chromophore-interacting side chains. The mGrape2 variant also provides an interesting example of the unpredictable results of altering the chromophore environment; it is reversibly photoactivatable with blue light excitation and relaxes spontaneously back to a nonfluorescent state when incubated in the dark. While no definitive evidence has yet been obtained for the mechanism, it seems somewhat likely that a light-induced *cis* to *trans* isomerization mechanism may be involved in this photoactivation process.

An unexpected mutant of the mTangerine lineage (Q66C) is mBanana, which contains the I197E mutation (Shaner et al. 2004). Unlike a tyrosine at this position, the glutamate apparently interacts with the chromophore phenolate, perhaps stabilizing the anionic form, and leads to an mOrange-like (likely thiazole-containing, as in mKO Kikuchi et al. 2008) chromophore with strongly blue-shifted excitation and emission. mBanana was resistant to further improvements by random and targeted mutagenesis and so remains one of the less useful mRFP1 variants. However, it does occupy a unique position in the color spectrum, one of the few true yellow emitters known to exist.

A unique approach was taken to develop alternative red-shifted mutants of mRFP1 utilizing somatic hypermutation (SHM) in B cells (Wang et al. 2004). Ramos cells with integrated copies of mRFP1.2 (an early precursor to mCherry), driven by an inducible promoter, were put through multiple rounds of induction followed by fluorescence-activated cell sorting (FACS)-based selection for long emission wavelength. It was hoped that this method would enable the identification of unexpected functional mutations that would not necessarily have been introduced by rational design techniques. After the first several rounds, a clone producing a mutant with emission red-shifted approximately 15 nm relative to mRFP1 was isolated among the population of B cells. This variant, eventually named mRaspberry, contains the mutations F65C and I161M, chromophore-interacting mutations that had not previously been identified for mRFP1 or its progeny. After several more rounds of selection, a clone with an unusually large Stokes shift was isolated from the library, giving red emission nearly 40 nm longer than mRFP1 but with a relatively similar excitation wavelength. This variant, named mPlum, contains two critical mutations, V16E and F65I, which are responsible for the vast majority of its spectral shift. Further analysis of mPlum and position 16 mutants combined with several x-ray crystal structures have revealed that the newly introduced Glu16 side chain interacts with the terminal acylimine of the chromophore, leading to the large observed red-shift. Combining the successful results of rational design and SHM-based screening, the major sites responsible for red-shift in mPlum were randomized in mGrape2, and the resulting library was screened for red-shifted mutants. The best clone from this library, mGrape3, achieved the longest excitation and emission wavelengths among the mRFP1 variants (Lin et al. 2009). Despite not being identical mutations, the F65M and V16T substitutions present in mGrape3 probably act by a similar mechanism to the F65I and V16E substitutions in mPlum. Unfortunately, none of the far-red mutants of mRFP1 are bright enough to be highly practical for imaging applications. However, along with the other mutants, they serve to illustrate the potential for spectral diversity from a single FP template. Figure 4.5 gives a photograph of several FPs spanning the visible spectrum, including the DsRed-derived mOrange2, mCherry, and mPlum.

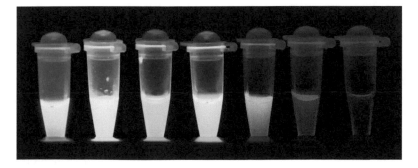

Figure 4.5 Photograph of seven different purified FPs under ultraviolet illumination, ordered by increasing emission wavelength. From left to right: EBFP2, mTurquoise, mNeonGreen (Shaner et al. 2013), mCitrine, mOrange2, mCherry, and mPlum. Note that mOrange2, mCherry, and mPlum are coral-derived; EBFP2, mTurquoise, and mCitrine are jellyfish-derived; mNeonGreen is derived from the cephalochordate *Branchiostoma lanceolatum*.

Photophysical properties of fluorescent proteins

4.4 THIRD-GENERATION MONOMERIC DsRED DERIVATIVES: OPTIMIZATION OF PHOTOSTABILITY

Of the mFruit series, only mCherry was sufficiently resistant to photobleaching to be practical in a wide range of imaging experiments, and while its brightness was sufficient for many purposes, brighter variants were still highly desirable. It was thus decided that the very bright but poorly photostable protein mOrange should be the basis for development of variants with higher photostability (Shaner et al. 2008). Initially, residue 163 was targeted for directed mutagenesis on the basis that the M163Q mutation had a large positive effect on photostability in mCherry. In principle, this approach was successful, producing M163Q and M163K variants of mOrange, which had substantially higher photostability. Unfortunately, these mutants had other drawbacks: mOrange-M163Q had a reduced QY, which was not recoverable through directed evolution, and mOrange-M163K was highly acid-sensitive, making it possibly useful as a pH sensor, but not practical for general use. Because the site-directed approach targeting the only known photostability-related side chain failed to yield useful photostable mOrange mutants, randomly mutagenized libraries were screened to identify other amino acid substitutions that improved photostability. Plates of library-expressing *E. coli* were exposed to high-intensity excitation light for varying lengths of time, and colonies that maintained bright emission relative to the original mOrange clone were selected and sequenced. After several rounds of random and site-directed library selection, the final variant, mOrange2, had substantially higher photostability than its parent, despite somewhat reduced brightness and longer maturation time. As with many other variants in the mFruit series, the exact set of mutations responsible for improving mOrange2's photostability were somewhat surprising. Two mutations, neither of which conferred a large enhancement of photostability alone, were responsible for a 25-fold increase in photostability, apparently by suppressing oxidative bleaching pathways. Neither of these mutations, Q64H and F99Y, are in direct contact with the chromophore, illustrating that more subtle packing interactions within the FP can produce substantial changes to its optical properties.

Despite the success of mCherry as a fusion tag, brighter and more photostable red-emitting FPs remained in high demand. The bright variant mOrange once again served as the starting material for evolving a brighter red variant (Shaner et al. 2008). Based on the observation that mCherry, like mStrawberry, could be converted to a high QY blue-shifted species at high pH, it was reasoned that it might be possible to use the acid tolerance already evolved in mOrange to produce a variant in which a high QY red species could exist at neutral pH. To test this hypothesis, the T66Q mutation was introduced into mOrange, eliminating the threonine necessary to form the oxazole ring in mOrange's unusual chromophore. Consistent with expectation, mOrange-T66Q formed a red chromophore similar to mCherry, which converted into a high QY species at high pH, but with a much lower pK_a than mCherry. This mutant was then used as the starting material for directed evolution with random and site-directed libraries, with selection for high QY and photostability. After several rounds of selection, the final variant, mApple, had a much higher QY and so was substantially brighter than mCherry. Though mApple was evolved using selection for photostability in the same manner as with mOrange2, it displays unusual photophysical behavior when used under standard microscopy conditions. Rather than fully bleaching, mApple reversibly photoswitches to a dark state when placed under strong excitation light and then recovers its fluorescence relatively quickly in the dark. Though this property may cause some complications when quantitation of fluorescence is necessary, it may not have such a negative impact on most imaging experiments.

4.5 OTHER CORAL-DERIVED FPs

Though the discovery of DsRed was reported along with several other potentially interesting coral FPs, it quickly became the singular focus of most research and development activity in the field. This is perhaps not entirely surprising, since it was not clear for several years if producing functional monomeric derivatives of any of these tetrameric proteins would be possible. Many novel FPs were reported in the literature or commercialized without thorough optimization or a substantial amount of testing, and the experimental performance from many of these novel proteins was often rather disappointing. Many coral FPs with seemingly promising optical properties have been apparently abandoned over the years since their appearance in the literature, perhaps due to a combination of poor performance in their wild-type forms, the time-consuming nature of optimization efforts, and the narrow focus of the field on a few "mainstream" FPs. Despite this trend, after the example of mRFP1, researchers became more adventurous in their FP efforts, and several other coral-derived FPs began to receive their deserved attention. Success in monomerization of any given tetrameric wild-type FP remains far from a sure bet, but these later efforts have nonetheless produced a large number of highly useful FPs, many with significant advantages over avGFP variants and the DsRed-derived monomers. With each generation of new FPs, the lessons of the past help avoid many of the pitfalls of earlier generations. This has led to increasingly high expectations for performance of novel FPs, as well as higher barriers for publication of new research in the field. Since such a broad range of colors have been observed in naturally occurring FPs, there is also less of a need for generation of color variants (apart from far-red) in these later-generation FPs, and none have been the subject of the same degree of wavelength-diversification effort as was applied to mRFP1. Among the notable later-generation FPs that have gained widespread use are proteins derived from *Galaxeidae* sp., *Fungia* sp., *Clavularia* sp., and *Entacmaea quadricolor*. Table 4.1 gives a list of the vast majority of nonphotoconvertible coral-derived FPs suitable for imaging experiments and is ordered by wavelength.

4.5.1 mAG AND mKO

The development of many monomeric FPs derived from corals other than *Discosoma* occurred more or less concurrently with development of the mRFP1-derived wavelength variants. After mRFP1, the next monomeric coral FP to be reported, mAG, was a derivative of a green tetramer from *Galaxeidae* coral and served as a comforting confirmation that tetramers other than DsRed could be successfully monomerized (Karasawa et al. 2003). This first coral-derived alternative to avGFP may in many cases be a superior alternative to older green FPs, though it has not been as thoroughly evaluated as the avGFP variants. In a follow-up to mAG, and still before the first generation mFruits were reported, Kusabira-Orange (KO), the first naturally occurring orange FP,

Photophysical properties of fluorescent proteins

Table 4.1 Comparison of excitation and emission maxima, EC, QY, quaternary structure, and brightness for selected anthozoan FPs

FP (ACRONYM)	EXCITATION MAX (NM)	EMISSION MAX (NM)	OLIGOMERIC STATE	EC (× 10⁻³ M⁻¹ CM⁻¹)	QY	BRIGHTNESS	SOURCE	REFERENCES
Blue FPs								
mTagBFP	399	456	Monomer	52	0.63	33 (97)	*E. quadricolor*	Subach et al. (2008)
mBlueberry1	398	452	Monomer	11	0.48	5 (15)	*Discosoma* sp.	Ai et al. (2007)
mBlueberry2	402	467	Monomer	51	0.48	24 (71)	*Discosoma* sp.	Ai et al. (2007)
Cyan FPs								
AmCyan	458	489	Tetramer	44	0.24	11 (32)	*Anemonia majano*	Matz et al. (1999)
MiCy	472	495	Dimer	27	0.90	25 (74)	*Acropora* sp.	Karasawa et al. (2004)
mTFP1	462	492	Monomer	64	0.85	54 (159)	*Clavularia* sp.	Ai et al. (2006)
Green FPs								
mAG	492	505	Monomer	55	0.74	41 (121)	*Galaxeidae* sp.	Karasawa et al. (2003)
mWasabi	493	509	Monomer	70	0.80	56 (165)	*Clavularia* sp.	Ai et al. (2008)
ZsGreen	493	505	Tetramer	43	0.91	39 (115)	*Zoanthus* sp.	Matz et al. (1999)
Yellow FPs								
ZsYellow	529	539	Tetramer	20	0.42	8 (24)	*Zoanthus* sp.	Matz et al. (1999)
Orange FPs								
mKO	548	559	Monomer	52	0.60	31 (91)	*Fungia concinna*	Karasawa et al. (2004)

mKO2	551	565	Monomer	64	0.62	40 (118)	*F. concinna*	Sakaue-Sawano et al. (2008)
mOrange	548	562	Monomer	71	0.69	49 (144)	*Discosoma* sp.	Shaner et al. (2004)
mOrange2	549	565	Monomer	58	0.60	35 (103)	*Discosoma* sp.	Shaner et al. (2008)
LSSmOrange	437	572	Monomer	52	0.45	23 (68)	*Discosoma* sp.	Shcherbakova et al. (2012)
dTomato	554	581	Dimer	69	0.69	48 (141)	*Discosoma* sp.	Shaner et al. (2004)
tdTomato	554	581	Tandem dimer	138	0.69	95 (279)	*Discosoma* sp.	Shaner et al. (2004)
DsRed	558	583	Tetramer	75	0.79	59 (174)	*Discosoma* sp.	Matz et al. (1999)
DsRed2	563	582	Tetramer	44	0.55	24 (71)	*Discosoma* sp.	Matz et al. (1999), Clontech (2001)
DsRed-Express	555	584	Tetramer	38	0.51	19 (56)	*Discosoma* sp.	Bevis and Glick (2002)
DsRed-Express2	554	586	Tetramer	36	0.42	15 (44)	*Discosoma* sp.	Strack et al. (2008)
DsRed-Monomer	556	586	Monomer	35	0.10	4 (12)	*Discosoma* sp.	Strongin et al. (2007)
DsRed-Max	560	589	Tetramer	48	0.41	20 (59)	*Discosoma* sp.	Strack et al. (2008)

(continued)

Photophysical properties of fluorescent proteins

Photophysical properties of fluorescent proteins

Table 4.1 (continued) Comparison of excitation and emission maxima, EC, QY, quartenary structure, and brightness for selected anthozoan FPs

FP (ACRONYM)	EXCITATION MAX (NM)	EMISSION MAX (NM)	OLIGOMERIC STATE	EC (× 10⁻³ M⁻¹ CM⁻¹)	QY	BRIGHTNESS	SOURCE	REFERENCES
TurboRFP	553	574	Dimer	92	0.67	62 (182)	E. quadricolor	Merzlyak et al. (2007)
TagRFP	555	584	Monomer	100	0.48	48 (142)	E. quadrico	Merzlyak et al. (2007)
TagRFP-T	555	584	Monomer	81	0.41	33 (97)	E. quadricolor	Shaner et al. (2008)
Red FPs								
mRuby	558	605	Monomer	112	0.35	39 (115)	E. quadricolor	Kredel et al. (2009)
mRuby2	559	600	Monomer	113	0.38	43 (127)	E. quadricolor	Lam et al. (2012)
mApple	568	592	Monomer	75	0.49	37 (109)	Discosoma sp.	Shaner et al. (2008)
mStrawberry	574	596	Monomer	90	0.29	26 (77)	Discosoma sp.	Shaner et al. (2004)
AsRed2	576	592	Tetramer	56	0.05	3 (9)	Anemonia sulcata	Wiedenmann et al. (2000), Clontech (2006)
mRFP1	584	607	Monomer	50	0.25	13 (38)	Discosoma sp.	Campbell et al. (2002)
mCherry	587	610	Monomer	72	0.22	16 (47)	Discosoma sp.	Shaner et al. (2004)

eqFP611	559	611	Tetramer	78	0.45	35 (103)	*E. quadricolor*	Wiedenmann et al. (2005)
tdRFP611	558	609	Tandem dimer	70	0.47	33 (97)	*E. quadricolor*	Kredel et al. (2008)
HcRed1	588	618	Dimer	20	0.02	0.4 (1)	*Heteractis crispa*	Gurskaya et al. (2001), Clontech (2003)
mRaspberry	598	625	Monomer	86	0.15	13 (38)	*Discosoma* sp.	Wang et al. (2004)
mKeima	440	620	Monomer	14	0.24	3 (9)	*Montipora* sp.	Kogure et al. (2006)
Far-red FPs								
tdRFP639	589	631	Tandem dimer	90	0.16	14 (41)	*E. quadricolor*	Kredel et al. (2008)
mKate	588	635	Monomer	32	0.28	9 (26)	*E. quadricolor*	Shcherbo et al. (2007)
mKate2	588	633	Monomer	63	0.40	25 (74)	*E. quadricolor*	Shcherbo et al. (2009)
Katushka	588	635	Dimer	65	0.34	22 (65)	*E. quadricolor*	Shcherbo et al. (2007)
tdKatushka	588	633	Tandem dimer	133	0.37	49 (144)	*E. quadricolor*	Shcherbo et al. (2009)
LSSmKate1	463	624	Monomer	31	0.08	2 (6)	*E. quadricolor*	Piatkevich et al. (2010)

(continued)

Photophysical properties of fluorescent proteins

Photophysical properties of fluorescent proteins

Table 4.1 (continued) Comparison of excitation and emission maxima, EC, QY, quartenary structure, and brightness for selected anthozoan FPs

FP (ACRONYM)	EXCITATION MAX (NM)	EMISSION MAX (NM)	OLIGOMERIC STATE	EC ($\times 10^{-3}$ M^{-1} CM^{-1})	QY	BRIGHTNESS	SOURCE	REFERENCES
LSSmKate2	460	605	Monomer	26	0.17	4 (12)	*E. quadricolor*	Piatkevich et al. (2010)
TagRFP675	598	675	Monomer	46	0.08	4 (12)	*E. quadricolor*	Piatkevich et al. (2013)
HcRed-Tandem	590	637	Tandem dimer	160	0.04	6 (18)	*H.crispa*	Fradkov et al. (2013)
mGrape1	595	625	Monomer	50	0.03	2 (6)	*Discosoma* sp.	Lin et al. (2009)
mGrape2	605	636	Monomer	33	0.03	1 (3)	*Discosoma* sp.	Lin et al. (2009)
mGrape3	608	646	Monomer	40	0.03	1 (3)	*Discosoma* sp.	Lin et al. (2009)
mPlum	590	649	Monomer	41	0.10	4 (12)	*Discosoma* sp.	Wang et al. (2004)
AQ143	595	655	Tetramer	90	0.04	4 (12)	*Actinia equina*	Shkrob et al. (2005)
TagRFP657	611	657	Monomer	34	0.10	3 (9)	*E. quadricolor*	Morozova et al. (2010)

Note: Brightness is the product of the EC and QY, while the value given in parenthesis is the percent brightness compared to EGFP, which has a brightness of 34. The table is ordered by emission wavelength and contains many of the highest performing coral-derived variants.

was engineered into a monomeric variant, mKO (Karasawa et al. 2004). Though not as bright as mOrange and somewhat slower to mature, mKO had several advantages, including acid tolerance and impressively high photostability. A faster-folding variant, mKO2, was later developed by traditional directed evolution techniques with selection for faster chromophore maturation and introduction of rational mutations to lower the protein pI (Sakaue-Sawano et al. 2008 and Atsushi Miyawaki, personal communication). mKO2 has shown promise as a Förster resonance energy transfer (FRET) acceptor from green and cyan FPs, including mAG and mTFP1 (Sun et al. 2009). As with mOrange, mKO houses an unusual three-ring chromophore, which accounts for its blue-shifted excitation and emission spectra relative to most red FPs. Rather than an oxazole ring, mKO contains a thiazole formed by cyclization of the chromophore acylimine with a cysteine side chain in the first chromophore position (Kikuchi et al. 2008).

4.5.2 mTFP1 AND VARIANTS

Coral FPs somewhat closely related to DsRed were also attractive early targets for monomerization efforts. Among the proteins discovered along with DsRed was cFP484 from *Clavularia* sp., a naturally occurring cyan FP with a tyrosine-containing chromophore (Matz et al. 1999). Because of its narrower excitation and emission spectra relative to avGFP tryptophan-based cyan variants and the availability of a high-quality crystal structure, this protein was chosen for monomerization (Ai et al. 2006). cFP484 was approached in much the same way as DsRed, with positively charged side chain substitutions placed to disrupt the two symmetrical dimer interfaces of the tetramer. As with DsRed, the evolution of a bright AC-interface dimer was a necessary intermediate step in creating the final monomeric form. The final variant had a somewhat red-shifted excitation and emission spectrum relative to the cFP484 precursor and hence was named monomeric "teal" FP or mTFP1. Inclusion of GFP-type termini in the mTFP1 design conferred reliable performance of this protein as a fusion tag, and it appears to be a good FRET partner to yellow FPs such as those derived from avGFP. Figure 4.6 shows a fluorescence image of a cell with five different FP fusion tags, including mTFP1 to the Golgi apparatus, demonstrating its utility in multiplex imaging experiments.

As with most known naturally occurring FPs with an avGFP-type chromophore, it is probably not possible to evolve mTFP1 variants with wavelengths beyond yellow-green. Nonetheless, mTFP1's impressive performance as a fusion tag warranted at least a limited amount of wavelength-diversification effort (Ai et al. 2008). For blue-shifted variants, aromatic substitutions to the mTFP1 chromophore produced shorter-wavelength equivalents to avGFP-derived CFPs and BFPs. While interesting in terms of modeling chromophore structure-wavelength relationships, the properties of these variants were not novel enough to garner further improvement efforts. For red-shifted variants, a structure-guided directed evolution process produced mWasabi, a bright green variant. mWasabi, like mAG, has the potential to be a direct substitute for traditional EGFP in many experiments and, in many cases, may perform substantially better. Of course, as with all later-generation FPs, mWasabi has not been vetted as thoroughly as EGFP.

4.5.3 PROTEINS DERIVED FROM *E. quadricolor*

Perhaps one of the most productive sources of useful FPs has been from the sea anemone *E. quadricolor*, which has yielded surprisingly bright variants over a broad range of red wavelengths. The first FP reported from *E. quadricolor* was eqFP611, a tetrameric protein

Photophysical properties of fluorescent proteins

Figure 4.6 Image of a cell labeled with five different FP fusion constructs. The histones, highlighting the nucleus in blue, are labeled with EBFP2 tagged to Histone 2b. The Golgi apparatus is labeled with the cyan FP mTFP1. Mitochondria, rendered in, are labeled with the yellow FP mCitrine. The peroxisomes, which are colored red, are labeled with mCherry. The focal adhesion protein Zyxin is labeled with the far-red FP mPlum and is represented in magenta. Note that EBFP2 and mCitrine are avGFP-derived.

that was shown to be much less tightly bound than the DsRed tetramer, raising hopes for successful monomerization (Wiedenmann et al. 2002). However, the first monomeric FP derived from *E. quadricolor* was not derived from eqFP611 but rather from eqFP578, an orange-red FP that shares approximately 76% identity at the amino acid level (Merzlyak et al. 2007). This protein, TagRFP, was substantially brighter than the best alternative at the time, mCherry, and performed quite well as a fusion partner. TagRFP was further improved through rational mutagenesis to produce a highly photostable variant, TagRFP-T, which remains one of the most photostable monomeric FPs (Shaner et al. 2008). Other variants of eqFP578 were evolved with selection for long emission wavelength, resulting in Katushka, which was then monomerized to produce mKate, and later the optimized variant mKate2, which was substantially brighter than existing far-red FPs (Shcherbo et al. 2007). mKate2 was targeted for further rational mutagenesis for additional red-shift, producing mNeptune, an unusually bright far-red FP (Lin et al. 2009). eqFP650 and eqFP670, dimeric far-red variants of Katushka, were later evolved and show promise for whole-animal imaging applications (Shcherbo et al. 2010). The TagRFP monomer has also shown a potential to yield highly diverse wavelength variants. After observations of a blue intermediate during red chromophore synthesis in TagRFP, a directed evolution strategy led to the creation of the bright blue FP TagBFP (Subach et al. 2008). Unlike previously known mechanisms to produce short-wavelength chromophores, the mutations introduced during TagBFP evolution apparently prevent the oxidation of the tyrosine side chain in an acylimine-containing chromophore, making TagBFP the lone representative of a new class of FPs (Subach et al. 2010). The excitation and emission spectrum of TagBFP compared to other coral-derived FPs spanning the visible spectrum is shown in Figure 4.7.

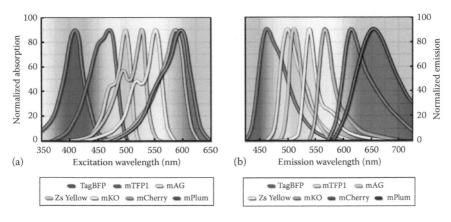

Figure 4.7 Wavelength diagram comparing the excitation and emission spectra of seven coral-derived FPs spanning the visible spectrum. (a) The excitation spectra of the selected FPs, showing their normalized absorption peaks. (b) The emission spectra of the selected FPs, showing their normalized emission peaks.

Ironically, despite its early promise, nearly seven years elapsed between the first report of eqFP611 and the publication of a bright monomeric variant (Kredel et al. 2009). In the intervening years, however, this protein was the focus of a similar wavelength-diversification effort to the development of Katushka from eqFP578 (Wiedenmann et al. 2005; Kredel et al. 2008). Several red-shifted (but still not monomeric) variants were created, but none have attracted a strong user base. Following the development of these red-shifted variants, a monomeric derivative of eqFP611, mRuby, was reported. Probably worth the wait, mRuby appears to be one of the brightest monomeric red fluorescent tags yet developed. In an interesting lesson about the unpredictable results of FP engineering, it was found that one dimeric variant of eqFP611 created during the process of evolving the monomer contained a peroxisome targeting signal at its C-terminus. Though not yet tested as thoroughly as other monomeric red FPs, mRuby will likely be a good starting point for engineering additional bright red and far-red variants. The utility of mRuby, as well as several other coral-derived FPs discussed, as a fusion partner is illustrated by Figure 4.8, which shows several fluorescence images of these proteins fused to proteins localized in a variety of subcellular compartments.

4.5.4 PHOTOCONVERTIBLE FPS

The pattern of serendipitous discoveries in the FP field continued with the identification of proteins that could convert from green to red with exposure to UV light. Identified when a tube of a green FP from *Trachyphyllia geoffroyi* left out on the lab bench turned red, Kaede was the first of a number of photoconvertible FPs to be discovered (Ando et al. 2002). The applications of these proteins are diverse and powerful, and thus much effort has gone into improving their utility as research tools. As with other FPs, strictly monomeric tags are the most generally useful, and the primary focus of most improvements to photoconvertible FPs has been producing high-performing monomeric variants of invariably tetrameric wild-type proteins. Among the naturally occurring photoconvertible proteins cloned, all but the original representative, Kaede, have been successfully monomerized.

Photophysical properties of fluorescent proteins

Development of new colors from coral fluorescent proteins

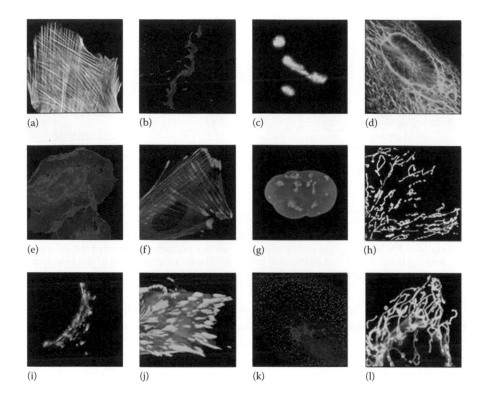

Figure 4.8 Anthozoan FPs from across the visible spectrum tagged to a variety of subcellular localization targets, all imaged with either a laser-scanning or spinning-disk confocal microscope. (a) mOrange2 fused to the F-actin-binding protein Lifeact showing the distribution of filamentous actin. (b) mApple fused to the gap junction protein connexin 43. (c) mTFP1 fused to fibrillarin. (d) Green fluorescent mWasabi fused to the intermediate filament cytokeratin. (e) mRuby fused to C-Src. (f) TagRFP fused to the focal adhesion protein zyxin. (g) TagBFP fused to the nuclear lamina protein Lamin B1. (h) TagRFP-T fused to a mitochondrial targeting signal. (i) The orange FP mKO2 fused to a Golgi targeting peptide. (j) mWasabi fused to the focal adhesion protein paxillin. (k) mCherry fused to clathrin light chain protein. (l) mTFP1 fused to a mitochondrial targeting sequence.

After the discovery of Kaede, it was also discovered that many other previously "red" coral FPs in fact shared this surprising ability to change from green to red upon UV light irradiation. Screening of novel FPs specifically for this optical property yielded several photoconvertible proteins from other corals. Among these were EosFP from *Lobophyllia hemprichii* (Wiedenmann et al. 2004) and Dendra (Gurskaya et al. 2006 or Dendra2, an improved variant later commercialized) from *Dendronephthya* sp., which has the unusual property of being photoconvertible with blue as well as UV illumination and is one of the first photoconvertible monomers to be developed. EosFP was initially engineered with two point mutations, each of which disrupted one of the dimer interfaces in the wild-type tetramer (Wiedenmann et al. 2004). Alone, these produced reasonably bright dimeric forms of EosFP, and together they yielded a monomeric variant. However, the original mEosFP was unable to fold at elevated temperatures required for mammalian cell experiments and was of limited utility. Later, a more concerted effort was applied to improve mEosFP stability at higher temperatures, and the resulting improved variant, mEos2, is one of the brighter photoconvertible probes available (McKinney et al. 2009). EosFP has since been further modified to produce IrisFP, which has the additional ability to reversibly photoswitch in both the green and red chromophore states, a property useful in superresolution imaging techniques (Adam et al. 2008).

To expand the range of photoconvertible proteins available as research tools, FPs with GFP-type chromophores were also targeted for rational engineering to produce this property, based on the hypothesis (supported by molecular evolution approaches) that naturally occurring photoconvertible proteins had evolved from a green ancestor. The green FP KikG, cloned from *Favia favus*, was the first example of successful application of this approach (Tsutsui et al. 2005). Because the mechanism of photoconversion in Kaede had already been determined, mutations were introduced that were expected to encourage this same activity in KikG based on its x-ray crystal structure. The first rationally designed mutant, which exhibited only a small amount of photoconversion activity, was selected from a large library of site-directed mutants targeting chromophore-interacting side chains. The final mutant, KikGR, was the result of 30 rounds of directed evolution. Later, this protein was engineered into a monomeric mutant, mKikGR, which has shown good utility for superresolution imaging (Habuchi et al. 2008).

A different strategy was taken in the design of mClavGR, a photoconvertible variant of the teal FP mTFP1 (Hoi et al. 2010). Rather than rely solely on predictions based on mTFP1's crystal structure, a bioinformatics-based approach was taken to engineer a consensus-based photoconvertible protein. The amino acid sequences of all known photoconvertible proteins were aligned, allowing the identification of a set of consensus amino acids with side chains internal to the beta-barrel (based on known photoconvertible protein crystal structures). The corresponding residues in mTFP1 were substituted with the consensus residues, and monomerization-related residues in mTFP1 were maintained. Surprisingly, the resulting variant was reasonably bright and already displayed strong photoconversion activity. Several rounds of directed evolution with selection for high photoconversion efficiency and brightness yielded mClavGR2, the first consensus-guided photoconvertible protein to be reported. While mClavGR2 is rather pH sensitive and not as bright as other monomeric photoconvertible proteins, as a variant of mTFP1, it is expected to perform well as a fusion partner in many contexts. More recently, an improved variant of mClavGR2 was developed, dubbed mMaple, which offers further improvements to photoconversion efficiency and fusion tag performance (McEvoy et al. 2012).

4.6 PERSPECTIVE AND SUMMARY

A great deal of progress has so far been made in engineering coral FPs into a useful variety of wavelengths, but this process is by no means complete. Brighter variants of all colors will always be welcome—in particular, variants with emission peaks in the yellow range between 530 and 550 nm remain rare, and the few existing proteins in this range are far from optimal. Further expansion of available wavelengths into the far-red and near-infrared is also likely to continue with vigor for the foreseeable future. Continued discovery of novel FPs from corals and other marine animals should help identify

Photophysical properties of fluorescent proteins

promising starting points for the evolution of monomeric variants with additional practical applications, and continued engineering of existing monomeric variants will undoubtedly yield many more useful tools in the future.

REFERENCES

Adam, V., M. Lelimousin et al. (2008). Structural characterization of IrisFP, an optical highlighter undergoing multiple photo-induced transformations. *Proc Natl Acad Sci USA* **105**(47): 18343–18348.

Ai, H. W., J. N. Henderson et al. (2006). Directed evolution of a monomeric, bright and photostable version of *Clavularia* cyan fluorescent protein: Structural characterization and applications in fluorescence imaging. *Biochem J* **400**(3): 531–540.

Ai, H. W., S. G. Olenych et al. (2008). Hue-shifted monomeric variants of *Clavularia* cyan fluorescent protein: Identification of the molecular determinants of color and applications in fluorescence imaging. *BMC Biol* **6**: 13.

Ai, H.-W., N. C. Shaner et al. (2007). Exploration of new chromophore structures leads to the identification of improved blue fluorescent proteins. *Biochemistry* **46**(20): 5904–5910.

Ando, R., H. Hama et al. (2002). An optical marker based on the UV-induced green-to-red photoconversion of a fluorescent protein. *Proc Natl Acad Sci USA* **99**(20): 12651–12656.

Baird , G. S. (2001). Designing fluorescent biosensors by exploiting structure–function relationships in fluorescent proteins. PhD thesis, University of California, San Diego, CA.

Baird, G. S., D. A. Zacharias et al. (2000). Biochemistry, mutagenesis, and oligomerization of DsRed, a red fluorescent protein from coral. *Proc Natl Acad Sci USA* **97**(22): 11984–11989.

Bevis, B. J. and B. S. Glick (2002). Rapidly maturing variants of the *Discosoma* red fluorescent protein (DsRed). *Nat Biotechnol* **20**(1): 83–87.

Campbell, R. E., O. Tour et al. (2002). A monomeric red fluorescent protein. *Proc Natl Acad Sci USA* **99**(12): 7877–7882.

Cubitt, A. B., R. Heim et al. (1995). Understanding, improving and using green fluorescent proteins. *Trends Biochem Sci* **20**(11): 448–455.

Fradkov, A. F., V. V. Verkhusha et al. (2002). Far-red fluorescent tag for protein labelling. *Biochem J* **368**(1): 17–21.

Griesbeck, O., G. S. Baird et al. (2001). Reducing the environmental sensitivity of yellow fluorescent protein. Mechanism and applications. *J Biol Chem* **276**(31): 29188–29194.

Gross, L. A., G. S. Baird et al. (2000). The structure of the chromophore within DsRed, a red fluorescent protein from coral. *Proc Natl Acad Sci USA* **97**(22): 11990–11995.

Gurskaya, N. G., A. F. Fradkov, A. Terskikh, M. V. Matz, Y. A. Labas, V. I. Martynov, Y. G. Yanushevich, K. A. Lukyanov, S. A. Lukyanov (2001). GFP-like chromoproteins as a source of far-red fluorescent proteins. *FEBS Lett* **507**(1):16–20.

Gurskaya, N. G., V. V. Verkhusha et al. (2006). Engineering of a monomeric green-to-red photoactivatable fluorescent protein induced by blue light. *Nat Biotechnol* **24**(4): 461–465.

Habuchi, S., H. Tsutsui et al. (2008). mKikGR, a monomeric photoswitchable fluorescent protein. *PLoS One* **3**(12): e3944.

Heim, R., A. B. Cubitt et al. (1995). Improved green fluorescence. *Nature* **373**(6516): 663–664.

Heim, R., D. C. Prasher et al. (1994). Wavelength mutations and posttranslational autoxidation of green fluorescent protein. *Proc Natl Acad Sci USA* **91**(26): 12501–12504.

Hoi, H., N. C. Shaner et al. (2010). A monomeric photoconvertible fluorescent protein for imaging of dynamic protein localization. *J Mol Biol* **401**(5): 776–791.

Karasawa, S., T. Araki et al. (2003). A green-emitting fluorescent protein from *Galaxeidae* coral and its monomeric version for use in fluorescent labeling. *J Biol Chem* **278**(36): 34167–34171.

Karasawa, S., T. Araki et al. (2004). Cyan-emitting and orange-emitting fluorescent proteins as a donor/acceptor pair for fluorescence resonance energy transfer. *Biochem J* **381**(Pt 1): 307–312.

Kikuchi, A., E. Fukumura et al. (2008). Structural characterization of a thiazoline-containing chromophore in an orange fluorescent protein, monomeric Kusabira Orange. *Biochemistry* **47**(44): 11573–11580.

Kogure, T., S. Karasawa et al. (2006). A fluorescent variant of a protein from the stony coral *Montipora* facilitates dual-color single-laser fluorescence cross-correlation spectroscopy. *Nat Biotechnol* **24**(5): 577–581.

Kredel, S., K. Nienhaus et al. (2008). Optimized and far-red-emitting variants of fluorescent protein eqFP611. *Chem Biol* **15**(3): 224–233.

Kredel, S., F. Oswald et al. (2009). mRuby, a bright monomeric red fluorescent protein for labeling of subcellular structures. *PLoS One* **4**(2): e4391.

Lam, A. J., F. St-Pierre et al. (2012). Improving FRET dynamic range with bright green and red fluorescent proteins. *Nat Methods* **9**(10): 1005–1012.

Lin, M. Z., M. R. Mckeown et al. (2009). Autofluorescent proteins with excitation in the optical window for intravital imaging in mammals. *Chem Biol* **16**(11): 1169–1179.

Matz, M. V., A. F. Fradkov et al. (1999). Fluorescent proteins from nonbioluminescent Anthozoa species. *Nat Biotechnol* **17**(10): 969–973.

McEvoy, A. L., H. Hoi et al. (2012). mMaple: A photoconvertible fluorescent protein for use in multiple imaging modalities. *PLoS One* **7**(12): e51314.

McKinney, S. A., C. S. Murphy et al. (2009). A bright and photostable photoconvertible fluorescent protein. *Nat Methods* **6**(2): 131–133.

Merzlyak, E. M., J. Goedhart et al. (2007). Bright monomeric red fluorescent protein with an extended fluorescence lifetime. *Nat Methods* **4**(7): 555–557.

Mishin, A. S., F. V. Subach et al. (2008). The first mutant of the *Aequorea victoria* green fluorescent protein that forms a red chromophore. *Biochemistry* **47**(16): 4666–4673.

Morozova, K. S., K. D. Piatkevich et al. (2010). Far-red fluorescent protein excitable with red lasers for flow cytometry and superresolution STED nanoscopy. *Biophys J* **99**(2): L13–L15.

Ormo, M., A. B. Cubitt et al. (1996). Crystal structure of the *Aequorea victoria* green fluorescent protein. *Science* **273**(5280): 1392–1395.

Pakhomov, A. A. and V. I. Martynov (2008). GFP family: Structural insights into spectral tuning. *Chem Biol* **15**(8): 755–764.

Piatkevich, K. D., J. Hulit et al. (2010). Monomeric red fluorescent proteins with a large Stokes shift. *Proc Natl Acad Sci USA* **107**(12): 5369–5374.

Piatkevich, K. D., V. N. Malashkevich et al. (2013). Extended stokes shift in fluorescent proteins: Chromophore-protein interactions in a near-infrared TagRFP675 variant. *Sci Rep* **3**: 1847.

Sakaue-Sawano, A., H. Kurokawa et al. (2008). Visualizing spatiotemporal dynamics of multicellular cell-cycle progression. *Cell* **132**(3): 487–498.

Shaner, N. C., R. E. Campbell et al. (2004). Improved monomeric red, orange and yellow fluorescent proteins derived from *Discosoma* sp. red fluorescent protein. *Nat Biotechnol* **22**(12): 1567–1572.

Shaner, N. C., G. G. Lambert et al. (2013). A bright monomeric green fluorescent protein derived from *Branchiostoma lanceolatum*. *Nat Methods* **10**(5): 407–409.

Shaner, N. C., M. Z. Lin et al. (2008). Improving the photostability of bright monomeric orange and red fluorescent proteins. *Nat Methods* **5**(6): 545–551.

Shcherbakova, D. M., M. A. Hink et al. (2012). An orange fluorescent protein with a large Stokes shift for single-excitation multicolor FCCS and FRET imaging. *J Am Chem Soc* **134**(18): 7913–7923.

Shcherbo, D., E. M. Merzlyak et al. (2007). Bright far-red fluorescent protein for whole-body imaging. *Nat Methods* **4**(9): 741–746.

Shcherbo, D., C. S. Murphy et al. (2009). Far-red fluorescent tags for protein imaging in living tissues. *Biochem J* **418**(3): 567–574.

Shcherbo, D., Shemiakina, II et al. (2010). Near-infrared fluorescent proteins. *Nat Methods* **7**(10): 827–829.

Shkrob, M. A., Y. G. Yanushevich et al. (2005). Far-red fluorescent proteins evolved from a blue chromoprotein from *Actinia equina*. *Biochem J* **393**(3): 649–654.

Strack, R. L., D. E. Strongin et al. (2008). A noncytotoxic DsRed variant for whole-cell labeling. *Nat Methods* **5**(11): 955–957.

Strongin, D. E., B. Bevis et al. (2007). Structural rearrangements near the chromophore influence the maturation speed and brightness of DsRed variants. *Protein Eng Des Sel* **20**(11): 525–534.

Subach, O. M., I. S. Gundorov et al. (2008). Conversion of red fluorescent protein into a bright blue probe. *Chem Biol* **15**(10): 1116–1124.

Subach, O. M., V. N. Malashkevich et al. (2010). Structural characterization of acylimine-containing blue and red chromophores in mTagBFP and TagRFP fluorescent proteins. *Chem Biol* **17**(4): 333–341.

Sun, Y., C. F. Booker et al. (2009). Characterization of an orange acceptor fluorescent protein for sensitized spectral fluorescence resonance energy transfer microscopy using a white-light laser. *J Biomed Opt* **14**(5): 054009.

Tsien, R. Y. (1998). The green fluorescent protein. *Annu Rev Biochem* **67**: 509–544.

Tsutsui, H., S. Karasawa et al. (2005). Semi-rational engineering of a coral fluorescent protein into an efficient highlighter. *EMBO Rep* **6**(3): 233–238.

Wang, L., W. C. Jackson et al. (2004). Evolution of new nonantibody proteins via iterative somatic hypermutation. *Proc Natl Acad Sci USA* **101**(48): 16745–16749.

Wiedenmann, J., C. Elke et al. (2000). Cracks in the beta-can: fluorescent proteins from *Anemonia sulcata* (Anthozoa, Actinaria). *Proc Natl Acad Sci USA* **97**(26): 14091–14096.

Wiedenmann, J., S. Ivanchenko et al. (2004). EosFP, a fluorescent marker protein with UV-inducible green-to-red fluorescence conversion. *Proc Natl Acad Sci USA* **101**(45): 15905–15910.

Wiedenmann, J., A. Schenk et al. (2002). A far-red fluorescent protein with fast maturation and reduced oligomerization tendency from *Entacmaea quadricolor* (Anthozoa, Actinaria). *Proc Natl Acad Sci USA* **99**(18): 11646–11651.

Wiedenmann, J., B. Vallone et al. (2005). Red fluorescent protein eqFP611 and its genetically engineered dimeric variants. *J Biomed Opt* **10**(1): 14003.

Yarbrough, D., R. M. Wachter et al. (2001). Refined crystal structure of DsRed, a red fluorescent protein from coral, at 2.0-A resolution. *Proc Natl Acad Sci USA* **98**(2): 462–467.

Red fluorescent proteins: Multipurpose markers for live-cell imaging

5

Jörg Wiedenmann
University of Southampton

G. Ulrich Nienhaus
Karlsruhe Institute of Technology

Contents

5.1 RED FLUORESCENT PROTEINS (RFPs): INTRODUCTION

5.1.1 WHY ARE RFPs DESIRABLE FOR LIVE-CELL IMAGING?

Fluorescence imaging of live cells, tissues, or even whole organisms poses several challenges that must be met by the fluorescent markers in order to be optimally detectable. In many cases, red fluorescent proteins (RFPs) or far-red fluorescent proteins (FPs) are the labels of choice to circumvent problems associated with the optical properties of the studied objects. In mammalian cells, compounds such as riboflavin, flavin coenzymes, and flavoproteins emit a green-yellow autofluorescence upon excitation with blue light. Coenzymes related to nicotinamide adenine dinucleotide phosphate (NADPH) contribute a bluish-green component to cellular autofluorescence (Billinton and Knight 2001). When using green and yellow fluorescent proteins (GFPs/YFPs) as markers, the overlap of their emission with the major cellular autofluorescence might hamper their quantitative detection (Figure 5.1). This is particularly of concern when the FPs are expressed at low levels or are not well localized in specific parts of the cells. Moreover, many cell culture media formulations contain substances that enhance the background fluorescence in the green-yellow region of the spectrum (Billinton and Knight 2001). This problem might be aggravated when pharmaceutical compounds added to the media are tested in cellular assays because many of these substances emit a bluish-green autofluorescence (Kredel et al. 2008). Choosing red- or far-red-emitting fluorescent markers can help to increase the signal-to-noise ratio. However, it is important to note that, in plant cells, the detection of far-red FPs might be comprised by the strong red autofluorescence of chlorophyll, which peaks around 685 nm. Moreover, application of both RFPs and far-red FPs in cyanobacteria is complicated by the presence of phycobiliproteins and chlorophyll, powerful emitters in orange-red spectral range.

RFPs and far-red FPs are, however, most promising labels for applications in living tissue and whole organisms, for example, to track the progression of fluorescently labeled tumors (Yang et al. 2005). Aside from decreased levels of autofluorescence, the application of red-shifted excitation light results in even further red-shifted emission, so that imaging benefits from reduced scattering of the longer wavelength light by

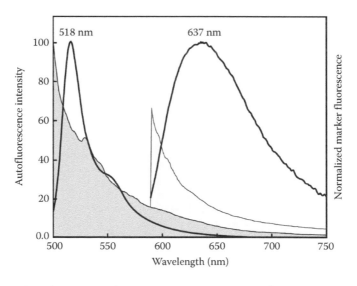

Figure 5.1 Overlap of cellular autofluorescence and marker protein fluorescence. The autofluorescence of human skin excited by 470 nm light (dark gray–shaded emission spectrum) is stronger than the autofluorescence induced by irradiation with 560 nm light (light gray–shaded emission spectrum). The overlap with the marker protein emission spectra in the same spectral range is higher for GFPs (here, EosFP [Wiedenmann et al. 2004], emission maximum at 518 nm) compared to RFPs (here, RFP637 [Kredel et al. 2008], emission maximum at 637 nm).

the tissue. Importantly, the absorption of hemoglobin is markedly reduced in the "optical window" at wavelengths above 600 nm. Therefore, RFPs with excitation and emission maxima at wavelengths >600 nm are desirable for whole-body or deep-tissue imaging applications using mammals (Lin et al. 2009; Shcherbo et al. 2010).

This chapter is dedicated to the RFPs representatives of the GFP family and the structural features responsible for their distinct spectroscopic properties. A detailed knowledge of structure–function relationships is a prerequisite to facilitate the generation of customized marker proteins for a large variety of live-cell imaging applications.

5.1.2 WHAT ARE THE DEMANDS ON RED FLUORESCENT MARKER PROTEINS IN LIVE-CELL IMAGING?

Live-cell imaging applications place specific demands on the biochemical properties of GFP-like FPs (Day and Davidson 2009; Wiedenmann et al. 2009). Some of them are specific for RFPs, whereas others will apply to GFP-like marker proteins in general (Nienhaus and Wiedenmann 2009). The ideal RFPs should be monomeric, feature fast and complete maturation, contain a highly photostable chromophore, emit bright red fluorescence with a large Stokes shift, and have excitation/emission peak wavelengths above 600 nm. However, different applications of RFPs sometimes have conflicting requirements and, therefore, it will become clear that no single optimal RFPs exists that is fit for all purposes. Rather, there are several variants that are optimally suited for certain applications.

5.1.3 RFPs IN NATURE

Several nonrelated groups of natural proteins show red fluorescence or have been engineered to obtain RFPs for imaging applications.

5.1.3.1 Phycobiliproteins

Phycobiliproteins are orange-red or bluish antenna pigments involved in light harvesting in the photosynthetic apparatus of red algae, cyanobacteria, and cryptomonads (Glazer 1985). The functional pigments are heteromultimers consisting of up to three different subunits (α, β, γ) and molecular masses typically ranging between 100 kDa (phycocyanins) and 240 kDa (phycoerythrins). Several linear tetrapyrrole chromophores covalently bound to the apoprotein are responsible for the outstanding fluorescence brightness of phycobiliproteins such as phycoerythrin. Their peak fluorescence ranges from 575 nm (phycoerythrins) to 660 nm (allophycocyanin). Their brightness and spectral properties make them attractive markers that are widely used for immunolabeling (Sun et al. 2003) in fixed cells. However, their complex biochemical synthesis prohibits their use as genetically encoded live-cell markers in most recombinant expression systems.

5.1.3.2 Phytochromes

Phytochromes are photoreceptors in plants and bacteria. They are dimeric proteins with a bilin chromophore attached to the protein moiety. The proteins can exist in two states that absorb light either around 660 nm or around 730 nm. A bacteriophytochrome from *Deinococcus radiodurans* that incorporates biliverdin as its chromophore was engineered into a monomeric, infrared fluorescent protein (IFP), with excitation and emission maxima of 684 and 708 nm, respectively. The IFP apoprotein expresses well in mammalian cells and mice; however, it becomes fluorescent once the expressing cells are supplied with biliverdin (Shu et al. 2009), which poses limits on its applicability as a fluorescence marker. The bacteriophytochrome RpBphP2 from the photosynthetic bacterium *Rhodopseudomonas palustris* has been engineered into a fluorescent variant with excitation and emission maxima at 690 and 713 nm, respectively; it is expressed in its functional form without biliverdin supplementation (Filonov et al. 2011). The low quantum yield of both phytochrome-derived marker proteins (~0.07) leaves room for optimization. However, their emission beyond 700 nm makes these FPs promising pilot structures for the development of advanced infrared marker proteins.

5.1.3.3 GFP-like RFPs

The GFP from the hydromedusa *Aequorea victoria* lent its name to a whole family of auto-FPs that exhibit emission colors ranging from cyan to red (Alieva et al. 2008; Nienhaus and Wiedenmann 2009). The β-can fold of GFP, in which the chromophore resides in the geometric center of an 11-stranded β-barrel, is conserved among all known red GFP homologues (Nienhaus and Wiedenmann 2009). In contrast to GFPs that are found occasionally in several metazoan phyla including cnidaria, ctenophora, arthropoda, and chordata, RFPs have been as yet discovered only in cnidarians, most often in the taxon anthozoa (Wiedenmann et al. 2011a,b) (Figure 5.2). Whereas GFP from *A. victoria* is monomeric in solution at concentrations relevant for marker applications, many of its red fluorescent homologues form tightly associated homodimers or homotetramers (Nienhaus and Wiedenmann 2009). As in GFP, the chromophore of red fluorescent variants forms autocatalytically, requiring nothing but oxygen and, in the case of photoactivatable RFPs, light of a distinct wavelength (Nienhaus and Wiedenmann 2009). The relative ease with which these single-gene encoded proteins can be customized through mutagenesis for various live-cell imaging applications makes them currently the most important pilot structures for the development of advanced marker proteins (Day and Davidson 2009; Wiedenmann et al. 2009, 2011a,b).

5.1.4 BRIEF HISTORICAL OVERVIEW OF THE DEVELOPMENT OF GFP-LIKE RED FLUORESCENT MARKER PROTEINS

Red fluorescence of sea anemones was described first in 1957 by L. Marden (Marden 1956). During a dive in the Red Sea, he observed the "unnatural" red appearances of the animals in depths where red light is mostly attenuated by the water column and explained this phenomenon by the presence of red fluorescent pigments (Marden 1956). In the following years, a considerable number of publications provided details about the optical properties of the fluorescent pigments of cnidarians, especially of reef corals (Wiedenmann et al. 2011a,b).

Despite these reports, red fluorescent pigments in cnidarians were not considered as possible candidates for marker gene applications until 1997 for two major reasons: (1) It was commonly believed that GFP-like proteins occur only as green secondary emitters in bioluminescent cnidarians, and (2) the red fluorescence of reef corals was believed to be based on biliproteins such as phycoerythrin.

The GFP-like nature of these proteins was independently discovered by Wiedenmann (1997) and Matz et al. (1999). Detailed studies of structure–function relations of RFPs enabled knowledge-based engineering approaches that resulted in a swift optimization of the marker proteins. Within about a decade, a stunning range of advanced red fluorescent marker proteins was created, and researchers succeeded in shifting their emission maxima by an incredible 100 nm toward longer wavelengths. An overview of key events in the RFPs discovery and marker protein development is given in Table 5.1 and Figure 5.3.

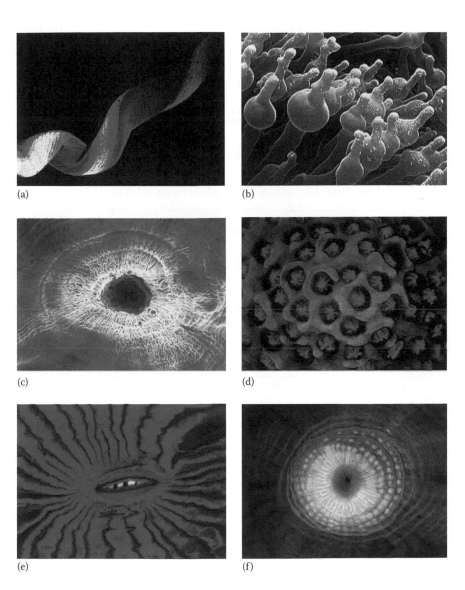

(a)

(b)

(c)

(d)

(e)

(f)

Figure 5.2 Origins of RFPs. (a) asFP595 expressed in the underside of tentacles of *Anemonia sulcata* (Wiedenmann et al. 2000). (b) eqFP583 and eqFP611 in the tentacles of *E. quadricolor* (Wiedenmann et al. 2002; Merzlyak et al. 2007). (c) dsFP586 (dsRed homologue) in *Discosoma* sp. (Nienhaus and Wiedenmann 2009). (d) amilFP597 in a colony of *Acropora millepora* (D'Angelo et al. 2008). (e) Green-to-red photoconvertible proteins in the oral disk of *Lobophyllia hemprichii* (EosFP) (Wiedenmann et al. 2004). (f) Green-to-red photoconvertible protein mcavRFP in a polyp of *Montastraea cavernosa* (Oswald et al. 2007). CFPs or GFPs present in the tissue appear blue (CFPs) or green (GFPs). Microscopic images (panels a, c–f) were acquired using a Leica MZ10 stereomicroscope (Leica, Wetzlar, Germany) using a CFP/DsRed double band-pass filter or the GFP plus filter (AHF Analysentechnik, Tübingen, Germany).

Photophysical properties of fluorescent proteins

Table 5.1 **Milestones of the discovery and development of GFP-like RFPs**

1956	Marden describes red fluorescent pigments in a sea anemone from the Red Sea (Marden 1956).
1997	Wiedenmann proposes that red fluorescent pigments from anthozoans are GFP-like proteins that can be used as marker proteins (Wiedenmann et al. 1997).
1999	Matz, Lukyanov, and coworkers clone the first RFPs (DsRed) (Matz et al. 1999).
2000	Gross, Tsien, and coworkers determine the chromophore structure of DsRed (Gross et al. 2000).
2000	Wall, Socolich, and Ranganathan determine the crystal structure of DsRed (Wall et al. 2000).
2001	Gurskaya, Lukyanov, and coworkers generate far-red fluorescing proteins from nonfluorescent anthozoan chromoproteins (Gurskaya et al. 2001).
2002	Campbell, Tsien, and coworkers generate the first monomeric RFPs (mRFP1) (Campbell et al. 2002).
2002	Wiedenmann, Nienhaus, and coworkers clone the eqFP611, the RFPs with the most red-shifted emission found among natural FPs (Wiedenmann et al. 2002).
2002	Ando, Miyawaki, and coworkers clone Kaede, the first photoactivatable RFPs (Ando et al. 2002).
2006	Kogure, Miyawaki, and coworkers generate Keima, an RFPs with an exceptionally large Stokes shift (Kogure et al. 2006).
2004–ﾠ2010	Generation of advanced RFPs (Figure 5.3).
2010	Morozova, Verkhusha, and coworkers generate TagRFP657: The monomeric marker protein with the most red-shifted emission (Morozova et al. 2010).
2010	Shcherbo, Chudakov, and coworkers generate eqFP670: The RFPs-variant with the most red-shifted emission maximum (Shcherbo et al. 2010).

5.1.5 ENGINEERING OF RFPs

Optimization of RFPs is greatly aided by the fact that the effects of mutations can be directly assessed via changes of their fluorescence properties (Figure 5.4). Mutations are commonly created by polymerase chain reaction (PCR)-based mutagenesis techniques. The considerable set of available crystal structures of GFP-like proteins enables knowledge-based engineering strategies that focus on defined key residues to alter protein properties in a predictable way. Site-directed mutagenesis was successfully applied to create monomeric RFPs and variants with red-shifted emission (Gurskaya et al. 2001; Campbell et al. 2002; Karasawa et al. 2004; Kredel et al. 2008, 2009). Randomization of distinct residues using mixtures of degenerate primers or multisite-directed mutagenesis techniques can help to select amino acid exchanges that alter the protein properties in the optimal manner (Kredel et al. 2009).

However, owing to the often unpredictable nature of mutations that improve folding and maturation, random mutagenesis is an irreplaceable tool to create optimized RFPs (Campbell et al. 2002). PCR-driven random mutagenesis relies on the increased error

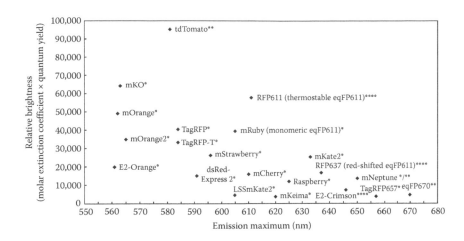

Figure 5.3 Relative brightness and emission peak positions of advanced red fluorescent marker proteins. The number of subunits involved in the formation of the functional marker is symbolized by asterisks next to the protein name. Values are from references (E2-Orange, Strack et al. 2009a; mOrange, Shaner et al. 2004; mKO, Tsutsui et al. 2008; mOrange2, Shaner et al. 2008; tdTomato, Shaner et al. 2004; TagRFP, Merzlyak et al. 2007; TagRFP-T, Shaner et al. 2008; DsRed-Express2, Strack et al. 2008; mStrawberry, Shaner et al. 2004; LSSmKate2, Piatkevich et al. 2010; mRuby, Kredel et al. 2009; mCherry, Shaner et al. 2004; RFP611, Kredel et al. 2008; mKeima, Kogure et al. 2006; mRaspberry, Wang et al. 2004; RFP637, Kredel et al. 2008; mKate2, Shcherbo et al. 2009; E2-Crimson, Strack et al. 2009b; mNeptune, Lin et al. 2009; TagRFP657, Morozova et al. 2010; eqFP670, Shcherbo et al. 2010).

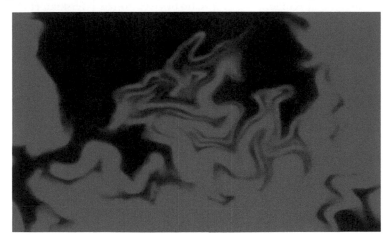

Figure 5.4 Growth patterns of *Escherichia coli* expressing a thermostable variant of eqFP611 in a confluent bacterial cover on an agar plate. The mutant strain can be clearly distinguished from bacteria producing nonfunctional variants (visible as dark background).

frequency of polymerases working under nonoptimal reaction conditions, resulting in randomly introduced mutations during amplification of the target sequence.

Fluorescence stereomicroscopes equipped with suitable filter sets allow for fast screening of mutants with desirable properties. Handheld blue or green light torches in combination with suitable filter goggles can be used for even faster screening, though at a lower level of sensitivity (Kredel et al. 2009). Automatized screening methods, such as fluorescence-activated cell sorting, have also been employed successfully; however, the majority of RFPs optimization was conducted using "manual" screening approaches (Wang et al. 2004). In many cases, the bacteria expressing the mutant proteins can be subjected to selective conditions to render the screening process more specific (Figure 5.4). Such selective conditions are, for instance, expression at elevated temperatures to select thermostable variants or light irradiation of the expressing colonies to identify photostable mutants (Kredel et al. 2008; Shaner et al. 2008). An overview of common mutagenesis and screening strategies is given in Table 5.2. In the following section, we will discuss the structure–function relations of RFPs that are relevant for the optimization of existing or newly discovered RFPs.

Table 5.2 **Common mutagenesis and screening strategies for RFPs optimization**

ALTERATION PROCESS	MUTAGENESIS STRATEGY	INITIAL SCREENING PARAMETER	REFERENCES (EXAMPLES)
Monomerization	Site-directed mutagenesis	Loss or reduction in fluorescence.	Campbell et al. (2002); Wiedenmann et al. (2005)
Maturation (speed)	Random mutagenesis	Detection of fluorescence during a time course of multiple screening rounds.	Bevis and Glick (2002); Terskikh et al. (2002); Kredel et al. (2008, 2009)
Maturation (completeness)	Random mutagenesis, site-directed mutagenesis	Ratio of green-to-red fluorescence.	Terskikh et al. (2000); Shaner et al. (2004)
Folding	Random mutagenesis	In the case of complete failure of folding, bacterial colonies should be screened after prolonged (weeks) incubation at 4°C.	Kredel et al. (2009)
Folding at higher temperatures	Random mutagenesis	Detection of fluorescence after expression at the desired temperature.	Kredel et al. (2008)
Red shift	Site-directed mutagenesis	Ratio of red to far-red fluorescence.	Gurskaya et al. (2001); Kredel et al. (2008); Shcherbo et al. (2010)
Photostability	Random mutagenesis, site-directed mutagenesis	Presence of fluorescence after prolonged light exposure.	Shaner et al. (2008)

Photophysical properties of fluorescent proteins

5.2 STRUCTURE–FUNCTION RELATIONS OF RFPs

5.2.1 OLIGOMERIZATION/AGGREGATION

Most anthozoan GFP-like proteins are obligate tetramers (Baird et al. 2000; Vrzheshch et al. 2000). Only a few dimeric FPs have been discovered, such as the cyan FPs (CFP) MiCy from a scleractinian coral and the orange-red FPs eqFP578 from the sea anemone *Entacmaea quadricolor*. Some RFPs tend to form even larger aggregates that have compromised their application as genetically encoded marker proteins. The RFPs known as eqFP611 from *E. quadricolor* is an obligate tetramer, with each monomer exhibiting the typical fold of a GFP-like protein (Figure 5.5a) (Nienhaus et al. 2003; Petersen et al. 2003; Wiedenmann et al. 2005). The overall structures of the two distinct A/B and A/C interfaces linking the monomers are similar to those found in the popular DsRed from *Discosoma* sp. (Figure 5.5b). The smaller A/B interface is formed by a cluster of hydrophobic amino acids encircled by polar amino acids. The hydrophilic A/C interface in eqFP611 is more extended than the A/B interface (Yarbrough et al. 2001; Verkhusha and Lukyanov 2004). The C-terminus of one protomer embraces the neighboring one, which likely has a stabilizing effect on the A/C interface. A large number of water molecules can be found between the amino acids forming the interface, many of which are involved in mediating hydrogen bonds between the monomers. With Tyr148, Tyr157, Tyr169, Phe192, Phe194, and Phe221, there are even more aromatic residues in the A/C interface of eqFP611 than in DsRed (Wiedenmann et al. 2005).

Monomerization of tetrameric reef coral and sea anemone proteins has proven to be rather difficult. To remove the interactions between a DsRed protomer and its two neighbors (Matz et al. 1999; Baird et al. 2000; Gross et al. 2000), 33 mutations were required (Campbell et al. 2002). Surprisingly, for eqFP611, functional dimers, denoted by d1eqFP611 and d2eqFP611, were readily obtained by the single mutations Val124Thr or Thr122Arg in the A/B interface, respectively (Wiedenmann et al. 2005). However, a functional monomer was only achieved after a total of seven rounds of random mutagenesis and four rounds of multisite-directed mutagenesis (Kredel et al. 2009). Compared with wild-type eqFP611, the bright red fluorescent monomer, denoted as mRuby, contains 28 amino acid replacements and has four amino acids removed (Figure 5.5c). All manipulations in the A/C interface resulted in an essentially complete loss of fluorescence. We note that the β-can of GFP-like proteins including eqFP611 is a thermodynamically very stable fold (Wiedenmann et al. 2002). Therefore, once they are properly folded, the functional monomers of eqFP611 are expected to form stable proteins. Presumably, the observed loss of fluorescence during mutagenesis of the A/C interface results from the inability of the polypeptide chain to assume its proper, functionally competent native structure.

Finally, we should note that, if functional monomers prove difficult or impossible to prepare, "pseudomonomeric" tandem dimers can be a viable alternative, consisting of two FPs domains connected by a flexible peptide linker (Campbell et al. 2002; Fradkov et al. 2002; Nienhaus et al. 2006).

5.2.2 MATURATION

Maturation of FPs consists of two major steps, folding of the polypeptide chain and autocatalytic formation of the chromophore. In GFP, folding occurs with a $t_{0.5} \sim 10$ min. The chemical reactions (cyclization, dehydration, and oxidation) that yield the

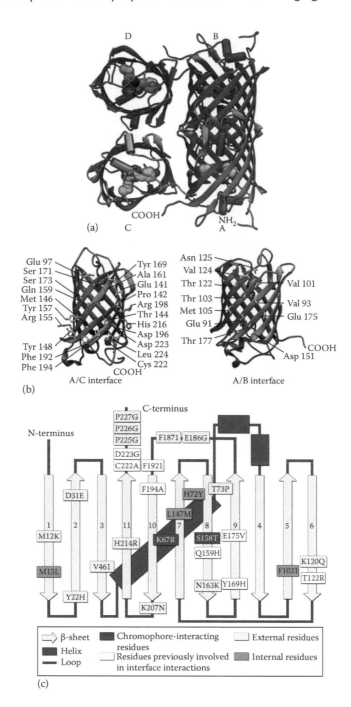

Figure 5.5 Oligomerization and mutagenesis of eqFP611. (a) Tetrameric assembly of eqFP611. (b) Residues involved in interface interactions in eqFP611. (Modified from Wiedenmann, J. et al., *J. Biomed. Opt.*, 10, 14003, 2005.) (c) Schematic display of the β-can structure of eqFP611 showing locations of the modifications introduced to generate the bright monomeric mRuby marker protein. (From Kredel, S. et al., *PLoS One*, 4, e4391, 2009.)

functional green chromophore, 4-(*p*-hydroxybenzylidene)-5-imidazolinone (*p*-HBI), are considerably slower ($t_{0.5}$ = 22–86 min) (Tsien 1998). Upon folding of the β-barrel, the central helix forms along its axis, which features a marked bend at the site of the chromogenic amino acid triad. The surrounding amino acids form a scaffold that positions the amide nitrogen of Gly67 next to the carbonyl carbon of Ser65 (in wild-type GFP). A nucleophilic attack results in formation of a heterocyclic intermediate, as depicted in Figure 5.6. Two amino acids, Arg96 and Glu222 (GFP numbering), located on opposite strands of the β-barrel, are highly conserved and provide a catalyst

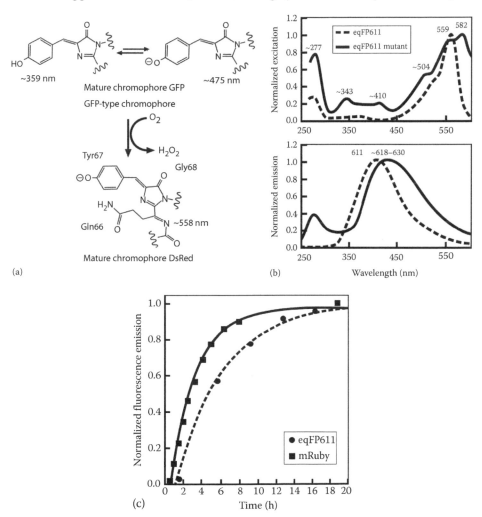

Figure 5.6 Maturation of RFPs. (a) Red fluorescent chromophores of the dsRed type are synthesized from GFP-type chromophores in an autocatalytic reaction. (Modified from Sniegowski, J.A. et al., *J. Biol. Chem.*, 280, 26248, 2005; Nienhaus, G.U. and Wiedenmann, J., *Chemphyschem*, 10, 1369, 2009.) (b) Excitation and emission spectra of the S171F, V184D, and N143S mutant of eqFP611 reveal the presence of green and red fluorescent species (Kredel et al. 2008). (c) The maturation speed can be accelerated by protein engineering as demonstrated for the optimized eqFP611 variant mRuby. (From Kredel, S. et al., *PLoS One*, 4, e4391, 2009.)

Photophysical properties of fluorescent proteins

for chromophore formation (Sniegowski et al. 2005; Wachter 2007). Glu222 has been suggested to function as a general base, facilitating proton abstraction from the heterocycle. Arg96 carries a positive charge and hydrogen bonds to the carbonyl oxygen of the imidazolinone ring, which apparently enhances the nucleophilic character of the glycine nitrogen. In ensuing steps, a water molecule is released, and the tyrosine Cα–Cβ bond is oxidized (Barondeau et al. 2003; Wachter 2007). This latter step is accompanied by hydrogen peroxide formation from molecular oxygen. In the planar p-HBI chromophore, the conjugated π-electron system extends from the p-hydroxybenzyl ring of the tyrosine to the imidazolinone ring. The surrounding residues and bound water molecules hold the two rings of the p-HBI chromophore tightly fixed in a coplanar arrangement, which is essential for a high quantum yield of fluorescence (Prescott et al. 2003; Henderson and Remington 2006). Indeed, the isolated chromophore in solution does not show any fluorescence because it rotates around its exocyclic bonds and undergoes fast radiationless decay by internal conversion upon reaching an avoided crossing of the S_0 and S_1 surfaces (Voityuk et al. 1998; Kummer et al. 2002).

In orange FPs and RFPs such as DsRed and eqFP611, the single bond between the amide nitrogen and Cα of the first of the three chromophore-forming amino acids is oxidized to an acylimine group residing in-plane with the p-HBI chromophore of GFPs (Verkhusha et al. 2004). In this 2-imino-5-(4-hydroxybenzylidene)-imidazolinone chromophore, the conjugated π-electron system is further extended, and thereby, the emission is shifted bathochromically to ~600 nm (Figure 5.5a). However, not all molecules form a red fluorescent chromophore; a certain fraction ends up as stably green fluorescent variants (Baird et al. 2000; Verkhusha et al. 2004) (Figure 5.6b). The fraction of green, incompletely matured species varies among RFPs from different species, from being minor as in eqFP611 from *E. quadricolor* to being significant as in DsRed (Wiedenmann et al. 2002). The residual green fluorescent chromophores reduce the brightness in the red spectral range of the protein bulk, for example, in a cell. Moreover, they prevent colocalization or fluorescence resonance energy transfer (FRET) studies that rely on dual-color labeling, typically with GFP as a second marker. Hence, the (residual) amount of green chromophores in RFPs should be kept to a minimum.

On the basis of the observation of green- and red-emitting species in DsRed (Baird et al. 2000; Gross et al. 2000), it was initially proposed that formation of the red chromophore involves a GFP-like intermediate (Verkhusha et al. 2004). Later, however, it was shown that the GFP-like chromophore is not an intermediate in the maturation pathway of the DsRed-like red chromophore but rather a dead-end product (Verkhusha et al. 2004). More recently, crystallographic and mass spectrometry data and the spectral properties of site-specific mutants of TagRFP allowed proposing another chemical pathway for the formation of the red chromophore. In a first step, the chromophore-forming triad undergoes cyclization. In contrast to the initial model, the next step is formation of the N-acylimine C=N bond, which yields a blue intermediate, with the imidazolinone ring and N-acylimine C=N bond forming the conjugated π-electron system (Subach et al. 2010). Finally, formation of the red chromophore is completed by oxidation of the Cα–Cβ bond of Tyr64 to include the tyrosine side chain into the π-electron system. In blue fluorescent proteins (BFPs), the chemical transformations are stopped right after the N-acylimine formation (Subach et al. 2008).

Maturation times of RFPs from various anthozoans differ considerably (Kredel et al. 2008). For eqFP611, most of the molecules have reached their fluorescent state within 12 h (Figure 5.6c). For DsRed, half-lives of maturation at room temperature beyond 24 h have been reported (Baird et al. 2000). This slow maturation has even been put to good use in a DsRed variant termed the "fluorescent timer" (Terskikh et al. 2000). At present, maturation times of engineered RFPs allow their application in most protein labeling experiments by using a standard overnight expression protocol. However, for experiments that require fast detection of the presence of the FPs, for instance, certain gene expression studies, short maturation times are indispensible.

Maturation times of FPs are often determined from experiments on purified proteins. However, if cells are transfected with DNA, a certain time is required until the DNA molecules migrate to the nucleus, where they are transcribed. This period can be shortened if the cells are injected with mRNA (Wacker et al. 2007). It is, however, important to note that the detectability threshold in imaging applications does not only depend on the number of functional chromophores but also on the number of emitted photons. Therefore, the time point at which the marker is detectable is not only dependent on the maturation time but also on the brightness of the marker (Wiedenmann et al. 2009). Hence, a very bright marker might be detectable earlier than a dim marker despite having a slower maturation speed. In addition, any other factor that affects the number of functional marker molecules in a cell, such as the strength of the promoter driving the expression, the stability of the mRNA, or the stability of the mature protein, can potentially influence the detectability threshold (Wiedenmann et al. 2009).

5.2.3 STOKES SHIFT

Peak excitation and emission wavelengths of presently known RFPs cover the ranges 559–605 nm and 605–670 nm, respectively (Nienhaus and Wiedenmann 2009; Shcherbo et al. 2010). The displacement between the excitation and emission maxima, that is, the Stokes shift, is important for the detectability of the fluorescent marker in devices depending on optical filter systems such as microscopes or fluorescence activated cell sorting (FACS) machines: The larger the Stokes shift, the easier is the separation of the excitation and emission light and, consequently, the signal-to-noise ratio. It can also be beneficial for cross-talk suppression in multicolor applications including FRET.

Natural RFPs typically show Stokes shifts ranging from 20 to 70 nm (Nienhaus and Wiedenmann 2009). The tetrameric eqFP611 displays an emission peak at 611 nm (Figure 5.7) that is Stokes-shifted by 52 nm from its excitation peak at 559 nm (Wiedenmann et al. 2002). In dsFP586, the Stokes shift is reduced to 27 nm. Boxer and coworkers have attributed the very large Stokes shift in the far-red-emitting DsRed variant mPlum to a picosecond solvation response (Abbyad et al. 2007). This time-dependent shift in emission was not observed in its parental proteins, implying that mPlum has a peculiar chromophore environment that allows such a pronounced reorganization. Mutational studies have identified a hydrogen bond between Glu16 and the chromophore as the major determinant of the red shift. The temporal shift of the fluorescence emission in mPlum was explained by a time-dependent interaction between Glu16 and the excited state of the chromophore, with the glutamic acid side chain changing its conformation in response to the modified charge distribution in the

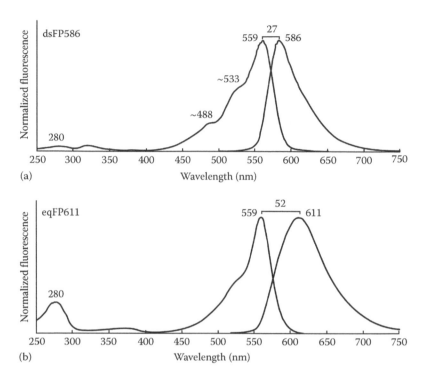

Figure 5.7 Natural RFPs show differences in the length of their Stokes shifts as exemplified by the (a) dsRed homologue dsFP586 from *Discosoma* sp. (Nienhaus and Wiedenmann 2009) and (b) eqFP611 from *E. quadricolor* (Wiedenmann et al. 2002). Numbers indicate fluorescence maxima and the distances between excitation and emission peaks in nanometer.

excited state. mKeima from the *Montipora* stony coral displays the largest Stokes shift among FPs reported to date, with a blue excitation peak at 440 nm and red emission peak at 620 nm (Kogure et al. 2006). Recently, it was confirmed that the mKeima chromophore is protonated and, upon excitation, undergoes excited-state proton transfer to yield an anionic species emitting in the red spectral range (Henderson et al. 2009).

5.2.4 RED SHIFT

In live-cell imaging, RFPs have received special attention because their emission is well separated from the green-yellow autofluorescence of cells, and moreover, the reduced light scattering and absorption at longer wavelengths facilitates imaging of thick tissues and even entire organisms (Shcherbo et al. 2010). RFPs are also desirable for multicolor labeling or FRET experiments (Goedhart et al. 2007; Wiedenmann et al. 2009). Multicolor applications require the separate detection of multiple fluorophores, for example, a green and a red marker protein. Hence, among RFPs, the more red-shifted species are preferable to allow for optimal separation of their emission from that of the green emitters. In combination with YFP, RFPs with an emission wavelength >600 nm are preferable. Orange emitters (560–580 nm) require far-red FPs with emission maxima beyond 630 nm as partners. One should bear in mind that the detectability threshold depends on the number of photons emitted by the marker molecules. Considering the

Photophysical properties of fluorescent proteins

fairly broad emission spectrum of FPs, a bright marker might produce a stronger signal at longer wavelengths than a more red-shifted emitter that is comparably dim.

In the red 2-imino-5-(4-hydroxybenzylidene)-imidazolinone chromophore, the conjugated π-electron system is further extended than in the green p-HBI chromophore (compare Figure 5.6a), and therefore, the emission is shifted bathochromically to ~600 nm. The delocalized π-electron system may be extended even further to include the carbonyl group of the preceding amino acid in FPs, resulting in an additional shift of the emission further into the red. In the monomeric DsRed variant mOrange (Shu et al. 2006) emitting at 562 nm, an oxazole heterocycle is created from the side chain of Thr66, the first amino acid of the chromogenic triad in this protein. Its hydroxyl group apparently attacks the preceding carbonyl carbon. In AsRed, the mutant A143S (Yanushevich et al. 2002) of asFP595 from *Anemonia sulcata*, the x-ray structure revealed that chromophore maturation is accompanied by a break in the polypeptide chain (Andresen et al. 2005; Quillin et al. 2005; Wilmann et al. 2005). Moreover, mass spectroscopy showed that the C-terminal fragment starts with a carbonyl, which is produced by hydrolysis of the intermediately formed acylimine group (Tretyakova et al. 2007). In addition to spontaneous chemical alterations of the green chromophore described earlier, a strictly photoinduced modification occurs in a particular class of photoconvertible FPs that includes Kaede (Ando et al. 2002), EosFP (Wiedenmann et al. 2004; Nienhaus et al. 2005), IrisFP (Adam et al. 2008; Fuchs et al. 2010), and a few other proteins (Shagin et al. 2004; Gurskaya et al. 2006; Oswald et al. 2007). By irradiation into the absorption band of the neutral green chromophore (~400 nm), the backbone is cleaved between the Nα and Cα atoms of the first amino acid in the chromophore triad, which is always a histidine in this class. Concomitantly, a double bond forms between its Cα and Cβ atoms, so that the p-HBI π-conjugation is extended via an all-*trans* ethenylene moiety into the histidine imidazole ring (Mizuno et al. 2003). The orange-colored emission at ~580 nm is similar to the one of DsRed and AsRed, which suggests that the increased delocalization is mainly due to the double bond extension by the ethenylene.

The chromophore environment may also induce bathochromic shifts of the fluorescence to the red region of the spectrum. In the YFP variant of *A. victoria* GFP, Thr203 is replaced by tyrosine. Moreover, residue 65 is Gly or Thr instead of Ser, which promotes ionization of the chromophore. The Tyr203 hydroxybenzyl side chain is stacked on top of the phenol ring of the chromophore, thereby effectively extending the delocalized π-electron system by π-stacking interactions (Wachter et al. 1998). In fact, any aromatic residue at position 203 (His, Trp, Phe, and Tyr) increases the excitation and emission wavelengths by up to 20 nm, with the extent of the shift increasing in the stated order (Tsien 1998).

In most orange FPs or RFPs hitherto known, the anionic fluorophore emits with high quantum yield in the *cis* conformation. The highly fluorescent eqFP611 (Wiedenmann et al. 2002; Nienhaus et al. 2008), with its chromophore in a planar *trans* conformation (Figure 5.8), is thus an exception. It displays the most red-shifted fluorescence emission of any unmodified FPs studied so far, with its emission peak at 611 nm.

In the dimeric variant d1eqFP611, a red-shifted species can be enhanced by irradiation with pulsed 532 nm light, and concomitant changes in the Raman spectrum suggested a *trans–cis* isomerization of the chromophore (Loos et al. 2006). In RFP630,

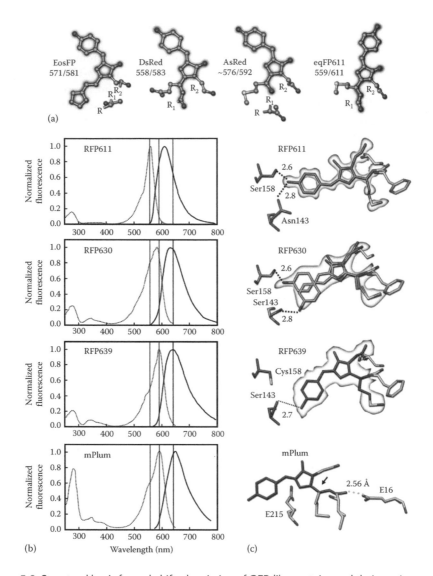

Figure 5.8 Structural basis for red-shifted emission of GFP-like proteins and their engineered variants. (a) Structural models showing differences among red-emitting chromophores. The conjugated π-electron systems are shown in color. Numbers below the FPs names represent the positions of the excitation/emission maxima given in nm. (b and c) X-ray structure models of the chromophores of eqFP611 and DsRed variants (c) show structural changes resulting in red-shifted emission spectra (b). (RFP611, RFP630, and RFP639 spectra and structural models adapted from *Chem. Biol.*, 15, Kredel, S., Nienhaus, K., Oswald, F. et al., Optimized and far-red-emitting variants of fluorescent protein eqFP611, 224–233, Copyright 2008, with permission from Elsevier; Nienhaus, K., Nar, H., Heilker, R., Wiedenmann, J., and Nienhaus, G.U., Trans-cis isomerization is responsible for the red-shifted fluorescence in variants of the red fluorescent protein eqFP611, *J. Am. Chem. Soc.*, 130, 12578–12579, Copyright 2008, American Chemical Society, respectively; The mPlum spectrum and structural model were modified from Wang, L. et al., *Proc. Natl. Acad. Sci. USA*, 101, 16745, 2004; Abbyad, P. et al., *Proc. Natl. Acad. Sci. USA*, 104, 20189, 2007, respectively, Copyright 2007, National Academy of Sciences, USA.) (Modified from Nienhaus, G.U. and Wiedenmann, J., *Chemphyschem*, 10, 1369, 2009.)

Photophysical properties of fluorescent proteins

the mutation N143S gives rise to a substantial red shift, and the x-ray structure of its dimeric version reveals a mixture of *cis* and *trans* chromophores (Kredel et al. 2008; Nienhaus et al. 2008). The *cis* form is stabilized by a hydrogen bond to the newly introduced Ser143. An additional modification, S158C, in the variant RFP639 (Figure 5.7) completely removes the hydrogen bond stabilization of the *trans* form and leads to a pure *cis* conformer, with a peak emission at 639 nm (Nienhaus et al. 2008). The same effect was also observed upon replacement of cysteine by alanine.

For chromophores, for which the dipole moment increases significantly upon optical excitation, the fluorescence emission often depends strongly on solvent polarity and shifts to longer wavelengths during the excited-state lifetime. This dynamic Stokes shift has been described earlier for mPlum (Abbyad et al. 2007).

5.2.5 PHOTOSTABILITY

Since practically all FPs applications rely on fluorescence measurements, photostability is a key parameter for the usefulness of any FPs marker. The photostability of a chromophore is quantified by its quantum yield of photobleaching, Φ_b, which, for an ensemble, is the ratio between the number of photobleached molecules and the total number of photons absorbed within a certain time interval (Figure 5.9). Typically, the fluorescent chromophore will emit 10^4–10^5 photons until it suffers permanent photodestruction (Wiedenmann et al. 2002; Schenk et al. 2004). Although insufficient photostability frequently limits the performance of a particular FPs, this particular property has often been neglected in FPs optimization. At present, the relation between structure and photostability is only poorly understood. Therefore, to improve photostability, multiple rounds of (directed or random) mutagenesis are typically performed, with screening for photostability after each round. Tsien and coworkers noticed the importance of residue 163 in influencing the photostability of mRFP1 variants (Shaner et al. 2008). For mTFP1, the N63T mutation resulted in a particularly large increase in photostability (Ai et al. 2006).

A key prerequisite for bright FPs fluorescence is the rigid enclosure of the chromophore within the protein cavity. However, proteins are inherently dynamic systems that can fluctuate among a vast number of conformational substates under physiological conditions (Frauenfelder et al. 1991). Structural fluctuations of the protein matrix give rise to chromophore motions, which may then cause variations in the fluorescence emission. Indeed, such fluctuations are universally observed in single-molecule experiments on FPs, clearly showing that thermally activated or light-driven transitions continuously occur between conformations with distinctly different emission properties (Schenk et al. 2004).

Fluorescence correlation spectroscopy (FCS) on protein solutions revealed emission fluctuations with characteristic times faster than milliseconds at room temperature (Haupts et al. 1998). For eqFP611 and DsRed, a pronounced sensitivity of flickering on the excitation rate was observed (Malvezzi-Campeggi et al. 2001; Schenk et al. 2004). The switching frequencies between bright and dark states increased from 3.5 kHz (at 100 kHz) to >15 kHz (at 10 MHz) for eqFP611, proving the photoinduced nature of these interconversions. A three-state analysis involving two bright and one dark state was able to quantitatively describe the power dependence of the FCS data (Schenk et al. 2004).

Flickering on longer time scales can be observed in single-molecule studies on immobilized FPs. Figure 5.9a shows a confocal laser scan image of wild-type eqFP611

Photophysical properties of fluorescent proteins

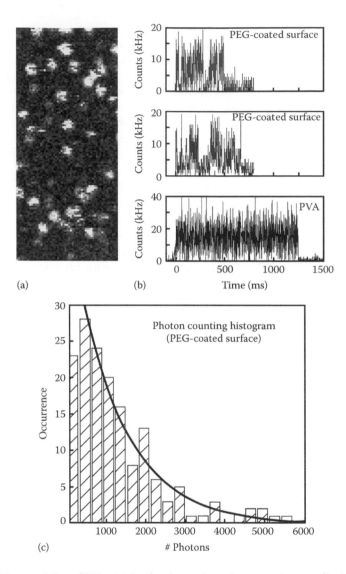

(a)   (b)

(c)

Figure 5.9 Photostability of RFPs. (a) Confocal scanning microscopy image of individual eqFP611 molecules attached to a PEG-coated surface. (b) Representative fluorescence time trajectories of individual eqFP611 molecules attached to a PEG-coated surface and embedded in PVA. (c) Histogram of the total number of photons collected from 160 individual time traces before photodestruction of eqFP611 molecules attached to a PEG-coated surface including an exponential fit. (Modified from Wiedenmann, J. et al., *Proc. Natl. Acad. Sci. USA*, 99, 11646, Copyright 2002, National Academy of Sciences USA; Schenk, A. et al., *Biophys. J.*, 86, 384, Copyright 2004, Biophysical Society.)

immobilized on a poly(ethylene glycol) (PEG)-coated surface, using 514 nm excitation from an Ar ion laser. For many such individual spots, the emission was recorded as a function of time. Typical emission time traces show that photons from individual PEG-conjugated eqFP611 molecules are detected at a rate of ~10 kHz for several hundred milliseconds until photobleaching occurs in a single step (Figure 5.9b, top, middle)

(Schenk et al. 2004). Thus, eqFP611 is monomeric when immobilized on the surface. Extended dark periods are clearly visible for time intervals >10 ms. Emission fluctuations on the millisecond timescale are markedly reduced for eqFP611 molecules embedded in poly(vinyl alcohol) (PVA) (Figure 5.9b, bottom) (Wiedenmann et al. 2002). Remarkably, flickering is completely absent when eqFP611 molecules are deposited on bare glass surfaces (Schenk et al. 2004). Apparently, fluctuations are suppressed by attachment of the proteins to rigid surfaces, supporting the notion that protein motions are crucially involved.

Stabilization of the chromophore within the protein scaffold might also explain the strongly decreased photobleaching yield of eqFP611 molecules embedded in PVA or deposited on bare glass surfaces compared to PEG-conjugated molecules (Wiedenmann et al. 2002; Schenk et al. 2004). Consequently, mutations that increase the rigidity of the protein structure or reduce motions of the chromophore can help to improve the photostability of the marker. In Figure 5.9c, we plot a photon counting histogram showing the total number of photons PEG-conjugated eqFP611 molecules collected from 160 individual time traces before photodestruction occurred.

5.2.6 PHOTOACTIVATION OF RFPs

The chromophore cage in eqFP611 is structured in such a way that a coplanar chromophore can be accommodated in the *cis* as well as in the *trans* isomeric states, both of which are highly fluorescent (Nienhaus et al. 2008). In anthozoan FPs, it is more often observed that the *trans* isomer is nonplanar and/or shows enhanced flexibility, so that it is nonfluorescent, as in the class of chromoproteins (Prescott et al. 2003). In a number of photoswitchable proteins, isomerization can be efficiently driven by light, so that FPs fluorescence can be switched on or off by external irradiation (Chudakov et al. 2003; Andresen et al. 2005; Habuchi et al. 2005; Loos et al. 2006) (Figure 5.10). Recently, Jakobs and coworkers reported two reversibly switching variants of the monomeric DsRed variant mCherry (Stiel et al. 2008). In rsCherry, the excitation light (absorbed by the anionic chromophore at 550 nm) turns its fluorescence on, whereas 450 nm light (absorbed by the neutral chromophore) switches it off. By contrast, rsCherryRev is switched off by the excitation light; 450 nm light switches it on again. Yet another mCherry variant, PAmCherry, is initially dark but becomes red fluorescent upon irradiation with violet light (Subach et al. 2009a,b). We have recently introduced IrisFP, a variant of EosFP that can be (irreversibly) photoconverted from a GFP to an RFPs by irradiation with ~400 nm light and, in addition, reversibly switched between a bright and a dark state in both the green and the red forms (Adam et al. 2008). For live-cell imaging applications, a monomeric form, mIrisFP, has also been generated (Fuchs et al. 2010). Figure 5.10a and b shows a sequence of switching events of the red mIrisFP chromophore and the corresponding chromophore models. Figure 5.10c illustrates an application of mIrisFP in ultrahigh-resolution dual-color total internal reflection fluorescence (TIRF) imaging, using photoactivation localization microscopy (PALM) (Betzig et al. 2006; Hess et al. 2006). This imaging technique (and related ones) is based on the detection of the emission from individual fluorophores, which allows their location to be determined very precisely. Indeed, with this and related approaches, an image resolution of >50 nm can be achieved routinely, which is far below the diffraction-limited lateral resolution of ~200 nm. Combining pulse–chase labeling with superresolution imaging, we have visualized focal adhesion points and their dynamics in live HeLa cells using the paxillin–mIrisFP fusion protein (Fuchs et al. 2010).

Photophysical properties of fluorescent proteins

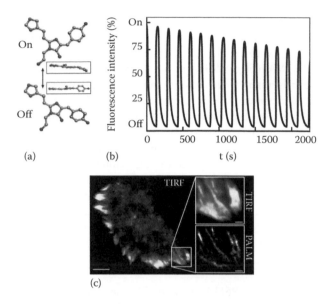

Figure 5.10 Photoactivation of RFPs. (a) Structural changes associated with the light-driven *cis–trans* isomerization result in (b) on/off switching of the red-emitting form of IrisFP. (c) Superresolution imaging of the distribution of paxillin in the focal adhesions of HeLa cells using the red form of mIrisFP. The red form of mIrisFP was produced by 405 nm laser irradiation for 30 s (50 W cm²). Image acquisition was performed by collecting 12,000 camera frames with a dwell time of 50 ms each, using 561 nm (200 W cm²) and 473 nm (30 mW cm²) laser illumination. The entire cell imaged in TIRF mode is shown in the left panel (scale bar, 5 µm). The area marked by the white frame is shown in an expanded view as a TIRF (upper right panel) and PALM image (lower right panel) (scale bar, 1 µm). (c: Modified from Fuchs, J. et al., *Nat. Methods*, 7, 627, 2010.)

Paxillin is a multidomain adaptor protein that localizes in focal adhesion complexes, which form at distinct sites where integrins are linked to the extracellular matrix.

RFPs with overlapping emission spectra can be identified by measuring the fluorescence emission intensity in two distinct spectral channels spanning ~100 nm of the visible spectrum (Gunewardene et al. 2011). This approach enables multiplexing with superresolution using three different photoactivatable RFPs in parallel.

5.3 CONCLUSION

GFP-like RFPs offer most favorable properties for live-cell imaging. Detailed insights into structure–function relations of RFPs have facilitated the development of advanced variants with optical and biochemical features customized for distinct imaging applications such as multiplexing, whole-body/deep-tissue imaging, or superresolution microscopy. The brightness of RFPs is of crucial importance for marker protein development as it affects the imaging threshold in the context of maturation speed, detectability at the red end of the emission spectrum, and photobleaching rate. Future challenges are to further improve the photostability and brightness of RFPs in the near-infrared spectral region and to reduce their oligomerization tendency.

Photophysical properties of fluorescent proteins

REFERENCES

Abbyad P., Childs W., Shi X.H., and Boxer S.G. (2007) Dynamic Stokes shift in green fluorescent protein variants. *Proceedings of the National Academy of Sciences of the United States of America* 104: 20189–20194.

Adam V., Lelimousin M., Boehme S. et al. (2008) Structural characterization of IrisFP, an optical highlighter undergoing multiple photo-induced transformations. *Proceedings of the National Academy of Sciences of the United States of America* 105: 18343–18348.

Ai H.W., Henderson J.N., Remington S.J., and Campbell R.E. (2006) Directed evolution of a monomeric, bright and photostable version of Clavularia cyan fluorescent protein: Structural characterization and applications in fluorescence imaging. *Biochemical Journal* 400: 531–540.

Alieva N.O., Konzen K.A., Field S.F. et al. (2008) Diversity and evolution of coral fluorescent proteins. *PLoS One* 3: e2680.

Ando R., Hama H., Yamamoto-Hino M., Mizuno H., and Miyawaki A. (2002) An optical marker based on the UV-induced green-to-red photoconversion of a fluorescent protein. *Proceedings of the National Academy of Sciences of the United States of America* 99: 12651–12656.

Andresen M., Wahl M.C., Stiel A.C. et al. (2005) Structure and mechanism of the reversible photoswitch of a fluorescent protein. *Proceedings of the National Academy of Sciences of the United States of America* 102: 13070–13074.

Baird G.S., Zacharias D.A., and Tsien R.Y. (2000) Biochemistry, mutagenesis, and oligomerization of DsRed, a red fluorescent protein from coral. *Proceedings of the National Academy of Sciences of the United States of America* 97: 11984–11989.

Barondeau D.P., Putnam C.D., Kassmann C.J., Tainer J.A., and Getzoff E.D. (2003) Mechanism and energetics of green fluorescent protein chromophore synthesis revealed by trapped intermediate structures. *Proceedings of the National Academy of Sciences of the United States of America* 100: 12111–12116.

Betzig E., Patterson G.H., Sougrat R. et al. (2006) Imaging intracellular fluorescent proteins at nanometer resolution. *Science* 313: 1642–1645.

Bevis B.J. and Glick B.S. (2002) Rapidly maturing variants of the *Discosoma* red fluorescent protein (DsRed). *Nature Biotechnology* 20: 83–87.

Billinton N. and Knight A.W. (2001) Seeing the wood through the trees: A review of techniques for distinguishing green fluorescent protein from endogenous autofluorescence. *Analytical Biochemistry* 291: 175–197.

Campbell R.E., Tour O., Palmer A.E. et al. (2002) A monomeric red fluorescent protein. *Proceedings of the National Academy of Sciences of the United States of America* 99: 7877–7882.

Chudakov D.M., Belousov V.V., Zaraisky A.G. et al. (2003) Kindling fluorescent proteins for precise in vivo photolabeling. *Nature Biotechnology* 21: 191–194.

D'Angelo C., Denzel A., Vogt A. et al. (2008) Blue light regulation of host pigment in reef-building corals. *Marine Ecology Progress Series* 364: 97–106.

Day R.N. and Davidson M.W. (2009) The fluorescent protein palette: Tools for cellular imaging. *Chemical Society Reviews* 38: 2887–2921.

Filonov G.S., Piatkevich K.D., Ting L.-M. et al. (2011) Bright and stable near-infrared fluorescent protein for in vivo imaging. *Nature Biotechnology* 29: 757–761.

Fradkov A.F., Verkhusha V.V., Staroverov D.B. et al. (2002) Far-red fluorescent tag for protein labelling. *Biochemical Journal* 368: 17–21.

Frauenfelder H., Nienhaus G.U., and Johnson J.B. (1991) Rate processes in proteins. *Berichte der Bunsengesellschaft fuer Physikalische Chemie* 95: 272–278.

Fuchs J., Böhme S., Oswald F. et al. (2010) Imaging protein movements in live cells with super-resolution using mIrisFP. *Nature Methods* 7: 627–630.

Glazer A.N. (1985) Light harvesting by phycobilisomes. *Annual Review of Biophysics and Biophysical Chemistry* 14: 47–77.

Goedhart J., Vermeer J.E., Adjobo-Hermans M.J., van Weeren L., and Gadella T.W. (2007) Sensitive detection of p65 homodimers using red-shifted and fluorescent protein-based FRET couples. *PLoS One* 2: e1011.

Gross L.A., Baird G.S., Hoffman R.C., Baldridge K.K., and Tsien R.Y. (2000) The structure of the chromophore within DsRed, a red fluorescent protein from coral. *Proceedings of the National Academy of Sciences of the United States of America* 97: 11990–11995.

Gunewardene Mudalige S., Subach Fedor V., Gould Travis J. et al. (2011) Superresolution imaging of multiple fluorescent proteins with highly overlapping emission spectra in living cells. *Biophysical Journal* 101: 1522–1528.

Gurskaya N.G., Fradkov A.F., Terskikh A. et al. (2001) GFP-like chromoproteins as a source of far-red fluorescent proteins. *FEBS Letters* 507: 16–20.

Gurskaya N.G., Verkhusha V.V., Shcheglov A.S. et al. (2006) Engineering of a monomeric green-to-red photoactivatable fluorescent protein induced by blue light. *Nature Biotechnology* 24: 461–465.

Habuchi S., Ando R., Dedecker P. et al. (2005) Reversible single-molecule photoswitching in the GFP-like fluorescent protein Dronpa. *Proceedings of the National Academy of Sciences of the United States of America* 102: 9511–9516.

Haupts U., Maiti S., Schwille P., and Webb W.W. (1998) Dynamics of fluorescence fluctuations in green fluorescent protein observed by fluorescence correlation spectroscopy. *Proceedings of the National Academy of Sciences of the United States of America* 95: 13573–13578.

Henderson J.N., Osborn M.F., Koon N. et al. (2009) Excited state proton transfer in the red fluorescent protein mKeima. *Journal of the American Chemical Society* 131: 13212–13213.

Henderson J.N. and Remington S.J. (2006) The kindling fluorescent protein: A transient photoswitchable marker. *Physiology* 21: 162–170.

Hess S.T., Girirajan T.P., and Mason M.D. (2006) Ultra-high resolution imaging by fluorescence photoactivation localization microscopy. *Biophysical Journal* 91: 4258–4272.

Karasawa S., Araki T., Nagai T., Mizuno H., and Miyawaki A. (2004) Cyan-emitting and orange-emitting fluorescent proteins as a donor/acceptor pair for fluorescence resonance energy transfer. *Biochemical Journal* 381: 307–312.

Kogure T., Karasawa S., Araki T. et al. (2006) A fluorescent variant of a protein from the stony coral *Montipora* facilitates dual-color single-laser fluorescence cross-correlation spectroscopy. *Nature Biotechnology* 24: 577–581.

Kredel S., Nienhaus K., Oswald F. et al. (2008) Optimized and far-red-emitting variants of fluorescent protein eqFP611. *Chemistry and Biology* 15: 224–233.

Kredel S., Oswald F., Nienhaus K. et al. (2009) mRuby, a bright monomeric red fluorescent protein for labeling of subcellular structures. *PLoS One* 4: e4391.

Kummer A.D., Kompa C., Niwa H. et al. (2002) Viscosity-dependent fluorescence decay of the GFP chromophore in solution due to fast internal conversion. *Journal of Physical Chemistry B* 106: 7554–7559.

Lin M.Z., McKeown M.R., Ng H.-L. et al. (2009) Autofluorescent proteins with excitation in the optical window for intravital imaging in mammals. *Chemistry and Biology* 16: 1169–1179.

Loos D.C., Habuchi S., Flors C. et al. (2006) Photoconversion in the red fluorescent protein from the sea anemone *Entacmaea quadricolor*: Is cis-trans isomerization involved? *Journal of the American Chemical Society* 128: 6270–6271.

Malvezzi-Campeggi F., Jahnz M., Heinze K.G., Dittrich P., and Schwille P. (2001) Light-induced flickering of DsRed provides evidence for distinct and interconvertible fluorescent states. *Biophysical Journal* 81: 1776–1785.

Marden L. (1956) Camera under the sea. *National Geographic* 109: 162–200.

Matz M.V., Fradkov A.F., Labas Y.A. et al. (1999) Fluorescent proteins from nonbioluminescent *Anthozoa* species. *Nature Biotechnology* 17: 969–973.

Merzlyak E.M., Goedhart J., Shcherbo D. et al. (2007) Bright monomeric red fluorescent protein with an extended fluorescence lifetime. *Nature Methods* 4: 555–557.

Mizuno H., Mal T.K., Tong K.I. et al. (2003) Photo-induced peptide cleavage in the green-to-red conversion of a fluorescent protein. *Molecular Cell* 12: 1051–1058.

Morozova K.S., Piatkevich K.D., Gould T.J. et al. (2010) Far-red fluorescent protein excitable with red lasers for flow cytometry and superresolution STED nanoscopy. *Biophysical Journal* 99: L13–L15.

Nienhaus G.U., Nienhaus K., Holzle A. et al. (2006) Photoconvertible fluorescent protein EosFP: Biophysical properties and cell biology applications. *Photochemistry and Photobiology* 82: 351–358.

Nienhaus G.U. and Wiedenmann J. (2009) Structure, dynamics and optical properties of fluorescent proteins: Perspectives for marker development. *Chemphyschem* 10: 1369–1379.

Nienhaus K., Nar H., Heilker R., Wiedenmann J., and Nienhaus G.U. (2008) Trans-cis isomerization is responsible for the red-shifted fluorescence in variants of the red fluorescent protein eqFP611. *Journal of the American Chemical Society* 130: 12578–12579.

Nienhaus K., Nienhaus G.U., Wiedenmann J., and Nar H. (2005) Structural basis for photo-induced protein cleavage and green-to-red conversion of fluorescent protein EosFP. *Proceedings of the National Academy of Sciences of the United States of America* 102: 9156–9159.

Nienhaus K., Vallone B., Renzi F., Wiedenmann J., and Nienhaus G.U. (2003) Crystallization and preliminary X-ray diffraction analysis of the red fluorescent protein eqFP611. *Acta Crystallographica Section D. Biological Crystallography* 59: 1253–1255.

Oswald F., Schmitt F., Leutenegger A. et al. (2007) Contributions of host and symbiont pigments to the coloration of reef corals. *FEBS Journal* 274: 1102–1109.

Petersen J., Wilmann P.G., Beddoe T. et al. (2003) The 2.0-angstrom crystal structure of eqFP611, a far red fluorescent protein from the sea anemone *Entacmaea quadricolor*. *Journal of Biological Chemistry* 278: 44626–44631.

Piatkevich K.D., Hulit J., Subach O.M. et al. (2010) Monomeric red fluorescent proteins with a large Stokes shift. *Proceedings of the National Academy of Sciences of the United States of America* 107: 5369–5374.

Prescott M., Ling M., Beddoe T. et al. (2003) The 2.2 Å crystal structure of a pocilloporin pigment reveals a nonplanar chromophore conformation. *Structure* 11: 275–284.

Quillin M.L., Anstrom D.M., Shu X. et al. (2005) Kindling fluorescent protein from *Anemonia sulcata*: Dark-state structure at 1.38 A resolution. *Biochemistry* 44: 5774–5787.

Schenk A., Ivanchenko S., Rocker C., Wiedenmann J., and Nienhaus G.U. (2004) Photodynamics of red fluorescent proteins studied by fluorescence correlation spectroscopy. *Biophysical Journal* 86: 384–394.

Shagin D.A., Barsova E.V., Yanushevich Y.G. et al. (2004) GFP-like proteins as ubiquitous metazoan superfamily: Evolution of functional features and structural complexity. *Molecular Biology and Evolution* 21: 841–850.

Shaner N.C., Campbell R.E., Steinbach P.A. et al. (2004) Improved monomeric red, orange and yellow fluorescent proteins derived from *Discosoma* sp. red fluorescent protein. *Nature Biotechnology* 22: 1567–1572.

Shaner N.C., Lin M.Z., McKeown M.R. et al. (2008) Improving the photostability of bright monomeric orange and red fluorescent proteins. *Nature Methods* 5: 545–551.

Shcherbo D., Murphy C.S., Ermakova G.V. et al. (2009) Far-red fluorescent tags for protein imaging in living tissues. *Biochemical Journal* 418: 567–574.

Shcherbo D., Shemiakina I.I., Ryabova A.V. et al. (2010) Near-infrared fluorescent proteins. *Nature Methods* 7: 827–829.

Shu X., Royant A., Lin M.Z. et al. (2009) Mammalian expression of infrared fluorescent proteins engineered from a bacterial phytochrome. *Science* 324: 804–807.

Shu X., Shaner N.C., Yarbrough C.A., Tsien R.Y., and Remington S.J. (2006) Novel chromophores and buried charges control color in mFruits. *Biochemistry* 45: 9639–9646.

Sniegowski J.A., Lappe J.W., Patel H.N., Huffman H.A., and Wachter R.M. (2005) Base catalysis of chromophore formation in Arg96 and Glu222 variants of green fluorescent protein. *Journal of Biological Chemistry* 280: 26248–26255.

Photophysical properties of fluorescent proteins

Stiel A.C., Andresen M., Bock H. et al. (2008) Generation of monomeric reversibly switchable red fluorescent proteins for far-field fluorescence nanoscopy. *Biophysical Journal* 95: 2989–2997.

Strack R.L., Bhattacharyya D., Glick B., and Keenan R. (2009a) Noncytotoxic orange and red/green derivatives of DsRed-Express2 for whole-cell labeling. *BMC Biotechnology* 9: 32.

Strack R.L., Hein B., Bhattacharyya D. et al. (2009b) A rapidly maturing far-red derivative of DsRed-Express2 for whole-cell labeling. *Biochemistry* 48: 8279–8281.

Strack R.L., Strongin D.E., Bhattacharyya D., Tao, W., Berman, A., Broxmeyer, H., Keenan, R., and Glick, B. (2008) A noncytotoxic DsRed variant for whole-cell labeling. *Nature Methods* 5: 955–957.

Subach F.V., Malashkevich V.N., Zencheck W.D. et al. (2009a) Photoactivation mechanism of PAmCherry based on crystal structures of the protein in the dark and fluorescent states. *Proceedings of the National Academy of Sciences of the United States of America* 106: 21097–21102.

Subach F.V., Patterson G.H., Manley S., Lippincott-Schwartz, J., and Verkhusha, V.V. (2009b) Photoactivatable mCherry for high-resolution two-color fluorescence microscopy. *Nature Methods* 6: 153–159.

Subach O.M., Gundorov I.S., Yoshimura M. et al. (2008) Conversion of red fluorescent protein into a bright blue probe. *Chemistry and Biology* 59: 1116–1124.

Subach O.M., Malashkevich V.N., Zencheck W.D. et al. (2010) Structural characterization of acylimine-containing blue and red chromophores in mTagBFP and TagRFP fluorescent proteins. *Chemistry and Biology* 17: 333–341.

Sun L., Wang S., Chen L., and Gong X. (2003) Promising fluorescent probes from phycobiliproteins. *IEEE Journal of Selected Topics in Quantum Electronics* 9: 177–188.

Terskikh A., Fradkov A., Ermakova G. et al. (2000) "Fluorescent timer": Protein that changes color with time. *Science* 290: 1585–1588.

Terskikh A.V., Fradkov A.F., Zaraisky A.G., Kajava A.V., and Angres B. (2002) Analysis of DsRed Mutants. Space around the fluorophore accelerates fluorescence development. *Journal of Biological Chemistry* 277: 7633–7636.

Tretyakova Y.A., Pakhomov A.A., and Martynov V.I. (2007) Chromophore structure of the kindling fluorescent protein asFP595 from *Anemonia sulcata*. *Journal of the American Chemical Society* 129: 7748–7749.

Tsien R.Y. (1998) The green fluorescent protein. *Annual Review of Biochemistry* 67: 509–544.

Tsutsui H., Karasawa S., Okamura Y., and Miyawaki A. (2008) Improving membrane voltage measurements using FRET with new fluorescent proteins. *Nature Methods* 5: 683–685.

Verkhusha V.V., Chudakov D.M., Gurskaya N.G., Lukyanov S., and Lukyanov K.A. (2004) Common pathway for the red chromophore formation in fluorescent proteins and chromoproteins. *Chemistry and Biology* 11: 845–854.

Verkhusha V.V. and Lukyanov K.A. (2004) The molecular properties and applications of *Anthozoa* fluorescent proteins and chromoproteins. *Nature Biotechnology* 22: 289–296.

Voityuk A.A., Michel-Beyerle M.E., and Rösch N. (1998) Quantum chemical modeling of structure and absorption spectra of the chromophore in green fluorescent proteins. *Chemical Physics Letters* 296: 269–276.

Vrzheshch P.V., Akovbian N.A., Varfolomeyev S.D., and Verkhusha V.V. (2000) Denaturation and partial renaturation of a tightly tetramerized DsRed protein under mildly acidic conditions. *FEBS Letters* 487: 203–208.

Wachter R.M. (2007) Chromogenic cross-link formation in green fluorescent protein. *Accounts of Chemical Research* 40: 120–127.

Wachter R.M., Elsliger M.A., Kallio K., Hanson G.T., and Remington S.J. (1998) Structural basis of spectral shifts in the yellow-emission variants of green fluorescent protein. *Structure* 6: 1267–1277.

Wacker S.A., Oswald F., Wiedenmann J., and Knochel W. (2007) A green to red photoconvertible protein as an analyzing tool for early vertebrate development. *Developmental Dynamics* 236: 473–480.

Wall M.A., Socolich M., and Ranganathan R. (2000) The structural basis for red fluorescence in the tetrameric GFP homolog DsRed. *Nature Structural and Molecular Biology* 7: 1133–1138.

Wang L., Jackson W.C., Steinbach P.A., and Tsien R.Y. (2004) Evolution of new nonantibody proteins via iterative somatic hypermutation. *Proceedings of the National Academy of Sciences of the United States of America* 101: 16745–16749.

Wiedenmann J. (1997). The application of an orange fluorescent protein and further colored proteins and the corresponding genes from the species group *Anemonia* sp. (*sulcata*) Pennant, (Cnidaria, Anthozoa, Actinaria) in gene technology and molecular biology. Patent DE 197 18 640. Deutsches Patent und Markenamt, Germany. Patent DE 197 18 640, pp. 1–18.

Wiedenmann J., D'Angelo C., and Nienhaus G.U. (2011a). Fluorescent proteins: Nature's colorful gifts for live cell imaging. In G. Jung (eds) *Fluorescent Proteins I—From Fundamental Research to Bioanalytics*, Springer Verlag, Berlin, Germany, pp. 1–31.

Wiedenmann J., Elke C., Spindler K.D., and Funke W. (2000) Cracks in the beta-can: Fluorescent proteins from *Anemonia sulcata* (Anthozoa, Actinaria). *Proceedings of the National Academy of Sciences of the United States of America* 97: 14091–14096.

Wiedenmann J., Gayda S., Adam V. et al. (2011b) From EosFP to mIrisFP: Structure-based development of advanced photoactivatable marker proteins of the GFP-family. *Journal of Biophotonics* 4: 377–390.

Wiedenmann J., Ivanchenko S., Oswald F. et al. (2004) EosFP, a fluorescent marker protein with UV-inducible green-to-red fluorescence conversion. *Proceedings of the National Academy of Sciences of the United States of America* 101: 15905–15910.

Wiedenmann J., Oswald F., and Nienhaus G.U. (2009) Fluorescent proteins for live cell imaging: Opportunities, limitations, and challenges. *IUBMB Life* 61: 1029–1042.

Wiedenmann J., Schenk A., Rocker C. et al. (2002) A far-red fluorescent protein with fast maturation and reduced oligomerization tendency from *Entacmaea quadricolor* (Anthozoa, Actinaria). *Proceedings of the National Academy of Sciences of the United States of America* 99: 11646–11651.

Wiedenmann J., Vallone B., Renzi F. et al. (2005) Red fluorescent protein eqFP611 and its genetically engineered dimeric variants. *Journal of Biomedical Optics* 10: 14003.

Wilmann P.G., Petersen J., Devenish R.J., Prescott M., and Rossjohn J. (2005) Variations on the GFP chromophore. *Journal of Biological Chemistry* 280: 2401–2404.

Yang M., Jiang P., Yamamoto N. et al. (2005) Real-time whole-body imaging of an orthotopic metastatic prostate cancer model expressing red fluorescent protein. *Prostate* 62: 374–379.

Yanushevich Y.G., Staroverov D.B., Savitsky A.P. et al. (2002) A strategy for the generation of non-aggregating mutants of Anthozoa fluorescent proteins. *FEBS Letters* 511: 11–14.

Yarbrough D., Wachter R.M., Kallio K., Matz M.V., and Remington S.J. (2001) Refined crystal structure of DsRed, a red fluorescent protein from coral, at 2.0-A resolution. *Proceedings of the National Academy of Sciences of the United States of America* 98: 462–467.

Optical highlighter photophysical properties

George H. Patterson
National Institutes of Health

Contents

6.1 INTRODUCTION

Other chapters in this book detail the fantastic diversity of fluorescent proteins (FP) from many different marine organisms, their development into various imaging tools, and their use in living cell biology studies. Here, the emphasis will be on the photophysical characteristics of a special class termed optical highlighter fluorescent proteins. These proteins are either nonfluorescent or can be made dark at the activated fluorescence wavelength, but display enhanced fluorescence after irradiation with the proper wavelength of light. Although the first examples of these FPs were reported shortly after the advent of green fluorescent protein (GFP), developments of superresolution microscopy using molecular localization (Patterson 2009), such as photoactivated localization microscopy (PALM) (Betzig et al. 2006) and fluorescence-photoactivated localization microscopy (F-PALM) (Hess et al. 2006), have heightened interest in these molecules. Optical highlighter fluorescent proteins also allow experimental alternatives to photobleaching when studying protein kinetics, gene expression, organelle dynamics, and cellular dynamics within a living specimen.

The molecules developed thus far are derived from several species of organisms and cover a broad wavelength range. They share many characteristics with their conventional counterparts, which is a helpful starting point in describing them.

First, they all have the same basic structure as GFP (Chalfie et al. 1994), which is an 11 strand β-barrel with an α-helix located in the interior of the barrel (Ormo et al. 1996, Yang et al. 1996). The chromophores are located within the interior α-helix and are made up of three amino acids, which undergo posttranslational modifications to form cyclized tripeptides. The positions of the amino acids in these tripeptides will be denoted as XYG for discussion here. The X position is the most N-terminal in the chromophore and can be almost any amino acid; the Y position is generally a tyrosine but is substituted by histidine, tryptophan, or phenylalanine to alter the excitation and emission spectra. The G position is invariably a glycine (Figure 6.1).

Chromophore formation generally proceeds in at least two sequential reactions. The cyclization step takes place when an amido nitrogen in the G position undergoes a nucleophilic attack on the peptidic carbonyl of the X position to produce a five-membered imidazolone ring (Heim et al. 1994). The next steps involve the dehydration of the carbonyl carbon in the X position amino acid and oxidation of the Y position C_α–C_β bond.

The chromophores for both conventional and optical highlighter fluorescent proteins generally come in three basic structures. One is exemplified in the GFP from *Aequorea victoria* (Chalfie et al. 1994). The *p*-hydroxybenzylidene-imidazolidinone chromophore consists of a cyclized tripeptide (XYG) made up of residues serine 65, tyrosine 66, and glycine 67 located within the interior α-helix. The second basic chromophore structure

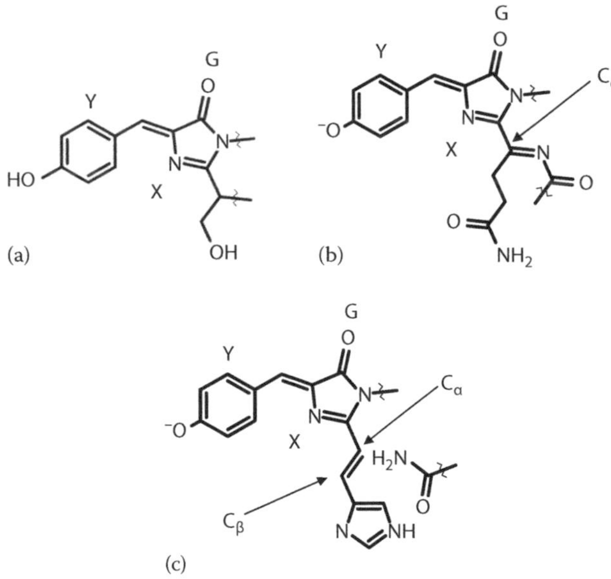

Figure 6.1 Basic chromophore structures. (a) *A. victoria* wtGFP, (b) DsRed, and (c) Kaede chromophores are shown here to represent three basic structures that make up the majority of FP chromophore structures. Since different amino acids can occupy the various positions, the lettering X, Y, and G (N-terminal to C-terminal) denote the three amino acids that form the core of the chromophores. The G position is invariably a glycine. The Y position is generally a tyrosine but can be substituted to alter the color. The X position can be one of several amino acids depending on the origin of the FP.

consists of a similar cyclized tripeptide, but it has the π-bonding system extended. The first of this type of chromophore was described for DsRed (Matz et al. 1999), which has extended the π-conjugation through the C_α–N bond of the X position residue (Wall et al. 2000, Yarbrough et al. 2001) (Figure 6.1b), whereas the third type extends the π-conjugation through the position X C_α–C_β bond (Mizuno et al. 2003) (Figure 6.1c). Remarkably, many varieties of both conventional fluorescent proteins and optical highlighter fluorescent proteins have been developed from the overall barrel shape and these three basic chromophores. The derivations rely on protein alterations within and around the chromophore, and these changes can alter the excitation and emission wavelengths, photostability, protein folding, chromophore formation, and molecular brightness.

6.2 CLASSES OF OPTICAL HIGHLIGHTERS

Most optical highlighter fluorescent proteins are subclassed based on the spectral characteristics of their activated and nonactivated states (green versus red fluorescence) and on the reversibility of the activated state (irreversible versus reversible) (Table 6.1). These proteins are often referred to as photoactivatable, photoconvertible, or photoswitchable fluorescent proteins depending on the mechanism of the change in the chromophore to elicit the fluorescence alterations. However, the naming of the different classes using these terms is still a little fluid, and unfortunately the terms are sometimes used interchangeably. In addition, some of the proteins have characteristics that can place them in more than one category. For the purposes of this chapter, most of the discussed proteins will fall into one of four designations and will grouped based on the characteristics of their proposed chromophore alterations and spectral changes during the highlighting step. Photoactivatable will usually denote molecules that display a change in fluorescence from an "off" state to an "on" state or a wavelength shift in their excitation spectrum, photoconvertible will indicate a shift in both fluorescence excitation and emission spectra, and photoswitchable will refer to molecules that can be repeatedly and controllably light driven between "on" and "off" states. Conventional will indicate FPs that are normally used in nonhighlighting experiments but have been found to display properties allowing their use as optical highlighters, although the mechanism of their spectral alterations (when determined) may put some of them in one of the aforementioned three categories.

6.2.1 PHOTOACTIVATION: GREEN FLUORESCENT PROTEINS

The original wild-type GFP from *A. victoria* was the first FP to be used as an optical highlighter. This was done by taking advantage of the photoinduced conversion (photoconversion), which occurs upon irradiation with ~400 nm light (Chattoraj et al. 1996, Yokoe and Meyer 1996) (Figure 6.2a). The wtGFP chromophore population exists as a mixture of neutral phenols (Y66 is protonated) and anionic phenolates (Y66 is deprotonated) giving rise to a major absorbance peak at ~397 nm and a minor absorbance peak at ~475 nm, respectively. Irradiation at ~400 nm causes the chromophore to undergo excited state proton transfer and convert into the anionic form (Chattoraj et al. 1996, Tsien 1998, Creemers et al. 1999) (Figure 6.2b). The transition to the form with the absorption peak at ~475 nm involves a transition through an excited state intermediate, which gives rise to green fluorescence (Figure 6.2c and d). The existence of the intermediate form was

Photophysical properties of fluorescent proteins

Table 6.1 Selected optical highlighter fluorescent proteins

PROTEIN	WAVELENGTHS (NM)[a]		CLASS	FOLD CONTRAST (POST/PRE)[b]	EXTINCTION COEFFICIENT ($M^{-1} cm^{-1}$)	QUANTUM YIELD	REFERENCE USED
	λ_{ex}	λ_{em}					
PA-GFP	400 (Pre)	515	Photoactivatable	70	20,700	0.13	Patterson and Lippincott-Schwartz (2002)
	504 (Post)	517			17,400	0.79	
PS-CFP	402 (Pre)	468	Photoactivatable	1500 green-to-cyan ratio	34,000	0.16	Chudakov et al. (2004)
	490 (Post)	511			27,000	0.19	
PS-CFP2	400 (Pre)	470	Photoactivatable		43,000	0.20	
	490 (Post)	511			47,000	0.23	
HbGFP1	389 (Pre)	512	Photoactivatable	130–160	39,420	0.83	Haddock et al. (2010)
	495 (Post)						
PAmRFP1-1	578 (Post)	605	Photoactivatable	70	10,000	0.08	Verkhusha and Sorkin (2005)
PAmCherry1	564 (Post)	595	Photoactivatable	4000	18,000	0.46	Subach et al. (2009b)
PATagRFP	562	595	Photoactivatable		66,000	0.38	Subach et al. (2010)
aceGFP-G222E	390 (Pre)	460	Photoactivatable	>1000	33,000 (pre)	0.07	Gurskaya et al. (2003)
	480 (Post)	505				0.45	
Kaede	508 (Pre)	518	Photoconvertible	2000 red-to-green ratio	98,800	0.80	Ando et al. (2002), Dittrich et al. (2005)
	572 (Post)	582			60,400	0.33	
Kikume green-red (KikGR)	507 (Pre)	517	Photoconvertible	2000 red-to-green ratio	28,200	0.70	Tsutsui et al. (2005)
	583 (Post)	593			32,600	0.65	

	Excitation (Pre/Post)	Emission (Pre/Post)	Class	Ratio	Extinction coefficient	Quantum yield	Reference
mKikGR	505 (Pre) / 580 (Post)	515 / 591	Photoconvertible		49,000 / 28,000	0.69 / 0.63	Habuchi et al. (2008)
EosFP	506 (Pre) / 571 (Post)	516 / 581	Photoconvertible		72,000 / 41,000	0.70 / 0.55	Wiedenmann et al. (2004)
mEosFP	505 (Pre) / 569 (Post)	516 / 581	Photoconvertible		67,200 / 37,000	0.64 / 0.62	Wiedenmann et al. (2004)
tdEosFP	506 (Pre) / 571 (Post)	516 / 581	Photoconvertible	200	84,000 (x2) / 33,000 (x2)	0.66 / 0.6	Wiedenmann et al. (2004)
mEos2	506 (Pre) / 573 (Post)	519 / 584	Photoconvertible		56,000 / 46,000	0.84 / 0.66	McKinney et al. (2009)
Dendra	486 (Pre) / 558 (Post)	505 / 575	Photoconvertible	4500 red-to-green ratio	21,000 / 20,000	0.72 / 0.70	Gurskaya et al. (2006)
Dendra2	490 (Pre) / 553 (Post)	507 / 573	Photoconvertible	300	45,000 / 35,000	0.50 / 0.55	
Dronpa	503 (Post)	518	Photoswitchable		95,000	0.85	Ando et al. (2004)
Dronpa-2	486	513	Photoswitchable		56,000	0.28	Ando et al. (2007)
Dronpa-3	487	514	Photoswitchable		58,000	0.33	Ando et al. (2007)
rsFastlime	496	518	Photoswitchable	67	39,100	0.77	Stiel et al. (2007)
Padron	503	522	Photoswitchable	143	43,000	0.64	Andresen et al. (2008)

(continued)

Photophysical properties of fluorescent proteins

Optical highlighter photophysical properties

Photophysical properties of fluorescent proteins

Table 6.1 (continued) Selected optical highlighter fluorescent proteins

PROTEIN	WAVELENGTHS (NM)[a] λ_{ex}	WAVELENGTHS (NM)[a] λ_{em}	CLASS	FOLD CONTRAST (POST/PRE)[b]	EXTINCTION COEFFICIENT (M^{-1} cm^{-1})	QUANTUM YIELD	REFERENCE USED
bsDronpa	460	504	Photoswitchable	17	45,000	0.50	Andresen et al. (2008)
rsCherry	572 (Post)	610	Photoswitchable	6.7	80,000	0.02	Stiel et al. (2008)
rsCherryRev	572 (Post)	608	Photoswitchable	20	84,000	0.005	Stiel et al. (2008)
asFP595	572 (Pre)	595 (Post)	Photoswitchable		56,200	<0.001	Lukyanov et al. (2000)
mTFP0.7	453	488	Photoswitchable		60,000	0.50	Henderson et al. (2007)
IrisFP	488 (Pre) 551 (Post)	516 580	Photoconvertible Photoswitchable		52,200 35,400	0.43 0.47	Adam et al. (2008)
KFP1	580	600	Photoactivatable and photoswitchable	30	1,23,000 59,000	<0.001 0.13	Chudakov et al. (2003)
Phamret	458 (Cyan) 458 (Green)	475 (Cyan) 517 (Green)	Fusion between CFP variant and PAGFP	15-fold green-to-cyan ratio			Matsuda et al. (2008)

[a] (Pre) and (post) indicate, respectively, the wavelengths of maximum intensity before and after photoactivation, photoconversion, or photoswitching.
[b] From original publication or calculated based on reported residual fluorescence in the pre-activated or "off" state.

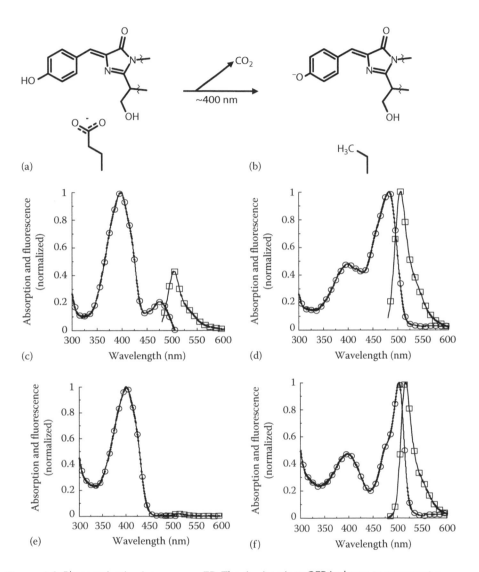

Figure 6.2 Photoactivation into a green FP. The *A. victoria* wtGFP is shown to represent a subset of proteins that initially form chromophores with the bulk of the population in the neutral form (a) and are altered to the anionic form (b) after irradiation with ~400 nm light. The neutral form absorbs light readily in the ~400 nm spectral region whereas the anionic form absorbs in the ~500 nm region. The highlighting event is accompanied by decarboxylation of a nearby glutamic acid (E222 in *A. victoria* wtGFP and PAGFP). The spectral alterations associated with the neutral to anionic changes are shown for wtGFP (c and d), PAGFP (e and f), and PSCFP2 (g and h). Absorption spectra are shown with open circles and are normalized to the wavelength of maximum chromophore absorption. Emission spectra are shown with open squares and are normalized to the wavelength of maximum fluorescence emission. The emission spectra for pre-activated wtGFP and PAGFP (c and e, open squares) are normalized to the wavelength of maximum fluorescence emission from the postactivated states (d and f, open squares).

(continued)

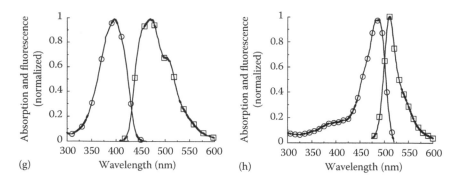

(g)   (h)   Wavelength (nm)

Figure 6.2 (continued) Photoactivation into a green FP. The *A. victoria* wtGFP is shown to represent a subset of proteins that initially form chromophores with the bulk of the population in the neutral form (a) and are altered to the anionic form (b) after irradiation with ~400 nm light. The neutral forms absorb light readily in the ~400 nm spectral region whereas the anionic form absorbs in the ~500 nm region. The highlighting event is accompanied by decarboxylation of a nearby glutamic acid (E222 in *A. victoria* wtGFP and PAGFP). The spectral alterations associated with the neutral to anionic changes are shown for wtGFP (c and d), PAGFP (e and f), and PSCFP2 (g and h). Absorption spectra are shown with open circles and are normalized to the wavelength of maximum chromophore absorption. Emission spectra are shown with open squares and are normalized to the wavelength of maximum fluorescence emission. The emission spectra for pre-activated wtGFP and PAGFP (c and e, open squares) are normalized to the wavelength of maximum fluorescence emission from the postactivated states (d and f, open squares).

confirmed by low-temperature spectral hole burning experiments (Creemers et al. 1999). Other structural changes taking place are the rotation of the threonine residue at 203 located close to the chromophore so it could hydrogen bond with the anionic Y66, but the most important alteration is the decarboxylation of glutamic acid at position 222 (E222) (van Thor et al. 2002). Before irradiation at ~400 nm, electrostatic repulsion with E222 precludes the formation of the anionic phenolate chromophore, but the loss of the negative carboxyl group alleviates this, and the result is a shift of the chromophore population from a mostly neutral species to a mostly anionic species. Alternatively, loss of the carboxyl group produces a large change in the intricate hydrogen-bonding network around the chromophore which may result in stabilization of the anionic form. Either or both of these scenarios would increase absorption at ~475 nm region and give a subsequent increase in the fluorescence intensity when excited at this wavelength.

Subsequent studies on a GFP T203I mutant that also has decreased absorbance in the minor peak region (Heim et al. 1994, Ehrig et al. 1995) indicated that its predominantly neutral phenol chromophore population could also be converted into the longer wavelength absorbing anionic form (Patterson and Lippincott-Schwartz 2002). This indicated that the structural change of the T203 residue probably plays a minor role in the wtGFP photoconversion. In addition, several other substitutions at the 203 position that reduced the absorption in the 488 nm spectral still resulted in this spectral change (Patterson and Lippincott-Schwartz 2002). Most of these mutants, including T203I, are able to undergo photoconversion. However, the T203H mutant (denoted photoactivatable GFP [PA-GFP]) drastically decreased the minor absorbance peak giving >60- and >100-fold fluorescence increases after photoactivation in cells and in purified protein form, respectively (Patterson and Lippincott-Schwartz 2002). Structural characterization of PAGFP photoactivation indicates that it has the E222 decarboxylation after irradiation, which is a similar structural alteration as the wtGFP (Henderson et al. 2009).

Photoswitchable cyan fluorescent protein (PS-CFP) shifts both excitation and emission spectra in a change from a cyan-to-green FP (Chudakov et al. 2004) and has been referred to as a photoactivatable FP. Derived from a nonfluorescent variant, acGFPL (Gurskaya et al. 2003), PS-CFP initially displays excitation at 402 nm and emission at 468 nm (Chudakov et al. 2004) (Figure 6.2d). With activation by ~400 nm light, new peaks are observed at 490 and 511 nm, respectively, leading to ~1500-fold increase in the green-to-cyan fluorescence ratio. An improved version, PS-CFP2, which develops fluorescence more efficiently and is thus brighter, is available through Evrogen. The mechanism for the shift from the cyan state to the green state is thought to occur via a similar change as wtGFP and PAGFP and is usually categorized as photoactivatable instead of photoconvertible. The chromophore population shifts from neutral chromophores to anionic chromophores are accompanied by a decarboxylation of a nearby glutamic acid. However, unlike the *A. victoria* GFPs, the PS-CFPs more efficiently produce cyan emission (468 nm) when the neutral chromophore is excited rather than emit from an excited state intermediate after excited state proton transfer.

Recently, a green fluorescent molecule, HbGFP (Haddock et al. 2010), was found in a member of the Ctenophora phylum and naturally has optical highlighting properties. Its spectra before irradiation are similar to that of PAGFP or "off" state Dronpa, but it can be switched on by irradiation with blue light instead the usual ~400 nm light. It is found to decay back to a nonfluorescent after being highlighted with a half-life of ~32 min, which differs from the wtGFP, PAGFP, and PS-CFP, but it is placed in this category since this appears to be a non-light-driven switching process.

6.2.2 PHOTOACTIVATION: RED FLUORESCENT PROTEINS

DsRed (Matz et al. 1999) has been the source of numerous FP variants, and some of these have been further altered to make highlighter molecules (Table 6.1). For example, the monomeric version of DsRed, mRFP1 (Campbell et al. 2002) was made into a series of photoactivatable fluorescent proteins, PAmRFP1-1, PAmRFP1-2, and PAmRFP1-3 (Verkhusha and Sorkin 2005). PAmRFP1-1 was the brightest and produced an ~70-fold increase in red fluorescence upon ultraviolet light excitation but has a very low quantum yield of only 0.08 (Verkhusha and Sorkin 2005). This was furthered by development of mCherry (a derivative of mRFP1) (Shaner et al. 2004) into a photoactivatable marker. PAmCherry1 (Subach et al. 2009b) also had little fluorescence before photoactivation, increased ~4000 fold and was ~10 times brighter than PAmRFP1-1 (Figure 6.3a). Based on the crystal structure (Subach et al. 2009a) and the mutations required to render it photoactivatable, the mechanism for PAmCherry1 photoactivation involves a decarboxylation of a nearby glutamic acid as for wtGFP, PAGFP, and PSCFP, but this reaction does not just rearrange the hydrogen-bonding network.

The PAmCherry1 chromophore forms similarly to most other FPs through the cyclized tripeptide intermediate. The oxidation step that follows normally leads to double-bond formation between the C_α and C_β of the tyrosine residue in the Y position (using the XYG notation for chromophore constituents introduced earlier), but for PAmCherry1, the first oxidation step takes place in the methionine in the X position to form a double bond between the C_α and N. This produces a nonplanar chromophore lacking absorption

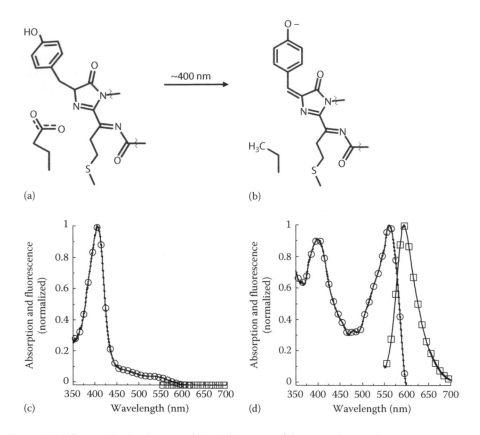

(a) (b)

(c) (d)

Wavelength (nm)

Figure 6.3 Photoactivation into a red FP. A depiction of the PAmCherry1 chromophore is shown as an example for this category. The model based on structural studies suggests that the chromophore is in a neutral form (a), which is activated into an anionic form (b). However, this model differs significantly from that proposed for wtGFP. Here, the first oxidation step takes place in the methionine in the X position to form a double bond between the C_α and N rather than between the C_α and C_β of the tyrosine residue in the Y position. This produces a nonplanar chromophore absorbing at 404 nm. Irradiation at ~400 nm results in the glutamic acid decarboxylation and a second oxidation reaction at the Y position C_α and C_β. Extension of the π-conjugation and formation of the *trans* isomer of the chromophore increases absorption at ~562 nm and produces red fluorescence when excited in this region. Absorption spectra are shown with open circles and are normalized to the wavelength of maximum chromophore absorption. The emission spectrum for pre-activated PAmCherry1 (c, open squares) is normalized to the wavelength of maximum fluorescence emission from the postactivated state (d, open squares).

except at 404 nm (Subach et al. 2009a). Irradiation at ~400 nm results in the glutamic acid decarboxylation, which is thought to initiate steps leading to a second oxidation reaction at the Y position C_α and C_β. This extends the π-conjugation and results in the formation of the *trans* isomer of the chromophore (Figure 6.3a). Although the *cis* isomer is the usual conformation for this family of red proteins, the *trans* isomer chromophore produces red fluorescence under ~561 nm excitation.

The brightness of the TagRFP made it an appealing target for developing a photoactivatable fluorescent protein. Indeed, PATagRFP (Subach et al. 2010) is ~3 times brighter than PAmCherry1 with extinction coefficient and quantum yield of 66,000 M^{-1} cm^{-1} and 0.38, respectively. The proposed mechanism for PATagRFP photoactivation resembles that for PAmCherry1 with an exception that it involves two separate photon absorption events. First, it is thought that the chromophore tripeptide cyclization step proceeds normally, but the first oxidation step taking place at the C_α–N bond of the methionine in the X position requires absorption of one photon (λ_{max} ~ 351 nm). This is thought to yield a nonfluorescent chromophore that can then absorb a second photon of light (λ_{max} ~ 412 nm) during an oxidation step taking place in the Y position tyrosine C_α–C_β bond (Subach et al. 2010). This results in extension of the π-conjugation from the tyrosine moiety in position Y to the carbonyl of the position X methionine and to produces the red fluorescent PATagRFP chromophore.

6.2.3 PHOTOCONVERSIONS: GREEN TO RED

At least four optical highlighters, not including their numerous derivatives (Table 6.1), undergo a green-to-red photoconversion. These molecules have the same basic cyclized tripeptide chromophore consisting of a histidine in the X position, tyrosine in the Y position, and glycine in the G position (using the XYG notation for chromophore constituents introduced earlier). Similar to the formation of most FP chromophores, newly synthesized proteins fold into the overall barrel shape, proceed through the cyclization step, and then a first oxidation step on position Y tyrosine C_α–C_β bond. This produces molecules with chromophores existing as a mixed population of neutral phenols (the tyrosine is protonated) and anionic phenolates (the tyrosine is deprotonated). The neutral form absorbs near 400 nm, and the anionic form absorbs near 500 nm. Excitation of the anionic form produces green fluorescence, whereas the neutral form does not seem to fluoresce under low-level excitation. However, intense irradiation at ~400 nm leads to a cleavage of the position X histidine C_α–N bond and to formation of a double bond between the histidine C_α and C_β. This again extends the π-conjugation of the chromophore and dramatically red-shifts the excitation and emission spectra (Figure 6.4a).

The first of these to be discovered was Kaede, which can be photoconverted by irradiation at ~400 nm (Ando et al. 2002). Before photoconversion, Kaede has a major absorbance peak at 508 nm and emission at 518 nm. After photoconversion, Kaede exhibits a new red-shifted absorbance peak at 572 nm, which upon excitation fluoresces with a new emission peak at 582 nm. This shift in both the excitation and emission peaks results in a >2000-fold increase in the red-to-green fluorescence ratio (Figure 6.4b).

The original KikG from *Favia favus* did not exhibit photoconvertible properties, but using their structural characterizations of Kaede as a guide, Miyawaki and colleagues engineered KikG into the KikGR variant (Tsutsui et al. 2005), which produced >2000-fold increase in fluorescence contrast during ratio imaging of the red and green components after photoactivation. Similar to Kaede, KikGR was also found to be a tetramer (Tsutsui et al. 2005) but was engineered into a monomeric form (Habuchi et al. 2008).

EosFP also shows a green-to-red fluorescence photoconversion upon ultraviolet or near-ultraviolet light irradiation (Wiedenmann et al. 2004). Similar to Kaede, EosFP has a pre-activated excitation maximum at 506 nm with emission at 516 nm. Upon irradiation at 405 nm, the photoconverted excitation peak is located at 571 nm with emission at 581 nm. Initially determined to be a tetramer, EosFP was engineered into two dimeric forms, d1EosFP and d2EosFP. The combination of the mutations produced a monomeric

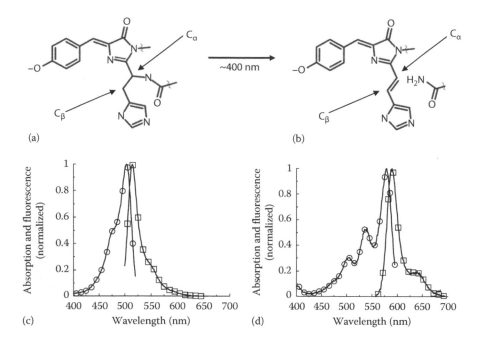

(a)   ~400 nm   (b)

(c)   Wavelength (nm)   (d)   Wavelength (nm)

Figure 6.4 Photoconversion from a green to a red FP. The chromophore of mKikGR is depicted in this cartoon to represent structural alterations observed for this category of optical highlighter proteins (a and b). These molecules are initially expressed as green FPs with core chromophores similar to that observed for *A. victoria* GFP with a major difference being the histidine in the X position. This chromophore population is made up of neutral forms (not shown), which absorb ~390 nm, and a large population of anionic forms (a), which absorb ~505 nm and produce green fluorescence (c). Irradiation in the 390 nm region results in extension of the π-conjugation through the histidine C_α and C_β (b) and a subsequent red-shift of the absorption and emission spectra (d). Absorption spectra are shown with open circles and are normalized to the wavelength of maximum chromophore absorption. Emission spectra are shown with open squares and are normalized to the wavelength of maximum fluorescence emission.

molecule with a K_d of ~0.1 mM, which has been named mEosFP. The emission maxima of these mutants remain constant while excitation maxima and brightness change slightly (Table 6.1). The mEosFP inefficiently forms a fluorescent molecule when expressed at 37°C (Wiedenmann et al. 2004) and has subsequently been developed into an improved optical highlighter, mEos2 (McKinney et al. 2009), and a photoswitchable molecule, IrisFP (Adam et al. 2008). Based on the extinction coefficient and quantum efficiency, mEos2 is one the brightest of the optical highlighters available.

Dendra exhibits up to 4500-fold photoconversion from its green-to-red fluorescent forms (Gurskaya et al. 2006). The wild-type dendGFP (Labas et al. 2002) was engineered into the monomeric Dendra with its photoconversion properties. Dendra can be activated with ~400 nm light, which is required by most of the other photoconvertible and photoactivatable fluorescent proteins but allows the option for activation with high levels of potentially less phototoxic wavelengths (~488 nm). Although Dendra is reported to efficiently develop fluorescence when expressed at 37°C, a mutated version, Dendra2, commercially available from Evrogen represents an improvement in this characteristic.

6.2.4 FLUORESCENT PROTEINS: PHOTOSWITCHABLE

This category of PA-FPs has the capability of being switched "on" and switched "off" by excitation at different wavelengths. This was first noted in single molecules of GFP T203 mutants yellow fluorescent protein [YFP] (Dickson et al. 1997), but most of these photoswitchable FPs were developed from proteins found in numerous marine species and display this behavior with high efficiency. In an early observation of this phenomenon, the red fluorescence (λ_{max} ~ 595 nm) of asFP595, a protein isolated from the sea anemone, *Anemonia sulcata*, was increased by green light and quenched by blue light (Lukyanov et al. 2000). This molecule has not found wide practical use as a marker since it matures slowly, exists as a tetramer, and has a low quantum yield. Nevertheless, as microscopy techniques advanced, the photoswitching capabilities of asFP595 have been utilized in a fluorescence microscopy technique, reversible saturable optical fluorescence transitions (RESOLFT), to break the diffraction barrier (Hofmann et al. 2005).

A variant of asFP595, kindling fluorescent protein (KFP1), was introduced as a FP having the capability to photoswitch or to be irreversibly photoactivated (Chudakov et al. 2003). KFP1 initially exhibits little fluorescence when excited at ~560 nm, but upon activation with 532 nm laser, light increases its red fluorescence ~30-fold. The reversion of KFP1 depends on the photoactivation excitation intensity. Lower-level excitation (~1 W cm^{-2} for 2 min) produced red KFP1 fluorescence that relaxed to the nonfluorescent state with a half-time of ~50 s, whereas 200 times this excitation (~20 W cm^{-2} for 20 min) irreversibly photoactivated ~50% of the population. With a quantum yield of 0.07, the photoactivated KFP1 lacks the brightness of many of the other PA-FPs. In addition, its tetrameric oligomerization state and slow maturation time ($t_{1/2}$ ~ 5 h) limit its use as a marker. However, the use of the potentially less phototoxic green activation light (532 nm) instead of ~400 nm light makes KFP1 live cell imaging amenable (Chudakov et al. 2003).

A protein derived from *Pectimiidae* named Dronpa (Ando et al. 2004) initially displays green fluorescence with an excitation maximum at 503 nm and emission maximum of 518 nm (Figure 6.5a). With moderate irradiation at 490 nm (0.4 W cm^{-2}), the 503 nm absorption and green fluorescence emission is lost. Irradiation at 400 nm (0.14 W cm^{-2}) restores the 503 nm absorbance and green fluorescence. Remarkably, the on–off cycling can occur 100 times with only a loss of 25% of the original fluorescence. Its extinction coefficient (95,000 M^{-1} cm^{-1}) and quantum yield (0.85) make it one of the brightest optical highlighters. Its capabilities make it useful in most highlighting experiments, but the reversible nature allows the same experiment to be repeated many times within the same areas (Ando et al. 2004). Dronpa and some of its derivatives, rsFastlime and Padron, have shown their utility in superresolution molecular localization (Andresen et al. 2008). Interestingly, the Padron variant photoswitches under opposite irradiation protocols by being turned "on" (Andresen et al. 2008) by 488 nm and "off" by 405 nm (Andresen et al. 2008), which allows it to be used in a double label experiment with rsFastlime.

Other photoswitchable FPs in the lineage of the DsRed include two versions of mCherry, rsCherry and rsCherryRev (Stiel et al. 2008). These molecules are switched back and forth between their "on" and "off" states using red and green/yellow light. A derivative of EosFP, IrisFP, can be irreversibly photoconverted just as EosFP but has also joined the photoswitchable category by having the capability to photoswitch between "on" and "off" states while in the unconverted green form as well as photoswitch between "on" and "off" red states after photoconversion (Adam et al. 2008).

Photophysical properties of fluorescent proteins

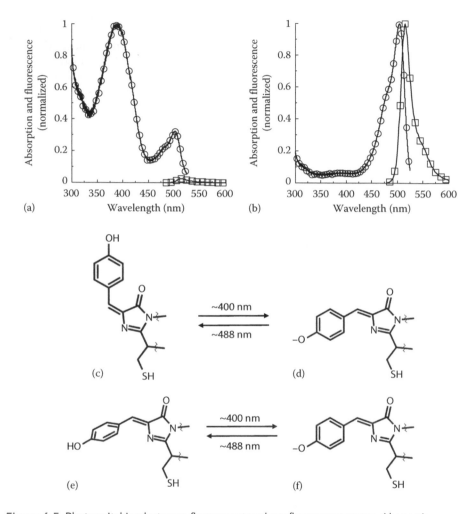

Figure 6.5 Photoswitching between fluorescent and nonfluorescent states. Absorption spectra for Dronpa in the "off" state (a) and "on" state (b) are shown with open circles and are normalized to the wavelength of maximum chromophore absorption. Emission spectra for Dronpa in the "off" state (a) and "on" state (b) are shown with open squares and are normalized to the wavelength of maximum fluorescence emission in the "on" state (b, open squares). Prior to irradiation with 488 nm laser light, (b) Dronpa absorbed light well at ~503 nm with strong emission at ~518 nm. After irradiation, absorption decreased in this region and increased at ~390 nm (a). Note for this example the Dronpa was not completely photoswitched "off" to maintain and demonstrate some residual "on" state. Proposed structural alterations associated with photoswitchable fluorescent proteins are depicted in cartoons (c–f) using the Dronpa chromophore as an example. One mechanism holds that the dark (a) and fluorescent (b) states can be explained by an alteration of the chromophore between *trans* (c) and *cis* (d) conformations, respectively. A modification of this suggests that rather than a switch into the *trans* state (c) to turn "off" the fluorescence, the chromophore could switch into a nonplanar nonfluorescent form (e) from the fluorescent *cis* state (f). Additional alterations are the protonation and deprotonation of the tyrosine in Y position to produce the neutral and anionic forms, respectively.

Photoswitching mechanisms in these proteins have been studied intensely by structure analyses. The models derived from the structures on the different variants are somewhat diverse, suggesting a common mechanism may not hold for this class. However, for at least one of the molecules, Dronpa, some of the proposed mechanisms are in conflict (Figure 6.5c and d). One model from a crystal structure of the light state of Dronpa holds that the difference between the light and dark states are the protonation states of the central tyrosine in the chromophore, which is altered by subtle changes in the chromophore environment upon irradiation (Wilmann et al. 2006). Another proposes the neutral state of the tyrosine in the Y position is accompanied by increased structural flexibility of the Dronpa chromophore and β-barrel in the dark state, which would diminish fluorescence (Mizuno et al. 2008, 2010). Alternatively, chromophores of asFP595 (a precursor of KFP1) (Andresen et al. 2005) and Dronpa (Andresen et al. 2007) are thought to switch between *cis* (fluorescent) and *trans* (dark) states (Figure 6.5c and d). However, it can be argued (Quillin et al. 2005) that simply the change to a *trans* chromophore probably does not explain the loss in fluorescence since other brightly fluorescent proteins, such as eqFP611 (Petersen et al. 2003) and PAmCherry1 (Subach et al. 2009a), have been observed with chromophores in the *trans* configuration. On the other hand, the tyrosine in the Y position would likely experience different local environments within the β-barrel of the protein when shifted from a *cis*-to-*trans* orientation, and interactions with different side chains could alter absorption characteristics or depopulate the excited state through nonradiative processes. Finally, from structural studies on the dark state of the photoswitchable fluorescent protein KFP1 (Chudakov et al. 2003), rather than a dependence on *cis* versus *trans* orientations, it is suggested that the dark state may be caused by a structural alteration of the chromophore from its normal coplanar conformation (Quillin et al. 2005) (Figure 6.5e and f).

6.2.5 CONVENTIONAL FLUORESCENT PROTEINS: NOT SO CONVENTIONAL

One of the first instances of optical highlighters in the red spectrum was actually discovered with proteins derived from *A. victoria* variants. This development relied on the phenomenon that under low oxygen conditions, several forms, such as wtGFP; GFPmut1, 2, and 3 (Cormack et al. 1996); S65T (Heim et al. 1995); I167T (Heim et al. 1994); and GFPuv (Crameri et al. 1996), convert into red fluorescent species upon irradiation with 488 nm light (Elowitz et al. 1997, Sawin and Nurse 1997). The high contrast with background fluorescence was utilized for monitoring protein diffusion in bacteria (Elowitz et al. 1997), but the low oxygen conditions required for the conversion has limited the use of these proteins in this technique. Recent studies on enhanced GFP (EGFP) under anaerobic conditions found that it could be converted even more efficiently (~9× better than oxygen scavenger alone) in the presence of reduced riboflavin using 488 nm light (Matsuda et al. 2010). The photoconversion resulted in a broad absorption spectrum with a peak at ~500 nm and extending to ~580 nm. Excitation at 532 nm produced an emission spectrum with a peak at ~561 nm. This worked efficiently enough that superresolution PALM imaging could be performed on the photoconverted molecules (Matsuda et al. 2010).

However, in the past few of years, studies have found that a similar "redding" phenomenon can be observed with EGFP under aerobic conditions when incubated in the presence of electron acceptors. These compounds include potassium ferricyanide, benzoquinone, 3-[4,5-dimethylthiazol-2-yl]-2,5-diphenyl tetrazolium bromide, cytochrome *c*, flavin adenine dinucleotide (FAD), flavin mononucleotide (FMN) glucose oxidase, and nicotinamide adenine dinucleotide (NAD) (Bogdanov et al. 2009). Irradiation with 488 nm light results in a shift of the excitation and emission peaks to ~575 and 607 nm, respectively. Similar results were obtained on other FPs in the green spectral range, such as AcGFP1, TagGFP, zFP506, amFP486, and ppluGFP2.

Several of the widely used red and orange variants have also displayed photoconversion into species fluorescing at wavelengths other than their original (Kremers et al. 2009). These occurred under usual live cell imaging conditions, and of 12 variants studied, 8 showed these behaviors. Katusha, mKate, HcRed1, mPlum, and mRaspberry had blue shifts of their spectra to varying degrees, whereas mOrange1, mOrange2, and Kusabira-Orange displayed red spectral shifts after irradiation. These suggest caution in using these molecules in photobleaching studies, but the authors showed that these molecules can be used in complementary highlighting experiments also.

Molecules of the DsRed tetramer exist as mixed populations of immature green fluorescent molecules and mature red fluorescent molecules, and this is thought to lead to another highlighting phenomenon. The model for this event holds that since the subunits are close together and the emission spectrum of the green species overlaps with that of the red species, the energy of the excited green molecules can transfer nonradiatively (fluorescence resonance energy transfer [FRET]) to the red molecules. Photobleaching the red molecule dequenches the green fluorescence and leads to an ~2.4-fold increase in green fluorescence emission (Marchant et al. 2001). However, the study by Kremers and colleagues (Kremers et al. 2009) on the spectral alterations associated with irradiation of several FP suggests that the greening may also be due in part to a photoconversion taking place in the DsRed.

In fact, it seems the DsRed has several different spectral alterations that can make it a highlighter molecule. Intense irradiation at 543 or 532 nm was found to slightly shift the peak from excitation and emission peaks from 559 and 593 nm to 575 and 595 nm, respectively (Cotlet et al. 2001). This alteration is proposed to be dependent on a *cis*-to-*trans* isomerization of the chromophore often associated with photoswitchable fluorescent proteins (see Section 6.2.4) but is also accompanied by a decarboxylation of a nearby glutamic acid (Habuchi et al. 2005b) similar to what occurs with wtGFP, PA-GFP, and PS-CFP photoactivation (see Section 6.2.1).

6.3 PHOTOBLEACHING CHARACTERISTICS

Just as with any fluorophore, photobleaching or photodestruction is a limiting factor. Previous sections discussed some of the processes that occur on subpopulations of molecules during normal imaging as well as during photobleaching studies. The introduction of new optical highlighters is generally accompanied by a photobleaching comparison with other FPs, although some report absolute photobleaching quantum yields. The photobleaching quantum yield is likely the best indicator for accomplishing the final objective of determining the photostabilities of the various proteins. However, to be meaningful, these data usually must be reported for proteins with which investigators may be most familiar. Thus, the most common method is to compare photobleaching rates, rate constants, or $t_{1/2}$ values with those of the familiar EGFP and mCherry. Unfortunately, this approach often makes difficult comparisons of data from several labs since the conditions for imaging and photobleaching may differ vastly based on lamp profiles, laser lines, filters, fluorophore brightness, and light detection efficiencies of different microscopes. Nevertheless, scaling the values based on comparisons with commonly used proteins should provide at least a rough approximation of relative optical highlighter photostability. This is attempted in Tables 6.2 and 6.3, but caution should be used in interpreting these values for the reasons listed earlier. The reported values are included in the table along with the author's attempt at normalizing the results. Short explanations for attaining the values are included in the table.

To normalize reported values across various studies, the paper from McKinney et al. (2009), which includes data from several optical highlighters, was chosen as a standard. The photobleaching values are reported for the green form before photoconversion and the red form after photoconversion. The interest in the green form photobleaching is that the green-to-red photoconverters are often used in ratio imaging experiments to produce very large contrasts between the pre- and postphotoconversion images of a time series.

For most comparisons, distinctions are made between imaging experiments performed with widefield versus confocal and these are separated here also. Compared with EGFP, all of the green components of the green-to-red photoconvertible proteins are less photostable (20%–39% under widefield imaging and 5%–20% under confocal imaging) (McKinney et al. 2009). Comparisons of the green optical highlighters with EGFP show an even wider range from ~12% up to 180% (Table 6.2). The PAGFP performs poorly by comparison (McKinney et al. 2009), whereas PSCFP (Chudakov et al. 2004), rsFastlime (Andresen et al. 2008), and bsDronpa (Andresen et al. 2008) show better photostability than EGFP. These data result from comparisons within separate reports and some differences in data collection do exist. For instance, data collected for the rsFastlime and bsDronpa required additional irradiation with 405 nm light to maintain the fluorescent "on" state (Andresen et al. 2008). Since the photoswitching to the dark state for many of the Dronpa derivatives tends to be much easier than photobleaching, this is one approach to get information of the photobleaching rate of the fluorescent state. In a subset of highlighting experiments, this may not represent a normal protocol since one would usually wish to observe the kinetics of a subpopulation of molecules after switching to the fluorescent state and irradiation with 405 nm would reduce contrast in the entire image through an increase the background fluorescence. However, this should not be a problem and is likely an advantage for imaging techniques such as reversible saturable optical fluorescence transitions (RESOLFT) (Hofmann et al. 2005) and optical lock in detection (OLID) (Mao et al. 2008).

For the red molecules, the comparisons with mCherry also produced a wide range of photostabilities and drastically different results for widefield versus confocal imaging (Table 6.3). For instance, McKinney et al. (2009) found that with the exception of mKiKGR, all of the green-to-red photoconverters were two to three times more photostable under widefield imaging whereas with the exception of mEos2, they were less stable than mCherry under confocal imaging. In fact, of the red optical highlighters for which reports could be found, only the Eos2 had photostability under confocal imaging similar to mCherry. Under widefield imaging, the photoactivated PAmCherry1 was observed to be as stable as mCherry, whereas the PAmRFP1-1 was significantly lower, and PATagRFP was 1.7 times as photostable as mCherry. Here is a point where caution is urged since these numbers were derived by comparing overlapping molecules

Photophysical properties of fluorescent proteins

Table 6.2 **Photostability behaviors of selected green FPs**

PROTEIN[a]	PHOTOSTABILITY (REPORTED VALUES $t_{1/2}$)[b]	PHOTOSTABILITY (COMPARED WITH EGFP $t_{1/2}$)[b]	PHOTOBLEACHING QUANTUM YIELD (QY_{PB})	REFERENCES USED
EGFP	w = 202 s c = 5010 s			McKinney et al. (2009)
EGFP	w = 115 s			Shaner et al. (2004)
EGFP	w = 53 s			Andresen et al. (2008)
EGFP	w = ~4 s			Chudakov et al. (2004)
EGFP			$QY_{PB} = 8.95 \times 10^{-6}$ Ensemble (calculated based on 4.7×10^{-3} s^{-1} [W cm^{-2}]$^{-1}$)	Chiu et al. (2001)
EGFP			$QY_{PB} = 8 \times 10^{-6}$ Single molecule	Peterman et al. (1999)
EGFP			$QY_{PB} = 6.9 \times 10^{-5}$ Single molecule	Harms et al. (2001)
EGFP	w = 174 s c = 5000 s			Shaner et al. (2008)
PA-GFP	w = 24 s c = 710 s	0.12 (w) 0.14 (c)		McKinney et al. (2009)
PS-CFP	w = ~6 s	1.5 (w) using EGFP(~6 s)		Chudakov et al. (2004)
Dronpa			Green $QY_{PB} = 3.2 \times 10^{-4}$	Ando et al. (2004)

rsFastlime	w = 65 s using 405 nm light to maintain "on" state	1.22 (w) using EGFP (53 s)	Stiel et al. (2007), Andresen et al. (2008)
Padron	w = 40 s	0.75 (w) using EGFP (53 s)	Andresen et al. (2008)
bsDronpa	w = 98 s using 405 nm light to maintain "on" state	1.8 (w) using EGFP (53 s)	Andresen et al. (2008)
Kaede	w = 53 s c = 990 s	0.26 (w) 0.2 (c)	Ando et al. (2002), Dittrich et al. (2005), McKinney et al. (2009)
mKikGR	w = 14 s c = 80 s	0.07 (w) 0.13 (c)	McKinney et al. (2009)
EosFP	w = 58 s c = 510 s	0.29 (w) 0.1 (c)	McKinney et al. (2009)
dEosFP (T158H)	w = 49 s c = 350 s	0.24 (w) 0.07 (c)	McKinney et al. (2009)
tdEosFP	w = 47 s c = 430 s	0.23 (w) 0.09 (c)	McKinney et al. (2009)
mEos2	w = 42 s c = 240 s	0.21 (w) 0.05 (c)	McKinney et al. (2009)
Dendra2	w = 45 s c = 260 s	0.22 (w) 0.05 (c)	McKinney et al. (2009)

[a] Multiple reports for the same protein are included in an attempt to normalize findings.
[b] (w) and (c) indicate values obtained from widefield and confocal imaging, respectively.

Photophysical properties of fluorescent proteins

Photophysical properties of fluorescent proteins

Table 6.3 Photostability behaviors of selected red FPs

PROTEIN[a]	PHOTOSTABILITY (REPORTED VALUES $t_{1/2}$)[b]	PHOTOSTABILITY (COMPARED WITH mCherry $t_{1/2}$)[b]	PHOTOBLEACHING QUANTUM YIELD (QY_{PB})	REFERENCES USED
mCherry	w = 168 s c = 2770 s			McKinney et al. (2009)
mCherry	w = 19 s			Subach et al. (2009b)
mCherry	w = 68 s			Shaner et al. (2004)
mCherry	w = 96 s c = 1800 s			Shaner et al. (2008)
DsRed	w = 326 s			Shaner et al. (2008)
DsRed			1. $QY_{PB} = 1.4 \times 10^{-6}$ 2. $QY_{PB} = 1.2 \times 10^{-7}$ Single molecule	Lounis et al. (2001)
DsRed			$QY_{PB} = 9.5 \times 10^{-6}$ FCS	Heikal et al. (2000)
DsRed			$QY_{PB} = 1.5 \times 10^{-5}$ Single molecule	Harms et al. (2001)
mRFP1	w = 6.2 s			Shaner et al. (2004)
mRFP1	w = 8.7 s c = 210 s			Shaner et al. (2008)
mRFP1			$QY_{PB} = 5.1 \times 10^{-5}$ Single molecule	Steinmeyer et al. (2005)

Kaede	w = 386 s c = 1660 s	2.3 (w) 0.6 (c)		Ando et al. (2002), Dittrich et al. (2005), McKinney et al. (2009)
mKikGR	w = 21 s c = 530 s	0.13 (w) 0.19 (c)	Red $QY_{PB} = 6.47 \times 10^{-6}$ Single molecule	Habuchi et al. (2008), McKinney et al. (2009)
EosFP	w = 489 s c = 1890 s	2.9 (w) 0.68 (c)		McKinney et al. (2009)
mEosFP			Red $QY_{PB} = 3.0 \times 10^{-5}$ Single molecule	Wiedenmann et al. (2004)
dEosFP (T158H)	w = 485 s c = 3130 s	2.9 (w) 1.1 (c)	Red $QY_{PB} = 2.4 \times 10^{-5}$ Single molecule	Wiedenmann et al. (2004), McKinney et al. (2009)
tdEosFP	w = 380 s c = 2730 s	2.3 (w) 0.99 (w)		Wiedenmann et al. (2004), McKinney et al. (2009)
mEos2	w = 323 s c = 2700 s	1.9 (w) 0.97 (c)		McKinney et al. (2009)
Dendra	Red (c; 3.3 times DsRed)			Gurskaya et al. (2006)
Dendra2	w = 378 s c = 2420 s	2.25 (w) 0.87 (c)		McKinney et al. (2009)

(continued)

Photophysical properties of fluorescent proteins

Table 6.3 (continued) Photostability behaviors of selected red FPs

PROTEIN[a]	PHOTOSTABILITY (REPORTED VALUES t$_{1/2}$)[b]	PHOTOSTABILITY (COMPARED WITH mCherry t$_{1/2}$)[b]	PHOTOBLEACHING QUANTUM YIELD (QY$_{PB}$)	REFERENCES USED
PAmRFP1-1	w = ~30 s	0.27(w) PAmRFP1-1 (~30 s)/mRFP1 (~10 s) normalized by Shaner et al. mRFP1 (6.2 s)/mCherry (68 s)		Verkhusha and Sorkin (2005), Shaner et al. (2004)
PAmCherry1	w = 18 s	0.95 (w) using mCherry = 19 s		Subach et al. (2009b)
PATagRFP	w = 180 s	1.7 (w) PATagRFP (180 s)/PAmCherry1 (102 s) normalize by Subach et al. PAmCherry (18 s)/mCherry (19 s)		Subach et al. (2010), Subach et al. (2009b)

[a] Multiple reports for the same protein are included in an attempt to normalize findings.
[b] (w) and (c) indicate values obtained from widefield and confocal imaging, respectively.

reported in different papers. For instance, the PAmRFP1-1 was compared with mRFP1, which was compared with mCherry in Shaner et al. (2004), and PATagRFP was compared with PAmCherry1 (Subach et al. 2010), which was previously compared with mCherry in Subach et al. (2009b).

Also listed in Tables 6.2 and 6.3 are the photobleaching quantum yields (QY_{PB}) for several proteins including EGFP and DsRed for comparison. As mentioned earlier, these are perhaps the best indicators of photostability since it simply represents the ratio of molecules photobleached per number of photons absorbed (Eggeling et al. 1998). The calculation requires the molecular cross sections of the molecules at the wavelength(s) of irradiation, the irradiation power levels, and the rate constant of the fluorescence decay. These three variables represent the major differences encountered when trying to compare photobleaching results across numerous imaging approaches. Differences in microscope optics (objective NA), excitation filters, laser lines, zoom factors, etc., affect irradiation power levels and excitation efficiencies, which in turn affect fluorescence decay kinetics. Inclusion of the precisely determined molecular cross sections, power levels, and rate constants accounts for differences in extinction coefficients at different wavelengths, differences in power levels, and photon energies resulting in a value that should be consistent across multiple imaging platforms in a variety of laboratories.

However, even these can lead to some discrepancies. For instance, mKiKGR QY_{PB} was found to be almost four times lower (Habuchi et al. 2008) (more photostable) than dEosFP (T158H) (Wiedenmann et al. 2004), whereas $t_{1/2}$ values indicated dEosFP (T158H) ranged from 5 to 22 times more photostable than mKiKGR (McKinney et al. 2009) (Table 6.3). It should be noted that the QY_{PB} in these examples were calculated based on single molecule imaging data, and it may be that photobleaching characteristics are also highly dependent on the imaging mode employed. Therefore, readers are again urged to use caution and compare the values in Tables 6.2 and 6.3 for rough estimations only since they are not meant as absolute comparisons. Nevertheless, comparisons within the same study should still give a good approximation of molecule photostability.

6.4 HIGHLIGHTING CHARACTERISTICS

Table 6.1 indicates one of the most important characteristics of the optical highlighters—contrast. Contrast plays an important role in tracking and diffusion experiments, since the goal is to mark a subpopulation of molecules to follow within the total population molecules. Likewise, for the superresolution molecular localization techniques, such as PALM, the contrast of activated molecules versus nonactivated molecule within a diffraction-limited spot presents limitations for the localization precision.

The highest contrasts are often obtained by using one of the molecules that convert from one color to another (PSCFP, Kaede, KikGR, EosFP variants). By ratio imaging using the highlighted image (green or red) and the nonhighlighted image (cyan or green), remarkable contrasts of >4500-fold can be obtained (Gurskaya et al. 2006). With the exception of PAmCherry1, the molecules that are not amenable to ratio imaging are generally much lower (~100–300-fold). For the conventional fluorescent proteins that are found to undergo photoconversion, these numbers are also much lower with contrasts generally <100-fold, although the mOrange and mOrange2 are notable exceptions with contrasts of ~160-fold (Kremers et al. 2009). The values in Table 6.1 are

Photophysical properties of fluorescent proteins

the values reported in the original reports or were calculated using the reported residual fluorescence observed in the "off" or nonactivated state.

Highlighting is also dependent on the wavelengths of light used. For most proteins discussed here, ultraviolet or near-ultraviolet light (~400 nm) provides the necessary energy to photoactivate, photoconvert, or photoswitch the molecules to their fluorescent states. Notable exceptions in the red spectrum include Dendra, which can be photoconverted with blue light (488 nm) (Gurskaya et al. 2006); KFP1, which can be irreversibly photoactivated with high levels of 532 nm light (Chudakov et al. 2003); and the mCherry derivatives, rsCherry and rsCherryRev, which are switched back and forth between their fluorescent and nonfluorescent states using 488 and 561 nm light (Stiel et al. 2008). Green exceptions include Padron, which is turned "on" by 488 nm light and "off" by 405 nm light (Andresen et al. 2008), the opposite behavior of other Dronpa derivatives and HbGFP, which can be turned on by blue light (Haddock et al. 2010). Many of the conventional fluorescent proteins, which behave as highlighters, also use longer wavelength light. Derivatives of the *A. victoria* GFP can be switched into red proteins under low oxygen conditions (Elowitz et al. 1997, Sawin and Nurse 1997, Matsuda et al. 2010) or in the presence of electron acceptors using 488 nm light (Bogdanov et al. 2009). And finally, the orange and red proteins use blue (488 nm) or yellow (561 nm) light to develop their altered spectra (Kremers et al. 2009).

Just as the highlighter fluorescent proteins differ in their photostabilities, their rates of photoactivation, photoconversion, and photoswitching vary considerably. Comparing reported values for light intensity dependence on the highlighting step runs into the same problems, different microscope platforms, filters, laser lines, etc., discussed earlier regarding comparisons of photobleaching. Often the kinetics of the highlighting process for a given protein are reported in comparison to other proteins, and these are listed in Table 6.4. Proteins are listed multiple times since different studies report highlighting kinetics in various formats. Here, an attempt is made to compile and discuss this critical characteristic for as many proteins as possible.

Fortunately, in many reports, the quantum efficiencies or the rate constants at specified power levels are defined, which makes comparisons easier. The easiest comparisons are the quantum yields (ratio of the number of molecules to the number of photons absorbed) associated with the highlighting mechanism. Most of the proteins for which this characteristic has been reported have QY of 10^{-3} to 10^{-4}. For the green-to-red photoconverters, one of the most easily photoconverted molecules is KiKGR or its derivative mKikGR (Table 6.4). This is confirmed in another study by Stark et al. (Stark and Kulesa 2007) in which they found that KikGR reached its t_{max} most quickly of the four proteins tested.

The efficiencies of photoactivation and photoconversion are markedly lower in comparison with photoswitching proteins. Some of these proteins have remarkably high efficiencies and photoswitch "on" quite easily (Dronpa has a $QY_{ON} = 0.37$ and the green state of IrisFP has a $QY_{ON} = 0.5$). On the other hand, photoswitching "off" tends to be much less efficient (10^{-2} to 10^{-4}). The lower efficiency in this case is likely an advantage, otherwise the molecules would probably switch "off" before emitting enough photons to provide a reasonable signal-to-noise ratio. Nevertheless, increased efficiencies in switching off will require less light irradiation and will allow the maximum contrast between "on/off" states to be reached more quickly, which should be beneficial for techniques such as RESOLFT (Hofmann et al. 2005) and OLID (Mao et al. 2008).

Table 6.4 Optical highlighting behaviors of selected FPs

PROTEIN[a]	"ON" PHOTOACTIVATION, PHOTOCONVERSION, PHOTOSWITCHING (REPORTED VALUES)[b]	"OFF" PHOTOACTIVATION, PHOTOCONVERSION, PHOTOSWITCHING (REPORTED VALUES)[b]	RELAXATION TO EQUILIBRIUM	REFERENCES USED
Dronpa	"On" $QY_{ps} = 0.37$	"Off" $QY_{ps} = 3.2 \times 10^{-4}$		Ando et al. (2004)
Dronpa	$k_{ON} = 6.7 \times 10^{-4}$ s^{-1} using 1 mW cm^{-2} 405 nm light	$k_{OFF} = 9.6 \times 10^{-3}$ s^{-1} using 10 mW cm^{-2} 488 nm light		Habuchi et al. (2005a)
rsFastlime	"On" $t_{1/2} = 0.11$ s using 0.2 W cm^{-2} 405 ± 10 nm light	"Off" $t_{1/2} = 5$ s using 0.3 W cm^{-2} 488 ± 10 nm light	"Off" to "on" $t_{1/2} = 8$ min	Stiel et al. (2007)
Dronpa	"On" $t_{1/2} = 0.10$ s using 0.2 W cm^{-2} 405 ± 10 nm light	"Off" $t_{1/2} = 263$ s using 0.3 W cm^{-2} 488 ± 10 nm light	"Off" to "on" $t_{1/2} = 840$ min	Stiel et al. (2007)
rsFastlime	"On" $t_{1/2} = 0.03$ s using 0.62 W cm^{-2} 405 ± 5 nm light	"Off" $t_{1/2} = 2.6$ s using 0.56 W cm^{-2} 488 ± 5 nm light	"Off" to "on" $t_{1/2} = 8$ min	Andresen et al. (2008)
Padron	"On" $t_{1/2} = 5.6$ s using 0.56 W cm^{-2} 488 ± 5 nm light	"Off" $t_{1/2} = 0.06$ s using 0.62 W cm^{-2} 405 ± 5 nm light	"On" to "off" $t_{1/2} = 150$ min	Andresen et al. (2008)
bsDronpa	"On" $t_{1/2} = 0.04$ s using 0.62 W cm^{-2} 405 ± 5 nm light	"Off" $t_{1/2} = 1.25$ s using 0.56 W cm^{-2} 488 ± 5 nm light	"Off" to "on" $t_{1/2} = 54$ min	Andresen et al. (2008)
Dronpa	"On" $t_{1/2} = 0.12$ s using 0.62 W cm^{-2} 405 ± 5 nm light	"Off" $t_{1/2} = 115$ s using 0.56 W cm^{-2} 488 ± 5 nm light	"Off" to "on" $t_{1/2} = 840$ min	Andresen et al. (2008)
Dronpa		"Off" $QY_{ps} = 3 \times 10^{-4}$		Ando et al. (2007)

(continued)

Photophysical properties of fluorescent proteins

Table 6.4 (continued) Optical highlighting behaviors of selected FPs

PROTEIN[a]	"ON" PHOTOACTIVATION, PHOTOCONVERSION, PHOTOSWITCHING (REPORTED VALUES)[b]	"OFF" PHOTOACTIVATION, PHOTOCONVERSION, PHOTOSWITCHING (REPORTED VALUES)[b]	RELAXATION TO EQUILIBRIUM	REFERENCES USED
Dronpa-2		"Off" $QY_{ps} = 4.7 \times 10^{-2}$ using 0.4 W cm^{-2} 490 ± 10 nm light		Ando et al. (2007)
Dronpa-3		"Off" $QY_{ps} = 5.3 \times 10^{-3}$ using 0.4 W cm^{-2} 490 ± 10 nm light		Ando et al. (2007)
Dronpa		"Off" $QY_{ps} = 3 \times 10^{-4}$		Flors et al. (2007)
Dronpa-2		"Off" $QY_{ps} = 5 \times 10^{-2}$		Flors et al. (2007)
Dronpa-3		"Off" $QY_{ps} = 5 \times 10^{-3}$		Flors et al. (2007)
Dronpa	"On" $QY_{ps} = 0.37$	"Off" $QY_{ps} = 3.2 \times 10^{-4}$		Dedecker et al. (2007)
Dronpa-2	"On" $QY_{ps} = 0.37$	"Off" $QY_{ps} = 1.5 \times 10^{-3}$		Dedecker et al. (2007)
Kaede	$t_{90\%} = 50$ s using 1.6 W cm^{-2} ~400 nm light $QY_{pc} = 2.4 \times 10^{-4}$			Ando et al. (2002)

	$k_{pc} = 38.6$ s^{-1} using 13.6 mW 405 nm before 60× NA1.2 objective		Dittrich et al. (2005)
Kikume green-red (KikGR)	$QY_{pc} = 4.7 \times 10^{-3}$		Tsutsui et al. (2005), Habuchi et al. (2008)
mKikGR	$k_{pc} = 6.8 \times 10^{-3}$ s^{-1} at pH 5.0 using 11.6 mW cm^{-2} 405 nm $k_{pc} = 7.5 \times 10^{-4}$ s^{-1} at pH 8.0 using 11.6 mW cm^{-2} 405 nm $QY_{pc} = 7.5 \times 10^{-3}$		Habuchi et al. (2008)
PAmRFP1-1	$t_{1/2} = 4.2$–4.5 min using 175 W xenon lamp, 340–380 filter, 63× objective		Verkhusha and Sorkin (2005)
rsCherry	"On" $t_{1/2} = 0.05$ s using 4 W cm^{-2} 550 ± 20 nm light	"Off" $t_{1/2} = 0.05$ s using 4 W cm^{-2} 450 ± 20 nm light	"On" to "off" $t_{1/2} = 40$ s	Stiel et al. (2008)
rsCherryRev	"On" $t_{1/2} = 3$ s using 4 W cm^{-2} 450 ± 20 nm light	"Off" $t_{1/2} = 0.7$ s using 4 W cm^{-2} 550 ± 20 nm light	"On" to "off" $t_{1/2} = 13$ s	Stiel et al. (2008)
PAmCherry1	$t_{1/2} = 2.5$ s		Subach et al. (2009b)
tdEosFP	$t_{1/2} = 4$ s		Subach et al. (2009b)

(continued)

Photophysical properties of fluorescent proteins

Photophysical properties of fluorescent proteins

Table 6.4 (continued) Optical highlighting behaviors of selected FPs

PROTEIN[a]	"ON" PHOTOACTIVATION, PHOTOCONVERSION, PHOTOSWITCHING (REPORTED VALUES)[b]	"OFF" PHOTOACTIVATION, PHOTOCONVERSION, PHOTOSWITCHING (REPORTED VALUES)[b]	RELAXATION TO EQUILIBRIUM	REFERENCES USED
IrisFP	Green to red $QY_{pc} = 1.8 \times 10^{-3}$ Green photoswitching $QY_{ON} = 0.5$ Red photoswitching $QY_{ON} = 4.7 \times 10^{-2}$	Green photoswitching $QY_{OFF} = 1.4 \times 10^{-2}$ Red photoswitching $QY_{OFF} = 2 \times 10^{-3}$	Green "Off" to "on" $t_{1/2} = 5.5$ h Red "Off" to "on" $t_{1/2} = 3.2$ h	Adam et al. (2008)
EosFP	Green to red $QY_{pc} = 8 \times 10^{-4}$			Adam et al. (2008)
Dendra	$t_{max} = 0.2$ s using 200 W cm^{-2} 488 nm light $t_{max} = 2$–5 s using 1.4 W cm^{-2} 488 nm light $t_{max} = 10$–20 s using 0.6 W cm^{-2} 405 nm light			Gurskaya et al. (2006)
PATagRFP	2.4 fold slower than PAmCherry1 5.1 fold slower than PAGFP (using 50 mW cm^{-2})			Subach et al. (2010)
mTFP0.7			Time constant = 7.7 min at pH 7.5	Henderson et al. (2007)
KiKGR	$t_{max} = 37.7$ s			Stark and Kulesa (2007)

Kaede	t_{max} = 107.02 s		Stark and Kulesa (2007)
PAGFP	t_{max} = 837.25 s		Stark and Kulesa (2007)
PSCFP2	t_{max} = 4008 s		Stark and Kulesa (2007)
HbGFP		"On" to "off" $t_{1/2}$ = 32 min	Haddock et al. (2010)
Phamret	cyan to green QY_{pc} = 2.7 × 10^{-2}		Matsuda et al. (2008)

[a] Multiple reports for the same protein are included in an attempt to normalize findings.

[b] QY_{pc} is the photoconversion quantum yield. QY_{ps} is the photoswitching quantum yield.

Photophysical properties of fluorescent proteins

Last, the stability of the highlighted state in the absence of any light-driven processes must be considered. For most of the photoactivatable and photoconvertible molecules, this seems to be less of a problem perhaps due to the covalent modifications occurring within or around the chromophores, such as the PAGFP E222 decarboxylation or the extension of the red protein π-conjugation. Photoswitchable proteins relax to equilibrium with a range of temporal characteristics (Table 6.4). For instance, Dronpa relaxes from the "off" state to the "on" state with a $t_{1/2}$ of ~14 h (Andresen et al. 2008), whereas rsCherryRev relaxes from the "on" state to the "off" state with a $t_{1/2}$ of ~13 s (Stiel et al. 2008). Thus, this critical characteristic should be noted in decisions concerning which protein to use.

6.5 SUMMARY AND OUTLOOK

Optical highlighter fluorescent proteins display a wide range of behaviors, some of which were engineered and some seem to be common to almost all molecules. Depending on the purpose of the experiment, these can advantageous or they can be annoying to the point of making experiments inconclusive. For instance, the light-driven blinking has become a useful characteristic for some of the superresolution molecular localization methods, but thermal relaxation into fluorescent state after high-intensity irradiation used in a fluorescence photobleaching recovery (FPR) (Axelrod et al. 1976) adds another recovery component to what is often a complicated protein diffusion, transport, or binding assay. If the dark states do not absorb the wavelengths of light used for imaging, these likely prolong the available fluorescence and might be considered a blessing. However, this comes at the expense of a subpopulation of nonfluorescing molecules at any given time, which imposes its own limitations on the experiment, such as the maximum attainable signal-to-noise ratio.

When considering the characteristics of optical highlighters, understanding of the mechanisms associated with the highlighting step for all of these molecules would be helpful in further design. Just as with any fluorophore, emphasis is placed on improving them in two manners, better and redder. Brighter molecules are always in demand as new approaches to imaging, such as the superresolution molecular localization method, demand more photons and lower backgrounds. Redder molecules, which move their excitation and emission farther away from the major cellular autofluorescence and provide better penetration of dense tissue, will benefit bold approaches to whole animal studies.

How far can the optical highlighters take us for imaging? Will they reach the common usage associated with the conventional fluorescent proteins? Not to be discounted, commercial interests may prove to be the major driving force in bringing optical highlighter use to any biologist with the interest (and funding) to observe their molecule dynamically or on the superresolution scale. Several companies have photoactivation modules incorporated into their confocal microscope systems, and at least three have developed or are developing instruments devoted to superresolution using molecular localization. In addition, given the pace of optical highlighter development, major improvements may be just a few months or years away. And given some of the amazing behaviors, purposely engineered or otherwise, displayed by these molecules, even more startling developments are likely to accompany these and may lead to new clever approaches to peek inside the environment of a cell or organism.

ACKNOWLEDGMENTS

This work was supported by the Intramural Research Program of the National Institutes of Health including the National Institute of Biomedical Imaging and Bioengineering.

REFERENCES

Adam, V., Lelimousin, M., Boehme, S. et al. 2008. Structural characterization of IrisFP, an optical highlighter undergoing multiple photo-induced transformations. *Proceedings of the National Academy of Sciences of the United States of America*, **105**, 18343–18348.

Ando, R., Flors, C., Mizuno, H., Hofkens, J., and Miyawaki, A. 2007. Highlighted generation of fluorescence signals using simultaneous two-color irradiation on Dronpa mutants. *Biophysical Journal*, **92**, L97–L99.

Ando, R., Hama, H., Yamamoto-Hino, M., Mizuno, H., and Miyawaki, A. 2002. An optical marker based on the UV-induced green-to-red photoconversion of a fluorescent protein. *Proceedings of the National Academy of Sciences of the United States of America*, **99**, 12651–12656.

Ando, R., Mizuno, H., and Miyawaki, A. 2004. Regulated fast nucleocytoplasmic shuttling observed by reversible protein highlighting. *Science*, **306**, 1370–1373.

Andresen, M., Stiel, A. C., Folling, J. et al. 2008. Photoswitchable fluorescent proteins enable monochromatic multilabel imaging and dual color fluorescence nanoscopy. *Nature Biotechnology*, **26**, 1035–1040.

Andresen, M., Stiel, A. C., Trowitzsch, S. et al. 2007. Structural basis for reversible photoswitching in Dronpa. *Proceedings of the National Academy of Sciences of the United States of America*, **104**, 13005–13009.

Andresen, M., Wahl, M. C., Stiel, A. C. et al. 2005. Structure and mechanism of the reversible photoswitch of a fluorescent protein. *Proceedings of the National Academy of Sciences of the United States of America*, **102**, 13070–13074.

Axelrod, D., Koppel, D. E., Schlessinger, J., Elson, E., and Webb, W. W. 1976. Mobility measurement by analysis of fluorescence photobleaching recovery kinetics. *Biophysical Journal*, **16**, 1055–1069.

Betzig, E., Patterson, G. H., Sougrat, R. et al. 2006. Imaging intracellular fluorescent proteins at nanometer resolution. *Science*, **313**, 1642–1645.

Bogdanov, A. M., Mishin, A. S., Yampolsky, I. V. et al. 2009. Green fluorescent proteins are light-induced electron donors. *Nature Chemical Biology*, **5**, 459–461.

Campbell, R. E., Tour, O., Palmer, A. E. et al. 2002. A monomeric red fluorescent protein. *Proceedings of the National Academy of Sciences of the United States of America*, **99**, 7877–7882.

Chalfie, M., Tu, Y., Euskirchen, G., Ward, W. W., and Prasher, D. C. 1994. Green fluorescent protein as a marker for gene expression. *Science*, **263**, 802–805.

Chattoraj, M., King, B. A., Bublitz, G. U., and Boxer, S. G. 1996. Ultra-fast excited state dynamics in green fluorescent protein: Multiple states and proton transfer. *Proceedings of the National Academy of Sciences of the United States of America*, **93**, 8362–8367.

Chiu, C. S., Kartalov, E., Unger, M., Quake, S., and Lester, H. A. 2001. Single-molecule measurements calibrate green fluorescent protein surface densities on transparent beads for use with 'knock-in' animals and other expression systems. *Journal of Neuroscience Methods*, **105**, 55–63.

Chudakov, D. M., Belousov, V. V., Zaraisky, A. G. et al. 2003. Kindling fluorescent proteins for precise in vivo photolabeling. *Nature Biotechnology*, **21**, 191–194.

Chudakov, D. M., Verkhusha, V. V., Staroverov, D. B. et al. 2004. Photoswitchable cyan fluorescent protein for protein tracking. *Nature Biotechnology*, **22**, 1435–1439.

Cormack, B. P., Valdivia, R. H., and Falkow, S. 1996. FACS-optimized mutants of the green fluorescent protein (GFP). *Gene*, **173**, 33–38.

Cotlet, M., Hofkens, J., Habuchi, S. et al. 2001. Identification of different emitting species in the red fluorescent protein DsRed by means of ensemble and single-molecule spectroscopy. *Proceedings of the National Academy of Sciences of the United States of America*, **98**, 14398–14403.

Crameri, A., Whitehorn, E. A., Tate, E., and Stemmer, W. P. 1996. Improved green fluorescent protein by molecular evolution using DNA shuffling. *Nature Biotechnology*, **14**, 315–319.

Creemers, T. M., Lock, A. J., Subramaniam, V., Jovin, T. M., and Volker, S. 1999. Three photoconvertible forms of green fluorescent protein identified by spectral hole-burning. *Nature Structural Biology*, **6**, 557–560.

Dedecker, P., Hotta, J., Flors, C. et al. 2007. Subdiffraction imaging through the selective donut-mode depletion of thermally stable photoswitchable fluorophores: Numerical analysis and application to the fluorescent protein Dronpa. *Journal of the American Chemical Society*, **129**, 16132–16141.

Dickson, R. M., Cubitt, A. B., Tsien, R. Y., and Moerner, W. E. 1997. On/off blinking and switching behaviour of single molecules of green fluorescent protein. *Nature*, **388**, 355–358.

Dittrich, P. S., Schafer, S. P., and Schwille, P. 2005. Characterization of the photoconversion on reaction of the fluorescent protein Kaede on the single-molecule level. *Biophysical Journal*, **89**, 3446–3455.

Eggeling, C., Widengren, J., Rigler, R., and Seidel, C. A. M. 1998. Photobleaching of fluorescent dyes under conditions used for single-molecule detection: Evidence of two-step photolysis. *Analytical Chemistry*, **70**, 2651–2659.

Ehrig, T., O'Kane, D. J., and Prendergast, F. G. 1995. Green-fluorescent protein mutants with altered fluorescence excitation spectra. *FEBS Letters*, **367**, 163–166.

Elowitz, M. B., Surette, M. G., Wolf, P. E., Stock, J., and Leibler, S. 1997. Photoactivation turns green fluorescent protein red. *Current Biology*, 7, 809–812.

Flors, C., Hotta, J., Uji-i, H. et al. 2007. A stroboscopic approach for fast photoactivation-localization microscopy with Dronpa mutants. *Journal of the American Chemical Society*, **129**, 13970–13977.

Gurskaya, N. G., Fradkov, A. F., Pounkova, N. I. et al. 2003. A colourless green fluorescent protein homologue from the non-fluorescent hydromedusa *Aequorea coerulescens* and its fluorescent mutants. *Biochemical Journal*, **373**, 403–408.

Gurskaya, N. G., Verkhusha, V. V., Shcheglov, A. S. et al. 2006. Engineering of a monomeric green-to-red photoactivatable fluorescent protein induced by blue light. *Nature Biotechnology*, **24**, 461–465.

Habuchi, S., Ando, R., Dedecker, P. et al. 2005a. Reversible single-molecule photoswitching in the GFP-like fluorescent protein Dronpa. *Proceedings of the National Academy of Sciences of the United States of America*, **102**, 9511–9516.

Habuchi, S., Cotlet, M., Gensch, T. et al. 2005b. Evidence for the isomerization and decarboxylation in the photoconversion of the red fluorescent protein DsRed. *Journal of the American Chemical Society*, **127**, 8977–8984.

Habuchi, S., Tsutsui, H., Kochaniak, A. B., Miyawaki, A., and van Oijen, A. M. 2008. mKikGR, a monomeric photoswitchable fluorescent protein. *PLoS One*, **3**, e3944.

Haddock, S. H., Mastroianni, N., and Christianson, L. M. 2010. A photoactivatable green-fluorescent protein from the phylum Ctenophora. *Proceedings of the Royal Society B*, **277**, 1155–1160.

Harms, G. S., Cognet, L., Lommerse, P. H., Blab, G. A., and Schmidt, T. 2001. Autofluorescent proteins in single-molecule research: Applications to live cell imaging microscopy. *Biophysical Journal*, **80**, 2396–2408.

Heikal, A. A., Hess, S. T., Baird, G. S., Tsien, R. Y., and Webb, W. W. 2000. Molecular spectroscopy and dynamics of intrinsically fluorescent proteins: Coral red (dsRed) and yellow (Citrine). *Proceedings of the National Academy of Sciences of the United States of America*, **97**, 11996–12001.

Heim, R., Cubitt, A. B., and Tsien, R. Y. 1995. Improved green fluorescence. *Nature*, **373**, 663–664.

Heim, R., Prasher, D. C., and Tsien, R. Y. 1994. Wavelength mutations and posttranslational autoxidation of green fluorescent protein. *Proceedings of the National Academy of Sciences of the United States of America*, **91**, 12501–12504.

Henderson, J. N., Ai, H. W., Campbell, R. E., and Remington, S. J. 2007. Structural basis for reversible photobleaching of a green fluorescent protein homologue. *Proceedings of the National Academy of Sciences of the United States of America*, **104**, 6672–6677.

Henderson, J. N., Gepshtein, R., Heenan, J. R. et al. 2009. Structure and mechanism of the photoactivatable green fluorescent protein. *Journal of the American Chemical Society*, **131**, 4176–4177.

Hess, S. T., Girirajan, T. P., and Mason, M. D. 2006. Ultra-high resolution imaging by fluorescence photoactivation localization microscopy. *Biophysical Journal*, **91**, 4258–4272.

Hofmann, M., Eggeling, C., Jakobs, S., and Hell, S. W. 2005. Breaking the diffraction barrier in fluorescence microscopy at low light intensities by using reversibly photoswitchable proteins. *Proceedings of the National Academy of Sciences of the United States of America*, **102**, 17565–17569.

Kremers, G. J., Hazelwood, K. L., Murphy, C. S., Davidson, M. W., and Piston, D. W. 2009. Photoconversion in orange and red fluorescent proteins. *Nature Methods*, **6**, 355–358.

Labas, Y. A., Gurskaya, N. G., Yanushevich, Y. G. et al. 2002. Diversity and evolution of the green fluorescent protein family. *Proceedings of the National Academy of Sciences of the United States of America*, **99**, 4256–4261.

Lounis, B., Deich, J., Rosell, F. I., Boxer, S. G., and Moerner, W. E. 2001. Photophysics of DsRed, a red fluorescent protein, from the ensemble to the single-molecule level. *Journal of Physical Chemistry B*, **105**, 5048–5054.

Lukyanov, K. A., Fradkov, A. F., Gurskaya, N. G. et al. 2000. Natural animal coloration can be determined by a nonfluorescent green fluorescent protein homolog. *Journal of Biological Chemistry*, **275**, 25879–25882.

Mao, S., Benninger, R. K., Yan, Y. et al. 2008. Optical lock-in detection of FRET using synthetic and genetically encoded optical switches. *Biophysical Journal*, **94**, 4515–4524.

Marchant, J. S., Stutzmann, G. E., Leissring, M. A., LaFerla, F. M., and Parker, I. 2001. Multiphoton-evoked color change of DsRed as an optical highlighter for cellular and subcellular labeling. *Nature Biotechnology*, **19**, 645–649.

Matsuda, A., Shao, L., Boulanger, J. et al. 2010. Condensed mitotic chromosome structure at nanometer resolution using PALM and EGFP-histones. *PLoS One*, **5**, e12768.

Matsuda, T., Miyawaki, A., and Nagai, T. 2008. Direct measurement of protein dynamics inside cells using a rationally designed photoconvertible protein. *Nature Methods*, **5**, 339–345.

Matz, M. V., Fradkov, A. F., Labas, Y. A. et al. 1999. Fluorescent proteins from nonbioluminescent *Anthozoa* species. *Nature Biotechnology*, **17**, 969–973.

McKinney, S. A., Murphy, C. S., Hazelwood, K. L., Davidson, M. W., and Looger, L. L. 2009. A bright and photostable photoconvertible fluorescent protein. *Nature Methods*, **6**, 131–133.

Mizuno, H., Mal, T. K., Tong, K. I. et al. 2003. Photo-induced peptide cleavage in the green-to-red conversion of a fluorescent protein. *Molecular Cell*, **12**, 1051–1058.

Mizuno, H., Mal, T. K., Walchli, M. et al. 2008. Light-dependent regulation of structural flexibility in a photochromic fluorescent protein. *Proceedings of the National Academy of Sciences of the United States of America*, **105**, 9227–9232.

Mizuno, H., Mal, T. K., Walchli, M. et al. 2010. Molecular basis of photochromism of a fluorescent protein revealed by direct 13C detection under laser illumination. *Journal of Biomolecular NMR*, **48**, 237–246.

Ormo, M., Cubitt, A. B., Kallio, K. et al. 1996. Crystal structure of the *Aequorea victoria* green fluorescent protein. *Science*, **273**, 1392–1395.

Patterson, G. H. 2009. Fluorescence microscopy below the diffraction limit. *Seminars in Cell & Developmental Biology*, **20**, 886–893.

Patterson, G. H. and Lippincott-Schwartz, J. 2002. A photoactivatable GFP for selective photolabeling of proteins and cells. *Science*, **297**, 1873–1877.

Peterman, E. J. G., Brasslet, S., and Moerner, W. E. 1999. The fluorescence dynamics of single molecules of green fluorescent protein. *Journal of Physical Chemistry A*, **103**, 10553–10560.

Petersen, J., Wilmann, P. G., Beddoe, T. et al. 2003. The 2.0-A crystal structure of eqFP611, a far red fluorescent protein from the sea anemone *Entacmaea quadricolor*. *Journal of Biological Chemistry*, **278**, 44626–44631.

Quillin, M. L., Anstrom, D. M., Shu, X. et al. 2005. Kindling fluorescent protein from *Anemonia sulcata*: Dark-state structure at 1.38 A resolution. *Biochemistry*, **44**, 5774–5787.

Sawin, K. E. and Nurse, P. 1997. Photoactivation of green fluorescent protein. *Current Biology*, **7**, R606–R607.

Shaner, N. C., Campbell, R. E., Steinbach, P. A. et al. 2004. Improved monomeric red, orange and yellow fluorescent proteins derived from *Discosoma* sp. red fluorescent protein. *Nature Biotechnology*, **22**, 1567–1572.

Shaner, N. C., Lin, M. Z., McKeown, M. R. et al. 2008. Improving the photostability of bright monomeric orange and red fluorescent proteins. *Nature Methods*, **5**, 545–551.

Stark, D. A. and Kulesa, P. M. 2007. An in vivo comparison of photoactivatable fluorescent proteins in an avian embryo model. *Developmental Dynamics*, **236**, 1583–1594.

Steinmeyer, R., Noskov, A., Krasel, C. et al. 2005. Improved fluorescent proteins for single-molecule research in molecular tracking and co-localization. *Journal of Fluorescence*, **15**, 707–721.

Stiel, A. C., Andresen, M., Bock, H. et al. 2008. Generation of monomeric reversibly switchable red fluorescent proteins for far-field fluorescence nanoscopy. *Biophysical Journal*, **95**, 2989–2997.

Stiel, A. C., Trowitzsch, S., Weber, G. et al. 2007. 1.8 A bright-state structure of the reversibly switchable fluorescent protein Dronpa guides the generation of fast switching variants. *Biochemical Journal*, **402**, 35–42.

Subach, F. V., Malashkevich, V. N., Zencheck, W. D. et al. 2009a. Photoactivation mechanism of PAmCherry based on crystal structures of the protein in the dark and fluorescent states. *Proceedings of the National Academy of Sciences of the United States of America*, **106**, 21097–21102.

Subach, F. V., Patterson, G. H., Manley, S. et al. 2009b. Photoactivatable mCherry for high-resolution two-color fluorescence microscopy. *Nature Methods*, **6**, 153–159.

Subach, F. V., Patterson, G. H., Renz, M., Lippincott-Schwartz, J., and Verkhusha, V. V. 2010. Bright monomeric photoactivatable red fluorescent protein for two-color super-resolution sptPALM of live cells. *Journal of the American Chemical Society*, **132**, 6481–6491.

Tsien, R. Y. 1998. The green fluorescent protein. *Annual Review of Biochemistry*, **67**, 509–544.

Tsutsui, H., Karasawa, S., Shimizu, H., Nukina, N., and Miyawaki, A. 2005. Semi-rational engineering of a coral fluorescent protein into an efficient highlighter. *EMBO Reports*, **6**, 233–238.

van Thor, J. J., Gensch, T., Hellingwerf, K. J., and Johnson, L. N. 2002. Phototransformation of green fluorescent protein with UV and visible light leads to decarboxylation of glutamate 222. *Nature Structural Biology*, **9**, 37–41.

Verkhusha, V. V. and Sorkin, A. 2005. Conversion of the monomeric red fluorescent protein into a photoactivatable probe. *Chemical Biology*, **12**, 279–285.

Wall, M. A., Socolich, M., and Ranganathan, R. 2000. The structural basis for red fluorescence in the tetrameric GFP homolog DsRed. *Nature Structural Biology*, **7**, 1133–1138.

Wiedenmann, J., Ivanchenko, S., Oswald, F. et al. 2004. EosFP, a fluorescent marker protein with UV-inducible green-to-red fluorescence conversion. *Proceedings of the National Academy of Sciences of the United States of America*, **101**, 15905–15910.

Wilmann, P. G., Turcic, K., Battad, J. M. et al. 2006. The 1.7 A crystal structure of Dronpa: A photoswitchable green fluorescent protein. *Journal of Molecular Biology*, **364**, 213–224.

Yang, F., Moss, L. G., and Phillips, G. N., Jr. 1996. The molecular structure of green fluorescent protein. *Nature Biotechnology*, **14**, 1246–1251.

Yarbrough, D., Wachter, R. M., Kallio, K., Matz, M. V., and Remington, S. J. 2001. Refined crystal structure of DsRed, a red fluorescent protein from coral, at 2.0-A resolution. *Proceedings of the National Academy of Sciences of the United States of America*, **98**, 462–467.

Yokoe, H. and Meyer, T. 1996. Spatial dynamics of GFP-tagged proteins investigated by local fluorescence enhancement. *Nature Biotechnology*, **14**, 1252–1256.

Far-red and near-infrared fluorescent proteins

Jun Chu, Yan Xing, and *Michael Z. Lin*
Stanford University School of Medicine

Contents

7.1 INTRODUCTION

Fluorescent proteins (FPs), originally found in the jellyfish *Aequorea victoria* (Shimomura et al., 1962) and later in corals (Matz et al., 1999) and anemones (Merzlyak et al., 2007; Wiedenmann et al., 2002), have revolutionized cell biology by providing genetically encoded fluorescent labels. Used alone, FPs can serve as real-time protein tags, reporters of gene expression, or cell lineage tracers in living cells (Sugiyama et al., 2009; van Roessel and Brand, 2002). FPs have also been engineered to have sensitivity to pH (Miesenbock et al., 1998), ions (Bregestovski et al., 2009; Dittmer et al., 2009; Palmer and Tsien, 2006), and other analytes (Mutoh et al., 2011; Rhee et al., 2010), either due to intrinsic sensitivity of fluorescence or via attachment of sensing domains that then allosterically modulate chromophore fluorescence. Forster resonance energy transfer (FRET) between FPs of two colors can be used to detect protein–protein interactions or biochemical events that lead to protein conformational changes such as ion influx or kinase activation. These incredibly wide-ranging capabilities of FPs have enabled cell biologists to visualize a wide array of biochemical events in real time in living cells using a fluorescence microscope. These experiments are now routine in cells cultured *in vitro*, in embryos, and in small organisms such as the worm *Caenorhabditis elegans*. Indeed the first use of FPs outside jellyfish was in labeling specific neuronal populations in *C. elegans* (Chalfie et al., 1994).

In contrast to their facile use in cells and model organisms, the application space of FPs in whole animals is more limited. High-resolution microscopy can be performed during surgery or through optical windows to a depth of several hundred microns with confocal methods or 1 mm with multiphoton methods (Helmchen and Denk, 2005; Kobat et al., 2009). Beyond that depth, however, scattering of light by tissue reduces resolution, and absorbance by tissue components limits sensitivity. Hemoglobin in blood vessels is the most abundant absorber of light in the visible wavelengths in mammalian tissue, with broad absorbance of light from blue wavelengths up to 600 nm (Weissleder, 2001). While the effects of scattering can be compensated to some degree using tomographic imaging methods, only the use of excitation light in the so-called optical window beyond 600 nm can overcome hemoglobin absorbance.

The absorbance spectra (usually identical to excitation spectra) of natural FPs all peak below 600 nm, and thus substantial efforts have been undertaken to engineer red-shifted FPs that can be imaged in the optical window (Lin et al., 2009; Morozova et al., 2010; Shcherbo et al., 2010; Strack et al., 2009). In addition to an advantage for intravital imaging, artificially evolved red-shifted FPs allow additional channels for multiwavelength experiments or for FRET. Here, we review the historical development of FPs, their mechanisms of wavelength tuning, and initial results from their uses in whole-animal imaging.

A note about color terminology is useful here. Some FPs that have mostly orange emission have been historically called red FPs (RFPs). To establish objective standards for color names, we will only use "red" in this chapter to describe FPs that emit at least 50% of their photons above 620 nm, a commonly accepted orange/red boundary (Bruno and Svoronos, 2006). We will denote as "far red" those FPs that emit at least 50% of their photons above 650 nm, an arbitrary cutoff chosen to be most consistent with historical far-red FP designations, as there is no standard definition of "far red." We will denote as infrared FPs (IFPs) those proteins that emit the majority of their photons above 700 nm, consistent with definitions for the red/infrared boundary from the International Commission on Illumination (CIE) and NASA Infrared Processing and Analysis Center (IPAC). Note that other conventions may be different; for instance, in the plant literature, 650–700 nm light is called red, and 700–750 nm light, which is still dimly visible to the eye at high intensity, is called far red (Sharrock, 2008).

7.2 EVOLUTIONARY HISTORY

7.2.1 FAR-RED FPs FROM CORALS

DsRed from the *Discosoma* sp. coral, also named drFP583, was the first orange-red FP identified (peak excitation/emission 558/583 nm) (Matz et al., 1999) and served as the starting point for the development of a set of true RFPs (Lin et al., 2009; Shaner et al., 2004; Strack et al., 2008, 2009; Wang et al., 2004). Native DsRed is brighter than enhanced green fluorescent protein (EGFP), the most widely used variant of *A. victoria* green FP (GFP), with brightness values of 59 versus 34 (calculated as the product of the extinction coefficient per chromophore in units of per mM per cm and quantum yield). However, native DsRed has several limitations: its maturation is slow, its tetrameric nature prevents its use as a protein tag, and it is not sufficiently red-shifted for deep-tissue imaging in mammals. To remove these limitations, DsRed was deoligomerized by mutating subunit interface residues with charged residues, followed by iterative cycles of evolution to restore brightness and maturation. This strategy first yielded the dimeric protein dimer2 (Shaner et al., 2004) that was then evolved into the tandem dimer tdTomato (Shaner et al., 2004), which exhibits similar spectra to DsRed and most of its brightness (peak excitation/emission 554/581 nm, brightness 48 per chromophore) but dramatically faster maturation. Several rounds of directed evolution aimed at monomerizing dimer2 yielded the monomeric mRFP1 (Campbell et al., 2002). The excitation and emission spectra of mRFP1 are significantly red-shifted from DsRed (584/607 nm), with more than 50% of emission photons above 620 nm, thus making this a true RFP. The brightness of mRFP1 is reduced compared to EGFP (13 vs. 34). For cell imaging, however, the lack of autofluorescence at red wavelengths usually results in signal/background ratios that are actually higher than with the intrinsically brighter GFP.

mRFP1 has served as a starting point for a series of FPs with even further red-shifted spectra. mCherry was derived from mRFP1 by a combination of random mutagenesis and saturation mutagenesis of sites near the chromophore in bacteria and features slightly red-shifted spectra and increased brightness (peak excitation/emission 587/610 nm, brightness 16), as well as faster maturation (Table 7.1) (Shaner et al., 2004). mRaspberry and mPlum were derived from mRFP1 by somatic hypermutation in lymphocytes and feature more dramatically red-shifted spectra (peak excitation/emission 598/625 nm and 590/649 nm, respectively, Table 7.1) (Wang et al., 2004). mPlum thus became the first monomeric far-red FP. However, this red shift comes at the cost of reduced brightness (13 and 4.1 for mRaspberry and mPlum, respectively).

In a case of convergent artificial evolution, mRFP1 and DsRed derivatives with excitation in the optical window were evolved with a common red-shifting mechanism. Rational placement of a tyrosine residue above the chromophore in mRFP1, repeating the red-shifting strategy used to create yellow FP (YFP) from GFP, followed by random mutagenesis yielded mGrape3 (Lin et al., 2009). mGrape3 demonstrates complex optical behavior: the dark-adapted protein absorbs maximally at 470 nm, but protein exposed to blue light absorbs maximally 608 nm (Table 7.1), within the optical window. Independently, a fast-maturing and nonaggregating variant of DsRed, DsRed-Express2 (Strack et al., 2008), was subjected to random and site-directed mutagenesis and selected for red-shifted fluorescence, yielding E2-Crimson (peak excitation/emission 611/646 nm) (Table 7.1) (Strack et al., 2009). One of the selected mutations in E2-Crimson places a

Photophysical properties of fluorescent proteins

Table 7.1 Characteristics of selected RFPs and IFPs

PROTEIN	EXCITATION PEAK[a]	EMISSION PEAK[a]	EXTINCTION COEFFICIENT[b]	QUANTUM YIELD[c]	PEAK BRIGHTNESS[d]	TERTIARY STRUCTURE	NATURAL PARENT
RFP630	583	630	50	0.35	18	Dimer	eqFP611
mRFP	584	607	50	0.25	13	Monomer	DsRed
mKate	585	635	42	0.30	13	Monomer	eqFP578
mKate2	586	630	50	0.36	18	Monomer	eqFP578
mCherry	587	610	72	0.22	16	Monomer	DsRed
tdRFP639	589	631	90	0.16	14	Tandem dimer	eqFP611
mPlum	590	649	41	0.10	4.1	Monomer	DsRed
eqFP650	592	650	65	0.24	16	Dimer	eqFP578
mRaspberry	598	625	86	0.15	13	Monomer	DsRed
mNeptune	600	650	67	0.20	13	Monomer	eqFP578
eqFP670	605	670	70	0.06	4.2	Dimer	eqFP578
mGrape3	608	646	40[e]	0.03	1.2	Monomer	DsRed
E2-Crimson	611	646	59	0.12	7.0	Tetramer	DsRed
TagRFP657	611	657	34	0.10	3.4	Monomer	eqFP578
IFP1.4	684	708	92	0.07	6.4	Monomer	DrBphP

[a] Excitation and emission maxima in nm.
[b] Maximum extinction coefficient per chromophore in $mM^{-1}\ cm^{-1}$ measured by the alkali denaturation method.
[c] Quantum yield of fluorescence.
[d] Calculated as peak extinction coefficient per chain in $mM^{-1}\ cm^{-1}$ multiplied by quantum yield.
[e] After photoactivation by 470 nm light. See text for references.

tyrosine in the same position as in mGrape3. mGrape3 and E2-Crimson have brightness values of 1.2 and 7.0 per chromophore, respectively. Thus, mGrape3 continues the trend in the mRFP1 derivatives of trading brightness for longer-wavelength emission profiles, while the severalfold higher brightness of E2-Crimson over mGrape3 is consistent with the relative dimness of monomers first seen with the transition of mRFP1 from dimer2.

In addition to far-red FPs derived from DsRed, a dimeric far-red was created from the coral chromoprotein hcriCP by a combination of random and site-directed mutagenesis. This protein, HcRed1, was the first far-red FP to be described (Gurskaya et al., 2001) but is fairly dim (peak excitation/emission 590/637 nm, brightness 3.2 per chromophore).

7.2.2 FAR-RED FPs FROM ANEMONE CORALS

More recently, orange-red FPs from the bulb-tip anemone *Entacmaea quadricolor*, eqFP578 (Merzlyak et al., 2007) and eqFP611 (Wiedenmann et al., 2002), have served as scaffolds for engineering red and far-red FPs, including some that combine excitation in the optical window with fairly high intrinsic brightness. Using random mutagenesis, the tetrameric orange FP eqFP578 was optimized for faster maturation to make TurboRFP and then monomerized to make TagRFP (Merzlyak et al., 2007). TagRFP is similar to tdTomato in the spectral profile (peak excitation/emission 555/584 nm, brightness 40) but is only half as bright.

Using the orange TurboRFP and TagRFP as precursors, dimeric and monomeric far-red FPs were then created. Saturation mutagenesis of three positions near the chromophore and random mutagenesis were performed on TurboRFP in a screen for red shifting, yielding Katushka (Shcherbo et al., 2007). Katushka has similar excitation wavelength and brightness as mCherry but relatively red-shifted emission (peak excitation/emission 588/635 nm, brightness 22). Transfer of the responsible mutations into TagRFP produced a monomeric variant, mKate (Shcherbo et al., 2007), with similar spectra and most of the brightness of Katushka (peak excitation/emission 588/635 nm, brightness 15). Improved variants were further developed by structure-guided mutagenesis to reduce unwanted chromophore conformations and random mutagenesis to promote maturation. The resulting tandem dimer tdKatushka2 and monomeric mKate2 are about 50% brighter than their predecessors (Shcherbo et al., 2009).

Further evolution of mKate resulted in the first bright FPs with excitation maxima in the optical window. mNeptune (peak excitation/emission 600/650 nm, brightness 13) was evolved from mKate by saturation mutagenesis of several sites on two ends of the chromophore, followed by random mutagenesis and further monomerization (Lin et al., 2009). TagRFP657 (peak excitation/emission 611/657 nm, brightness 3.4) was developed from monomeric mKate by a combination of random mutagenesis and site-directed mutagenesis at sites identified to be important in mNeptune and mGrape3 (Morozova et al., 2010). The dimeric eqFP650 (peak excitation/emission 592/650 nm, brightness 16) and eqFP670 (peak excitation/emission 605/670 nm, brightness 4.2) were evolved from Katushka-9-5, a low-cytotoxicity variant of Katushka, by a similar strategy (Shcherbo et al., 2010).

In parallel, the tetrameric orange FP eqFP611, closely related to eqFP578, has been dimerized and then further evolved to yield the far-red dimer RFP630 (peak excitation/emission 583/630 nm, brightness 18) and tandem dimer tdRFP639 (peak excitation/emission 589/631 nm, brightness 14) (Kredel et al., 2008). A monomeric orange FP, mRuby, has also been created from eqFP611 (Kredel et al., 2009). mRuby has similar optical characteristics as TagRFP (peak excitation/emission 558/605 nm, brightness 39) and demonstrates remarkable structural stability and excellent performance in fusions. mRuby thus could be another starting point for evolution of additional monomeric far-red FPs.

7.2.3 IFPs FROM PHYTOCHROMES

Phytochromes are a superfamily of proteins present in plants, cyanobacteria, and eubacteria that utilize tetrapyrrole cofactors synthesized by the host organism to allow absorption of far-red light (Hughes, 2010). Most phytochromes exist in two interconvertible states: the dark-adapted far-red-absorbing Pr state and an infrared-absorbing Pfr state induced by far-red light absorption (note that in the phytochrome literature, these are described as absorbing red and far-red light, respectively). Many phytochromes have C-terminal domains with presumed functions in light-dependent signal transduction or demonstrate light-dependent protein–protein interactions (Hughes, 2010).

The first phytochrome to be converted to an IFP was *Synechococcus* sp. cyanobacterial phytochrome 1 (Cph1) (Fischer and Lagarias, 2004). Random mutants of the PAS–GAF–PHY domain of the protein, lacking the histidine kinase domain, were expressed in bacteria engineered to synthesize the cofactor phycocyanobilin. Screening by fluorescence-activated cell sorting found that a single variant, named PR-1 and containing mutation of Tyr176 to His, conferred bright far-red fluorescence. PR-1 has peak excitation/emission of 644/672 nm and quantum yield of 0.145 (Fischer et al., 2005). The extinction coefficient has not been measured but is likely to be similar to that of free phycocyanobilin at 36 per mM per cm (Stadnichuk, 1995). Using the more extended plant tetrapyrrole phytochromobilin instead of phycocyanobilin further red shifts the excitation and emission peaks of PR-1 by about 15 nm.

The ability to create nonphotoconvertible and fluorescent mutants of cyanobacterial phytochromes extends to mutations at other sites in various phytochromes. In *Synechococcus* sp. OS-B' Cph1, mutation of Asp86 to His allows far-red fluorescence, although quantum yields have not been measured (Ulijasz et al., 2008). In the *Thermosynechococcus elongatus* phytochrome Tlr0924, mutation of Cys499 to Asp creates a far-red FP with quantum yield of 0.13 (Rockwell et al., 2008). Mutation of one of two Cys attachment sites for phycocyanobilin in the cyanobacterial cyanochrome Te-PixJ, which is closely related to phytochromes but natively absorbs at blue-green wavelengths, creates a red-absorbing far-red-emitting FP of unknown brightness (Ulijasz et al., 2009).

The first fluorescent phytochrome shown to be expressible in mammalian cells without toxicity was a derivative of a eubacterial phytochrome, DrBphP from *Deinococcus radiodurans* (Shu et al., 2009a). Unlike cyanobacterial phytochromes, these bacteriophytochromes use biliverdin as the chromophore. The first step toward making DrBphP fluorescent was the discovery that mutation of Asp207 to His in the chromophore-binding pocket reduced Pr–Pfr conversion and increased fluorescence (Wagner et al., 2008). The mutant PAS–GAF domain, the minimal chromophore-binding domain, was then used as a template for site-directed mutagenesis. Sites contacting the biliverdin cofactor and mediating dimerization were mutated and brighter variants selected. The brightest variant, designated IFP1.4, has excitation and emission maxima of 686 and 713 nm. IFP1.4 has an extinction coefficient of 90 per mM per cm and a quantum yield of 0.07 when saturated with the biliverdin cofactor. Although small amounts of biliverdin are present in mammalian cells, exogenous biliverdin must be added to achieve optimal brightness when IFP1.4 is expressed in mammalian cells.

Native bacteriophytochromes from *Pseudomonas aeruginosa* (Yang et al., 2008) and *Rhodopseudomonas palustris* (Toh et al., 2010) have been found to be fluorescent when bound to their natural tetrapyrrole cofactors as well. In the case of the *R. palustris* protein RpBph3, the fluorescence quantum yield is 0.045. This raises the possibility that many bacteriophytochromes can serve as starting points for the development of brighter IFPs (Toh et al., 2010).

7.3 STRUCTURES AND MECHANISMS

7.3.1 COMMON CHROMOPHORE COVALENT STRUCTURE

Crystal structures of representative members of the above FPs reveal that they conserve the covalent chromophore structure of DsRed. The chromophore of DsRed, formed from the tripeptide sequence of Gln66–Tyr67–Gly68, contains the same *p*-hydroxybenzylideneimidazolinone (HBDI) group as the GFP chromophore but also features an additional double bond between the backbone nitrogen and alpha carbon of Gln66 due to an additional oxidation reaction (Wall et al., 2000; Yarbrough et al., 2001) (Figure 7.1). This double bond is in conjugation with the HBDI group and the backbone

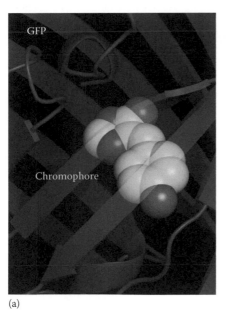
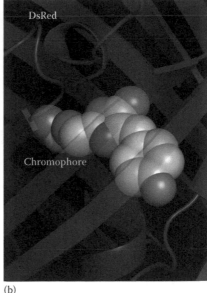

(a)   (b)

Figure 7.1 Covalent structures of GFP and DsRed chromophores. (a) The GFP chromophore is shown with carbon atoms in green, nitrogen in blue, and oxygen in red. The GFP chromophore is a π-conjugated system consisting of a phenolate ring (front) connected to an imidazolinone ring (rear) by a methylene bridge and is formed by a cyclization, dehydration, and oxidation reaction of a Ser–Tyr–Gly tripeptide sequence. (b) The DsRed chromophore is shown with carbon atoms in pink, nitrogen in blue, and oxygen in red. The DsRed chromophore consists of the same groups formed from cyclization of a Gln–Tyr–Gly sequence but with conjugation extended through an acylimine group formed by additional oxidation of the backbone N–Cα bond of the Gln residue.

Photophysical properties of fluorescent proteins

carbonyl group of amino acid 65. The extended π-conjugated chain created by the additional double bonds leads to more delocalized electron density and a decrease in the energy gap between the highest occupied molecular orbital and the lowest unoccupied molecular orbital, which represents the ground and excited states for the most common electron transition. The differences in the wavelengths of the proteins discussed here result from differences in the conformation of the chromophore and its interactions with the chromophore-binding pocket. As the chromophore is extended relative to GFP, some of the mechanisms of wavelength tuning that have evolved were not previously observed in GFP family proteins.

7.3.2 ALTERED ELECTRONIC INTERACTIONS WITH THE CHROMOPHORE RINGS IN mCherry

The crystal structure of mCherry shows several features that may account for the substantial red shift with respect to the progenitor DsRed (Shu et al., 2006). First, mCherry contains a substitution of the positively charged Lys163 with an uncharged Gln. The side chain is positioned directly underneath the phenol ring of the chromophore (when the N- and C-termini of the FP are oriented up), so its positive charge would be expected to stabilize electron density within the phenol ring (Figure 7.2). Its mutation to Gln would be expected to lead to a redistribution of electron density from the phenolate ring in the ground state. As negative charge is believed to

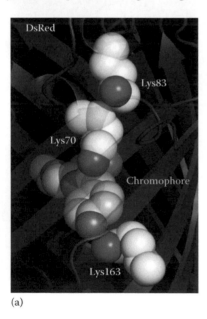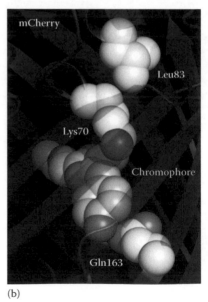

(a) (b)

Figure 7.2 Altered electronic interactions of the chromophore with its environment in mCherry. (a) In DsRed, the positive charges of Lys70 and Lys163 stabilize the ground state, where electron density preferentially resides in the phenol ring. (b) In mCherry, movement of Lys70 away from the chromophore and mutation of Lys163 to Gln reduces this electrostatic stabilization of the ground state. Movement of Lys70 in mCherry is due to loss of electrostatic repulsion by Leu83. The chromophore and side chains of these important amino acids are shown in space-filling representation, with carbon, nitrogen, and oxygen atoms shown in white, blue, and red, respectively.

move out of the phenolate ring upon excitation (Hasegawa et al., 2010), a reduction of electron density in the phenolate ring in the ground state would be expected to reduce the energy gap between ground and excited states, leading to a red shift.

Another important change is mutation of a polar Lys83 to a nonpolar Leu (a K83L mutation). A similar K83M mutation also confers a significant red shift of excitation and emission maxima in DsRed (Shu et al., 2006). The DsRed K83M crystal structure reveals movement of the amino group of a conserved Lys70 away from its original position between the methylene group and the phenol ring. A similar movement of Lys70 away from the phenol ring was also observed in mCherry (Figure 7.2). Mutation of Lys83 is likely to cause movement of Lys70 due to loss of electrostatic repulsion, as the direction of Lys70 movement is toward the new uncharged side chain at position 83. Because K83M, in the absence of other mutations, is sufficient to cause this movement in DsRed, K83L rather than other mutations is likely also sufficient for the similar shift in mCherry. The fact that Lys70 movement away from the phenol group causes a red shift in mCherry is consistent with this group functioning to stabilize the ground-state electron distribution in DsRed. Interestingly, quantum mechanics/molecular mechanics modeling suggests that the bond between the methylene group and the phenol ring has higher electron density in the ground versus the excited state.

7.3.3 EXCITED-STATE HYDROGEN BONDING TO THE ACYLIMINE IN mPlum

The further red-shifted mPlum has a similar loss of Lys163 as in mCherry, as well as a similar placement of Lys70, but it also has some unique features contributing to the additional red-shifted emission (Shu et al., 2009b). The neutral protonated carboxylic acid group of Glu16 directly hydrogen bonds to the carbonyl oxygen in the acylimine group of the chromophore (Figure 7.3). Time-resolved emission spectroscopy suggests that energetic relaxation of the chromophore occurs during the excited state (Abbyad et al., 2007). This may represent rotation of the side chain of Glu16 (Abbyad et al., 2007; Shu et al., 2009b). Electron density shift in the chromophore upon excitation may also polarize the acylimine carbonyl, strengthening the hydrogen bond with Glu16. Because conformational and/or hydrogen bond changes occur in the excited state, it preferentially reduces the energy of the excited state and thus the emitted photon energy (Shu et al., 2009b). This explains the large Stokes shift in mPlum and the lack of a major red shift in the excitation spectrum compared to mRFP1. Mutation of Leu65, the residue whose backbone contributes the carbonyl group to the chromophore acylimine, to Ile is required for these effects. In mPlum, the carboxylic acid group of Glu16 in mPlum assumes a conformation perpendicular to the chromophore plane (Shu et al., 2009b). The crystal structures of the Leu65 revertant and a Gln16 mutant of mPlum, which do not exhibit the large Stokes shift, show that the respective carboxylic acid and amide groups of these mutants are in a plane parallel to the chromophore, which presumably disallows the excited-state changes seen in mPlum (Shu et al., 2009b).

7.3.4 CHROMOPHORE ISOMERIZATION AND ALTERED ELECTRONIC DISTRIBUTIONS IN mKate

The crystal structure of mKate suggests new structural features that contribute to the red shift compared to its progenitor TagRFP (Pletnev et al., 2008). Three key mutations in the vicinity of the chromophore, Asn143 to Ser, Phe174 to Leu, and His197 to Arg,

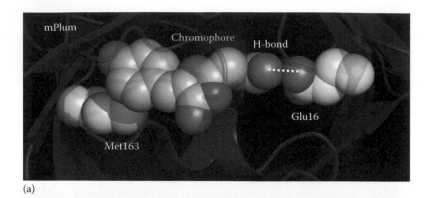

(a)

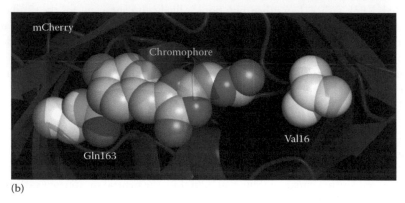

(b)

Figure 7.3 Energetic relaxations near the acylimine in the excited state of mPlum (a). Like mCherry (b), mPlum has lost the positive charge of Lys at position 163. It also contains a new hydrogen bond from Glu16 to the acylimine oxygen of the chromophore. Conformational changes to the chromophore or to Glu16 occur in the excited state, causing a large Stokes shift.

are believed to be primarily responsible for the red shift. The crystal structure of mKate reveals that the chromophore is in a *cis* configuration, similar to DsRed (Figure 7.4). Its precursor TagRFP is assumed to be in the *trans* configuration similar to the spectrally similar and highly homologous eqFP611, which has been crystallized (Petersen et al., 2003). The change from *trans* to *cis* chromophores in mKate reduces hydrogen bonding to the chromophore phenolate. In the *trans* conformation, the chromophore phenolate group of TagRFP makes two hydrogen bonds to Asn143 and Ser158. The chromophore of mKate in the *cis* conformation is only able to make a hydrogen bond to Ser143. The loss of a hydrogen bond would be expected to destabilize electron distribution on the phenolate, thereby increasing the energy of the ground state and causing a red shift.

A second red-shifting mechanism in mKate is the removal of a positive charge above the chromophore. In eqFP611 and TagRFP, a His197 residue extends above the chromophore, stabilizing electron density in the phenolate ring. This would tend to increase the energy required to transition an electron to the excited state (Ai et al., 2006). In mKate, this electronic interaction is altered; replacement of His197 by Arg leads to a repositioning of positive charge from directly above the phenolate ring to alongside it (Figure 7.4).

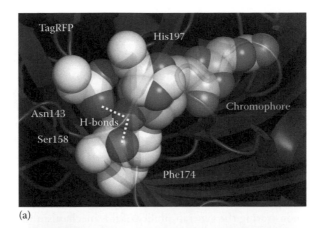

(a)

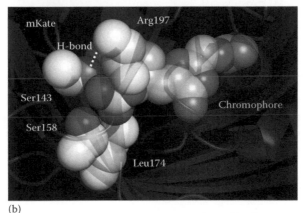

(b)

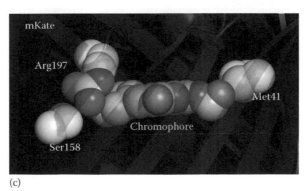

(c)

Figure 7.4 Chromophore isomerization, reduced hydrogen bonding, and altered electrostatic interactions in mKate. (a) The chromophore rings in TagRFP are in a *trans* conformation, where the phenolate oxygen makes two hydrogen bonds and the His197 undergoes a cation–π interaction with the phenolate ring. (b,c) Simultaneous mutation of Asn143 to Ser, Phe174 to Leu, and His197 to Arg to create mKate causes the chromophore to assume a *cis* conformation, losing one hydrogen bond and the cation–π interaction.

Photophysical properties of fluorescent proteins

7.3.5 INCREASED CHROMOPHORE PLANARITY AND A NEW HYDROGEN BOND IN NEPTUNE

The dimeric Neptune, and presumably its monomeric version mNeptune, has a combination of structural features that contribute to the substantial red shift of both excitation and emission maxima (Lin et al., 2009). First, a mutation of Met41 to Gly creates a cavity near the acylimine oxygen, which is filled by a water molecule. This water molecule creates a novel hydrogen bond to the acylimine oxygen of the chromophore (Figure 7.5). Reversion of position 41 to Met causes a substantial blue shift to the excitation spectrum, indicating that this hydrogen bond is preferentially stabilizing the excited state, which can be explained by the excited state having increased electron distribution on the acylimine oxygen relative to the ground state (Hasegawa et al., 2010). Similar mutations at position 41 contribute to red shifting in E2-Crimson, TagRFP657, and eqFP670, demonstrating the generalizability of this mechanism (Morozova et al., 2010; Shcherbo et al., 2010; Strack et al., 2009).

Second, Neptune exhibits increased coplanarity of the chromophore rings in comparison with its progenitor mKate. The side chain of Arg197, instead of extending across and to the side chain of the phenolate ring, extends parallel above the two rings of the chromophore (Figure 7.5). This change may be due to the mutation of Ser158 to Cys (S158C), which places the larger Cys side chain in a space adjacent to the phenolate ring originally occupied by the guanidinium group of Arg197. The new position of Arg197 allows the reduction of the twist angle between the phenolate and the imidazolinone rings of the chromophore. Coplanarity facilitates red shifting as it promotes more efficient electron delocalization and reduces the energy of the lowest unoccupied molecular orbital representing the excited state. Interestingly, the combination of S158C and M41G produces a 7 nm red shift, while S158C alone has negligible effect on wavelength. Apparently the repositioning of Arg197 may only be energetically favorable

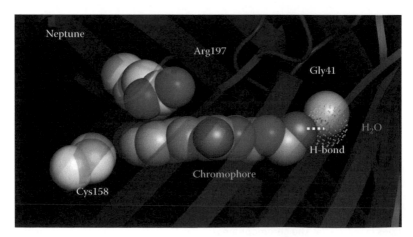

Figure 7.5 Increased chromophore planarity and a new hydrogen bond in Neptune. Mutation of Ser158 to Cys pushes Arg197 to a new position above the chromophore, allowing for a more coplanar arrangement of the two chromophore rings. Mutation of Met41 to Gly creates a new cavity for a water molecule to reside in near the acylimine. Hydrogen bonding between the water molecule and the acylimine oxygen selectively stabilizes the excited state and leads to an excitation red shift.

with loss of Met41 due to coordinated conformational changes. Finally, a mutation of Ser62 to Cys may alter the positioning of the internal alpha helix to allow better hydrogen bonding between the water molecule and the acylimine oxygen.

7.3.6 INCREASED ENVIRONMENT POLARIZABILITY IN mGrape, TagRFP657, AND E2-Crimson

mGrape3, E2-Crimson, and TagRFP657 all have a tyrosine residue at position 197 (DsRed numbering) and are the currently most red-shifted FPs in terms of excitation wavelength, with peaks at 608–611 nm. No crystal structures have been solved for these proteins, but they presumably utilize the same red-shifting mechanism as YFP. To create YFP, a tyrosine was rationally introduced at position 203 of EGFP (homologous to 197 of DsRed), leading to a red shift of both excitation and emission spectra by about 20 nm. Tyr was chosen because it can introduce a π–π stacking interaction between its phenol group and the phenolate group of the chromophore (Ormo et al., 1996). The highly polarizable phenol group in effect increases the polarizability of the protein solvent with which the chromophore interacts, and solvent polarizability is known to cause excitation red shifts in π-conjugated fluorophores. This is because the excited state of a π-conjugated fluorophore usually has a larger dipole moment than the ground state, but if it forms within a polarizable solvent, the dipole moment is partially counteracted by corresponding charge shifts in the solvent (Wachter et al., 1998).

7.3.7 REDUCED NONRADIATIVE DECAY IN FLUORESCENT BACTERIOPHYTOCHROMES

Unlike autocatalytic FPs, engineered fluorescent bacteriophytochromes differ from their native precursors primarily in being brighter rather than being red shifted. Structural studies would aid in understanding how higher brightness has evolved in these engineered fluorescent bacteriophytochromes. Structures of PR-1 and IFP1 have not been solved, but their fluorescence enhancements can be inferred from available phytochrome structures. Spectroscopic studies have suggested the bond between the C and D rings of phycocyanobilin in cyanobacterial phytochromes and of biliverdin in eubacterial phytochromes undergoes a *cis–trans* isomerization upon light absorption, leading to the Pfr conformation in the full-length protein. The structures of the PAS–GAF–PHY domain of PR-1 and the PAS–GAF domain of the IFP1 precursor DrBphP show that the chromophores are situated in a large binding pocket and are tethered to the protein by a covalent bond between a Cys side chain and ring A (Essen et al., 2008; Wagner et al., 2005). In both Cph1 and DrBphP, the chromophore-binding pocket is enlarged near the D ring, allowing space for rotation (Figure 7.6). The critical Y176H mutation in PR-1 that accounts for its brightness and the brightness-enhancing mutations in IFP1 is located within the D-ring-binding pocket.

Spectroscopic and molecular dynamics studies suggest that the effect of these fluorescence-enhancing mutations may be to suppress both *cis–trans* isomerization and other nonradiative mechanisms of relaxation from the excited state. The higher fluorescence of RpBphP3 relative to RpBph2 is associated with an extended excited-state lifetime, attributed to constraints placed on the rotation of the pyrrole D ring by hydrogen bonds with the surrounding protein matrix (Toh et al., 2010). This decreased motion in the excited state would be expected to lead to less efficient *cis–trans* isomerization, but it also reduces occurrence of an excited-state proton transfer

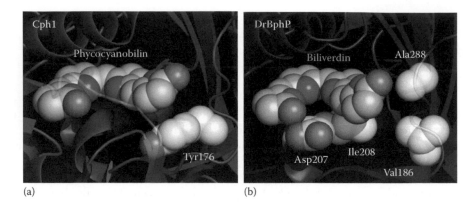

(a)   (b)

Figure 7.6 Sites of mutations that increase fluorescence of phytochrome family proteins. (a) In the creation PR-1 from Cph1, mutation of Tyr176 to His is sufficient to eliminate photoisomerization and increase fluorescence quantum yield. (b) In the creation of IFP1 from DrBphP, mutations of Asp207 to His, Val186 to Met, Ile208 to Thr, and Ala288 to Val likely tightened the chromophore-binding pocket near the fourth ring of biliverdin, increasing fluorescence.

(ESPT) event, which otherwise allows nonfluorescent decay from the Pr excited state to the Pr ground state. Still, the vast majority of excited states are terminated by ESPT, with only 4.5% of excited states yielding fluorescence and 6% yielding *cis–trans* isomerization of the C–D bond. A molecular dynamics simulation of IFP1.4 also suggests that suppression of ESPT contributes to IFP1.4 fluorescence (Samma et al., 2010). The simulation suggests that the IFP1.4 ground state experiences strain imposed by a noncomplementary chromophore cavity, which is released in the excited state as it adopts the more optimal geometry, leading to reduced ESPT and increased fluorescence. A more detailed understanding of this biophysical phenomenon together with the crystallization of PR-1 and IFP1.4 should help inform the future development of improved IFP tags.

7.4 APPLICATIONS

In theory, FPs that emit in the optical window allow for more sensitive wide-field imaging in deep tissues of mammals than GFPs, due to less absorption of the emission light by hemoglobin. Likewise, FPs that both excite and emit in the optical window will further improve sensitivity and accuracy by allowing excitation light to reach deeper and more uniformly into tissue. A few studies have directly compared detection of FPs at different wavelengths. One examined the influence of red emission alone on detectability of purified FP placed inside the abdomen of mice. In this comparison of EGFP, dTomato, mCherry, Katushka, mRaspberry, and mPlum, contrast was correlated with the extent of emission in the optical window (Deliolanis et al., 2008). Another study directly compared EGFP, tdTomato, and mCherry in tumor cells implanted in animals. This study found that tdTomato gave the highest signal and EGFP the lowest, using excitation light of 470 nm for EGFP and 560 nm for tdTomato and mCherry. The superior sensitivity of tdTomato is presumably due to its higher brightness than mCherry (Winnard et al., 2006).

However, the studies mentioned earlier used proteins that still require light of wavelengths below 600 nm for excitation, and so their visualization was still hindered by hemoglobin absorbance in deep tissue. The importance of excitation light was recently confirmed in animals expressing the far-red FPs mKate and Neptune in the liver, a large, deeply situated vascularized tissue (Lin et al., 2009). In a comparison, using two bands of light centered at 610 and 580 nm, excitation with light in the optical window was found to yield better signal/noise ratios, as expected (Lin et al., 2009). Imaging of the IFP1.4 precursor IFP1.1 was also performed using 660 nm excitation light, and the observed contrast was lower than seen with Neptune at 610 nm excitation, likely due to the lower intrinsic brightness of IFP1.4. Furthermore, with 660 nm excitation, fluorescence of food in the gut becomes significant, so reduced-fluorescence food should be used to improve contrast at this wavelength. Newer FPs have since been tested at multiple wavelengths but only in subcutaneous locations where even GFP is quite visible (Shcherbo et al., 2010).

These new far-red FPs that are excitable in the optical window can be utilized in all applications where bluer FPs have been adequate. Cells expressing GFPs and orange-red FPs can be visualized in subcutaneous locations with blue and green excitation light (Hoffman, 2005) or in deeper locations with surgical flaps, implanted windows, or fiber-optic microcopy. Far-red FPs are capable of the same uses but should also allow for more sensitive noninvasive imaging of deeper locations, especially if tomographic 3D reconstruction methods are used (Ntziachristos et al., 2002; Turchin et al., 2008). Far-red FPs also have the potential to serve as components for biosensors of specific pathways that can then be imaged in the context of a live animal noninvasively. Finally, far-red FPs provide an additional color in *in vitro* microscopy experiments that can be excited by 594 or 633 nm HeNe laser lines and are easily separable from orange FPs such as tdTomato, TagRFP, or mRuby.

7.5 CONCLUSIONS

How much redder and brighter can we make far-red FPs? The dsRed chromophore is constrained by space and the need to be encoded within a turn of an alpha helix. There are strains still present along the acylimine region of the chromophore even in the most red-shifted FPs such as Neptune. It is thus possible that relief of these strains by the appropriate mutations will lead to further red shifts. Further mutagenesis of bacteriophytochromes to generate brighter IFPs is likely to be productive as well, given the small amount of *in vitro* evolution attempted so far. Increasing the quantum yield of fluorescent bacteriophytochromes would be very desirable to make them more useful research tools. A challenge for IFPs in whole-animal imaging, however, is ensuring adequate supply of the chromophores to cells throughout the body.

REFERENCES

Abbyad, P., Childs, W., Shi, X. et al. (2007). Dynamic Stokes shift in green fluorescent protein variants. *Proc Natl Acad Sci USA 104*, 20189–20194.

Ai, H. W., Henderson, J. N., Remington, S. J. et al. (2006). Directed evolution of a monomeric, bright and photostable version of *Clavularia* cyan fluorescent protein: Structural characterization and applications in fluorescence imaging. *Biochem J 400*, 531–540.

Bregestovski, P., Waseem, T., and Mukhtarov, M. (2009). Genetically encoded optical sensors for monitoring of intracellular chloride and chloride-selective channel activity. *Front Mol Neurosci 2*, 15.

Bruno, T.J. and Svoronos, P.D.N. (2006). *CRC Handbook of Fundamental Spectroscopic Correlation Charts*. Boca Raton, FL: CRC Press.

Campbell, R. E., Tour, O., Palmer, A. E. et al. (2002). A monomeric red fluorescent protein. *Proc Natl Acad Sci USA 99*, 7877–7882.

Chalfie, M., Tu, Y., Euskirchen, G. et al. (1994). Green fluorescent protein as a marker for gene expression. *Science 263*, 802–805.

Deliolanis, N. C., Kasmieh, R., Wurdinger, T. et al. (2008). Performance of the red-shifted fluorescent proteins in deep-tissue molecular imaging applications. *J Biomed Opt 13*, 044008.

Dittmer, P. J., Miranda, J. G., Gorski, J. A. et al. (2009). Genetically encoded sensors to elucidate spatial distribution of cellular zinc. *J Biol Chem 284*, 16289–16297.

Essen, L. O., Mailliet, J., and Hughes, J. (2008). The structure of a complete phytochrome sensory module in the Pr ground state. *Proc Natl Acad Sci USA 105*, 14709–14714.

Fischer, A. J. and Lagarias, J. C. (2004). Harnessing phytochrome's glowing potential. *Proc Natl Acad Sci USA 101*, 17334–17339.

Fischer, A. J., Rockwell, N. C., Jang, A. Y. et al. (2005). Multiple roles of a conserved GAF domain tyrosine residue in cyanobacterial and plant phytochromes. *Biochemistry 44*, 15203–15215.

Gurskaya, N. G., Fradkov, A. F., Terskikh, A. et al. (2001). GFP-like chromoproteins as a source of far-red fluorescent proteins. *FEBS Lett 507*, 16–20.

Hasegawa, J. Y., Ise, T., Fujimoto, K. J. et al. (2010). Excited states of fluorescent proteins, mKO and DsRed: Chromophore-protein electrostatic interaction behind the color variations. *J Phys Chem B 114*, 2971–2979.

Helmchen, F. and Denk, W. (2005). Deep tissue two-photon microscopy. *Nat Methods 2*, 932–940.

Hoffman, R. M. (2005). The multiple uses of fluorescent proteins to visualize cancer in vivo. *Nat Rev Cancer 5*, 796–806.

Hughes, J. (2010). Phytochrome three-dimensional structures and functions. *Biochem Soc Trans 38*, 710–716.

Kobat, D., Durst, M. E., Nishimura, N. et al. (2009). Deep tissue multiphoton microscopy using longer wavelength excitation. *Opt Express 17*, 13354–13364.

Kredel, S., Nienhaus, K., Oswald, F. et al. (2008). Optimized and far-red-emitting variants of fluorescent protein eqFP611. *Chem Biol 15*, 224–233.

Kredel, S., Oswald, F., Nienhaus, K. et al. (2009). mRuby, a bright monomeric red fluorescent protein for labeling of subcellular structures. *PLoS One 4*, e4391.

Lin, M. Z., McKeown, M. R., Ng, H. L. et al. (2009). Autofluorescent proteins with excitation in the optical window for intravital imaging in mammals. *Chem Biol 16*, 1169–1179.

Matz, M. V., Fradkov, A. F., Labas, Y. A. et al. (1999). Fluorescent proteins from nonbioluminescent Anthozoa species. *Nat Biotechnol 17*, 969–973.

Merzlyak, E. M., Goedhart, J., Shcherbo, D. et al. (2007). Bright monomeric red fluorescent protein with an extended fluorescence lifetime. *Nat Methods 4*, 555–557.

Miesenbock, G., De Angelis, D. A., and Rothman, J. E. (1998). Visualizing secretion and synaptic transmission with pH-sensitive green fluorescent proteins. *Nature 394*, 192–195.

Morozova, K. S., Piatkevich, K. D., Gould, T. J. et al. (2010). Far-red fluorescent protein excitable with red lasers for flow cytometry and superresolution STED nanoscopy. *Biophys J 99*, L13–L15.

Mutoh, H., Perron, A., Akemann, W. et al. (2011). Optogenetic monitoring of membrane potentials. *Exp Physiol 96*, 13–18.

Ntziachristos, V., Bremer, C., Graves, E. E. et al. (2002). In vivo tomographic imaging of near-infrared fluorescent probes. *Mol Imaging 1*, 82–88.

Ormo, M., Cubitt, A. B., Kallio, K. et al. (1996). Crystal structure of the *Aequorea victoria* green fluorescent protein. *Science 273*, 1392–1395.

Palmer, A. E. and Tsien, R. Y. (2006). Measuring calcium signaling using genetically targetable fluorescent indicators. *Nat Protoc 1*, 1057–1065.

Petersen, J., Wilmann, P. G., Beddoe, T. et al. (2003). The 2.0-A crystal structure of eqFP611, a far red fluorescent protein from the sea anemone *Entacmaea quadricolor*. *J Biol Chem 278*, 44626–44631.

Pletnev, S., Shcherbo, D., Chudakov, D. M. et al. (2008). A crystallographic study of bright far-red fluorescent protein mKate reveals pH-induced *cis–trans* isomerization of the chromophore. *J Biol Chem 283*, 28980–28987.

Rhee, S. G., Chang, T. S., Jeong, W. et al. (2010). Methods for detection and measurement of hydrogen peroxide inside and outside of cells. *Mol Cells 29*, 539–549.

Rockwell, N. C., Njuguna, S. L., Roberts, L. et al. (2008). A second conserved GAF domain cysteine is required for the blue/green photoreversibility of cyanobacteriochrome Tlr0924 from *Thermosynechococcus elongatus*. *Biochemistry 47*, 7304–7316.

Samma, A. A., Johnson, C. K., Song, S. et al. (2010). On the origin of fluorescence in bacteriophytochrome infrared fluorescent proteins. *J Phys Chem B 114*, 15362–15369.

Shaner, N. C., Campbell, R. E., Steinbach, P. A. et al. (2004). Improved monomeric red, orange and yellow fluorescent proteins derived from *Discosoma* sp. red fluorescent protein. *Nat Biotechnol 22*, 1567–1572.

Sharrock, R. A. (2008). The phytochrome red/far-red photoreceptor superfamily. *Genome Biol 9*, 230.

Shcherbo, D., Merzlyak, E. M., Chepurnykh, T. V. et al. (2007). Bright far-red fluorescent protein for whole-body imaging. *Nat Methods 4*, 741–746.

Shcherbo, D., Murphy, C. S., Ermakova, G. V. et al. (2009). Far-red fluorescent tags for protein imaging in living tissues. *Biochem J 418*, 567–574.

Shcherbo, D., Shemiakina, I. I., Ryabova, A. V. et al. (2010). Near-infrared fluorescent proteins. *Nat Methods 7*, 827–829.

Shimomura, O., Johnson, F. H., and Saiga, Y. (1962). Extraction, purification and properties of aequorin, a bioluminescent protein from the luminous hydromedusan, *Aequorea*. *J Cell Comp Physiol 59*, 223–239.

Shu, X., Royant, A., Lin, M. Z. et al. (2009a). Mammalian expression of infrared fluorescent proteins engineered from a bacterial phytochrome. *Science 324*, 804–807.

Shu, X., Shaner, N. C., Yarbrough, C. A. et al. (2006). Novel chromophores and buried charges control color in mFruits. *Biochemistry 45*, 9639–9647.

Shu, X., Wang, L., Colip, L. et al. (2009b). Unique interactions between the chromophore and glutamate 16 lead to far-red emission in a red fluorescent protein. *Protein Sci 18*, 460–466.

Stadnichuk, I. N. (1995). Phycobiliproteins: Determination of chromophore composition and content. *Phytochem Anal 6*, 281–288.

Strack, R. L., Hein, B., Bhattacharyya, D. et al. (2009). A rapidly maturing far-red derivative of DsRed-Express2 for whole-cell labeling. *Biochemistry 48*, 8279–8281.

Strack, R. L., Strongin, D. E., Bhattacharyya, D. et al. (2008). A noncytotoxic DsRed variant for whole-cell labeling. *Nat Methods 5*, 955–957.

Sugiyama, M., Sakaue-Sawano, A., Iimura, T. et al. (2009). Illuminating cell-cycle progression in the developing zebrafish embryo. *Proc Natl Acad Sci USA 106*, 20812–20817.

Toh, K. C., Stojkovic, E. A., van Stokkum, I. H. et al. (2010). Proton-transfer and hydrogen-bond interactions determine fluorescence quantum yield and photochemical efficiency of bacteriophytochrome. *Proc Natl Acad Sci USA 107*, 9170–9175.

Turchin, I. V., Kamensky, V. A., Plehanov, V. I. et al. (2008). Fluorescence diffuse tomography for detection of red fluorescent protein expressed tumors in small animals. *J Biomed Opt 13*, 041310.

Ulijasz, A. T., Cornilescu, G., von Stetten, D. et al. (2008). Characterization of two thermostable cyanobacterial phytochromes reveals global movements in the chromophore-binding domain during photoconversion. *J Biol Chem 283*, 21251–21266.

Ulijasz, A. T., Cornilescu, G., von Stetten, D. et al. (2009). Cyanochromes are blue/green light photoreversible photoreceptors defined by a stable double cysteine linkage to a phycoviolobilin-type chromophore. *J Biol Chem 284*, 29757–29772.

van Roessel, P. and Brand, A. H. (2002). Imaging into the future: Visualizing gene expression and protein interactions with fluorescent proteins. *Nat Cell Biol 4*, E15–E20.

Wachter, R. M., Elsliger, M. A., Kallio, K. et al. (1998). Structural basis of spectral shifts in the yellow-emission variants of green fluorescent protein. *Structure 6*, 1267–1277.

Wagner, J. R., Brunzelle, J. S., Forest, K. T. et al. (2005). A light-sensing knot revealed by the structure of the chromophore-binding domain of phytochrome. *Nature 438*, 325–331.

Wagner, J. R., Zhang, J., von Stetten, D. et al. (2008). Mutational analysis of *Deinococcus radiodurans* bacteriophytochrome reveals key amino acids necessary for the photochromicity and proton exchange cycle of phytochromes. *J Biol Chem 283*, 12212–12226.

Wall, M. A., Socolich, M., and Ranganathan, R. (2000). The structural basis for red fluorescence in the tetrameric GFP homolog DsRed. *Nat Struct Biol 7*, 1133–1138.

Wang, L., Jackson, W. C., Steinbach, P. A. et al. (2004). Evolution of new nonantibody proteins via iterative somatic hypermutation. *Proc Natl Acad Sci USA 101*, 16745–16749.

Weissleder, R. (2001). A clearer vision for in vivo imaging. *Nat Biotechnol 19*, 316–317.

Wiedenmann, J., Schenk, A., Rocker, C. et al. (2002). A far-red fluorescent protein with fast maturation and reduced oligomerization tendency from *Entacmaea quadricolor* (Anthozoa, Actinaria). *Proc Natl Acad Sci USA 99*, 11646–11651.

Winnard, P. T. J., Kluth, J. B., and Raman, V. (2006). Noninvasive optical tracking of red fluorescent protein-expressing cancer cells in a model of metastatic breast cancer. *Neoplasia 8*, 796–806.

Yang, X., Kuk, J., and Moffat, K. (2008). Crystal structure of *Pseudomonas aeruginosa* bacteriophytochrome: Photoconversion and signal transduction. *Proc Natl Acad Sci USA 105*, 14715–14720.

Yarbrough, D., Wachter, R. M., Kallio, K. et al. (2001). Refined crystal structure of DsRed, a red fluorescent protein from coral, at 2.0-A resolution. *Proc Natl Acad Sci USA 98*, 462–467.

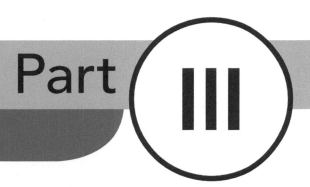

Part III

Applications

Genetically encoded fluorescent proteins and FRAP

Davide Mazza and *James G. McNally*
National Institutes of Health

Contents

8.1 INTRODUCTION

Fluorescence microscopy generates more than just structural information about a specimen. It quantifies macromolecular dynamics within live cells or tissues by examining how molecules move and interact with other cellular components. One of the first techniques developed to study macromolecular dynamics was fluorescence recovery after photobleaching (FRAP). FRAP was introduced in the mid-1970s to quantify the lateral diffusion of lipids and proteins in spatially confined biological membranes (Axelrod et al. 1976). Initially, FRAP was technically demanding, as specialized procedures were required to fluorescently label a molecule of interest and then introduce it into a specimen and specialized instruments were required to perform the FRAP measurement. This has changed with the advent of the fluorescent proteins (FPs) as genetically encoded markers and with the availability of commercial laser scanning confocal microscopes. FRAP is now much more accessible, and as a result, the use of FRAP has skyrocketed (Figure 8.1).

Throughout its history, FRAP has contributed to various fields, ranging from cell biology (Reits and Neefjes 2001; Snapp et al. 2003; Chen et al. 2006) to drug delivery (Meyvis et al. 1999; De Smedt et al. 2005) and materials science (Lorén et al. 2009). It is the cell biology field, however, that has most benefited from the introduction of the FPs. Here, it has been possible to tag cellular proteins of interest and then investigate their dynamics within live cells, often generating completely new insights about how proteins move and interact with other proteins in living cells. For example, FRAP powered by FP fusions has revised our view of DNA transcription and repair, demonstrating that these processes are far more dynamic than predicted from *in vitro* biochemistry (Hager et al. 2009).

Although many FRAP analyses of cellular proteins have revealed surprisingly dynamic behavior, protein mobility is often slower than would be expected from simple diffusion. This slower mobility arises from an assortment of other processes and effects

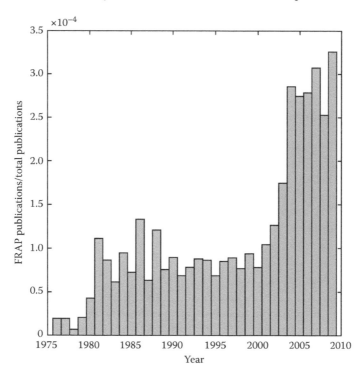

Figure 8.1 FRAP usage. FRAP applications have increased dramatically in the last decade. The plot shows the number of publications indexed in PubMed citing the words "fluorescence recovery photobleaching" in either the title or the abstract, normalized to the total number of PubMed publications in that year.

occurring within a living cell. For example, proteins may interact with each other and form larger complexes, or they may interact with more immobile structural elements in the cell. Furthermore, proteins may be transported directionally, or they may move in a highly crowded environment and be impeded by obstacles.

All the preceding cellular processes and properties will alter the mobility of a protein, and this can often be detected by FRAP. Thus, FRAP data on fluorescently tagged proteins almost always includes more information than just the protein's rate of simple diffusion. For example, FRAP can detect binding interactions of key proteins or probe subcellular architecture that might constrain diffusion. Extracting the biologically relevant information beyond the simple diffusion of proteins has been the focus of most of the recent work with FRAP and the FPs (Carrero et al. 2003; Bancaud et al. 2009; van Royen et al. 2009; Mueller et al. 2010).

The explosion in new applications for FRAP over the past decade has also stimulated the development of alternative approaches to assay protein dynamics *in vivo*. These include alternate forms of photobleaching microscopy (Dundr and Misteli 2003), such as fluorescence loss in photobleaching (FLIP), inverse FRAP (iFRAP) and its cousin photoactivation FRAPa, and continuous photobleaching FRAP (cpFRAP). In addition to photobleaching, other fluorescence microscopy methods, including fluorescence correlation spectroscopy (FCS) and single-molecule tracking (SMT), are also emerging as critical tools to probe protein movement and interaction within live cells (Dange et al. 2011; Erdel et al. 2011). All of these approaches for assaying cellular protein dynamics have relied heavily on labeling proteins inside living cells using the genetically encoded FPs.

In this chapter, we describe the basic principles of FRAP and what can be learned from applying the technique. We also discuss the various alternative and complementary approaches for assessing cellular protein dynamics *in vivo*. Then we consider the suitability of the FPs for both FRAP and other forms of photobleaching microscopy, as well as for FCS and SMT. We conclude with a plea for further testing and development of FPs designed specifically for either photobleaching microscopy, FCS, or SMT, highlighting the key features of an ideal FP for each of these techniques.

8.2 *IN VIVO* METHODS FOR MEASURING PROTEIN DYNAMICS

8.2.1 FLUORESCENCE RECOVERY AFTER PHOTOBLEACHING: AN OVERVIEW

FRAP quantifies the local mobility of fluorescently tagged proteins by photobleaching a selected region of a sample. Ideally, the photobleach is instantaneous and irreversible. This is followed by measurement of the increase in fluorescence as labeled molecules repopulate the bleached region from its surroundings.

A typical FRAP experiment can be divided into three phases (Figure 8.2a). First, in the prebleach phase, fluorescence in the cell is monitored over time to ensure that the labeled population is at equilibrium (corresponding to no significant fluctuations in the average fluorescence intensity). Second, in the photobleaching phase, the photobleach is applied by scanning a high-intensity laser beam over the region of interest. This photobleach is typically short (<1 s) to minimize the effects of diffusion during the photobleach, which can adversely affect subsequent quantitative analysis. Third, in

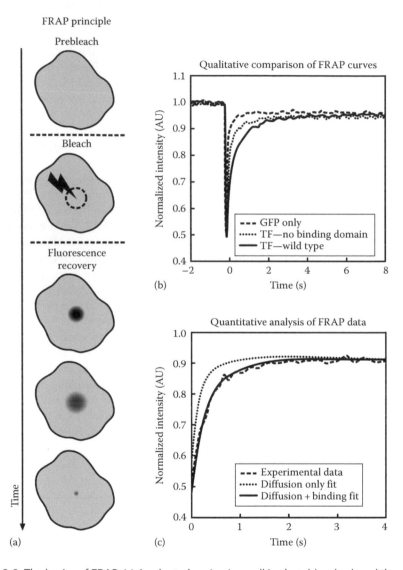

Figure 8.2 The basics of FRAP. (a) A selected region in a cell is photobleached, and then fluorescence from outside the bleach zone diffuses in. (b) Measurement of the fluorescent intensity as a function of time in the bleach zone yields the FRAP curve. Different molecules yield different curves that can be compared qualitatively. Shown are representative curves for a molecule like GFP that exhibits diffusion compared to a wild-type TF that exhibits diffusion and binding. Note that the removal of the TF's DNA-binding domain leads to a faster recovery. (c) Mathematical models can be devised to describe FRAP recoveries. Shown here are fits of two models to experimental data (dashed line). In this illustration, a simple diffusion model is insufficient to explain the FRAP recovery, whereas a diffusion and binding model provides a satisfactory fit.

Applications

the recovery phase, the laser power is reduced to the prebleach value and fluorescence intensity within the bleached area is again monitored over time until equilibrium is reached. During the recovery phase, additional photobleaching is often inadvertently induced over the entire imaged region. This "observational photobleaching" is typically less than the intentional photobleach, but it must be corrected before analyzing the experimental data. Observational photobleaching occurs not only in FRAP, but in all forms of photobleaching microscopy described in subsequent sections, and so in all cases, a correction method is essential to undue this effect.

FRAP can be performed on a wide range of fluorescence microscopes. Its simplest implementation uses a widefield microscope equipped with a laser that is focused at a selected location to induce photobleaching in a nearly diffraction-limited spot. The use of confocal microscopes equipped with acousto-optical illumination control increases the flexibility of the technique, allowing the user to define both the size and the shape of the bleached area, for example, to overlay a cellular structure of interest. With conventional one-photon excitation, however, it is not possible to confine the bleach volume along the optical axis, as all the molecules in an extended double cone above and below the focal plane will be photobleached. Also, the high power delivered on the sample to induce photobleaching causes the bleach volume to expand due to fluorescence saturation (Braeckmans et al. 2006). A solution to these limitations is provided by two-photon FRAP, where infrared pulsed light is used to photobleach only those molecules in a limited volume (about 1 μm thick) above and below the focal plane (Brown et al. 1999; Mazza et al. 2008).

8.2.2 QUALITATIVE FRAP

A qualitative or simple semiquantitative analysis of FRAP data can provide a wealth of information. Key features of the FRAP curve include the rate and the final level of recovery. The rate of fluorescence recovery in a FRAP can vary from less than a second to many hours, depending on the molecule under study. For example, in living cells, GFP alone recovers in a less than a second for a typical bleach spot of ~1 μm in diameter (Sprague et al. 2004). This has been used to estimate the cellular diffusion rates for GFP of about 25 μm²/s. In contrast, the histone H2B tagged with GFP takes more than 8 h to fully recover (Kimura and Cook 2001). This probably reflects very stable binding of the histone to the relatively immobile chromatin structure.

In addition to the rate of recovery, the level of recovery achieved relative to the initial level of fluorescence is also frequently of interest. Some molecules exhibit a fraction that recovers relatively rapidly, while another fraction may recover more slowly or not at all. In the latter case, this fraction is often referred to as the immobile fraction, presumably reflecting the proportion of the labeled molecule that is either tightly bound to an immobile cellular structure or potentially trapped by surrounding obstacles to diffusion. For example, membrane proteins exhibit different mobile fractions at different temperatures, and these different features have been used to construct models of lipid domains (Schram and Thompson 1997).

Comparison of recovery rates or mobile fractions can be very powerful (Figure 8.2b). For example, comparing the FRAP results for labeled RNA polymerase II in cells treated with or without the transcriptional inhibitor, DRB, permits the identification of the FRAP recovery phase that reflects transcript elongation, providing an estimation of the *in vivo*

Applications

elongation rate (Darzacq et al. 2007). Analogously, comparing the FRAP results obtained from cells with the wild-type p53 transcription factor (TF) to cells expressing mutants known to affect either specific site or nonspecific site DNA binding suggests that most of p53's interactions in the nucleus are at nonspecific sites, since only the FRAP of the mutant that affects nonspecific site binding shows a change in recovery rate compared to wild type (Hinow et al. 2006; Sauer et al. 2008). Another example of the power of semiquantitative FRAP is the comparison of recovery curves at different phases of the cell cycle. By performing FRAP on inner centromere proteins, Hemmerich et al. (2008) have dissected the assembly and maintenance of centromeres throughout the cell cycle, identifying components such as Centromeric protein A (CENP-A) that stably binds the centromere once loaded and others such as Centromeric protein H (CENP-H) that exhibit a more dynamic behavior. These and many other examples show how comparison of different FRAP curves can be used to extract biologically relevant information. Importantly, such comparisons must be made under identical conditions including the same size and shape of the bleached region, the same duration, and illumination intensity of the photobleach, as well as other parameters that are discussed in the succeeding text.

8.2.3 QUANTITATIVE FRAP

Quantitative analysis of FRAP data can be used to extract estimates of the biological parameters that influence protein mobility. These parameters include the cellular diffusion constant, a diffusion anomaly parameter that quantifies the extent to which cellular diffusion differs from a simple random walk, in addition to the association, and dissociation rates of binding to an immobile cellular substrate. These various estimates are achieved by mathematical or computational modeling of the measured FRAP recovery (van Royen et al. 2009; Mueller et al. 2010a,b). Such models are created by solving the equations describing how the molecules diffuse and potentially bind in the cellular region of interest. The models always include an initial condition that describes the distribution of fluorescence at the end of the intentional photobleach. Sometimes, the models also include boundary conditions such as the size of the intracellular compartment where the molecules can diffuse.

The simplest model for quantitative FRAP is for Brownian diffusion, which occurs if the molecule can execute an unconstrained random walk. Brownian diffusion models were the first models developed to analyze FRAP data (Axelrod et al. 1976; Soumpasis 1983). Over the years, these models have been extensively refined (Braeckmans et al. 2003; Braga et al. 2004; Kang et al. 2009) and applied to the measurement of diffusion in solutions (Hou et al. 1990), drug delivery systems (Alvarez-Manceñido et al. 2006), and cellular membranes (Yguerabide et al. 1982). However, simple diffusion models often fail to fit the FRAP data obtained from living cells because other factors beyond simple diffusion influence the molecule's motion. These additional factors include directed transport, intracellular flows, transient binding to immobilized scaffolds, or obstruction by other cellular components.

Most of the attempts to model these additional complicating features inside living cells have involved adjustments to the model of simple diffusion. Typically, it has been presumed that the molecule's mean square displacement is no longer proportional to time as in simple diffusion but rather proportional to time raised to the power α. Thus, for simple diffusion, $\alpha = 1$, whereas when $\alpha > 1$, the molecule moves faster than would be expected from simple diffusion. This super-diffusion rate might arise from flow or directed transport. When $\alpha < 1$, the molecule moves more slowly than would be expected from simple diffusion. This subdiffusion rate might arise from binding to immobile scaffolds or from impediments created by subcellular obstacles. The exponent alpha is known as the anomaly parameter, and the resultant behavior when $\alpha \neq 1$ is referred to as anomalous diffusion (Saxton 2001).

Adding an anomaly parameter to the FRAP diffusion model introduces a second free parameter in addition to the diffusion constant and thereby allows better fitting of the FRAP curves obtained from live cells (Lubelski and Klafter 2008). Unfortunately, a good fit with an anomalous diffusion model does not identify which of the possible underlying causes are responsible for the deviation. Furthermore, the estimated value of the anomaly parameter α has no direct connection with an underlying cellular variable. Thus, in recent years, more specific FRAP models have been developed, which explicitly include features that could cause anomalous diffusion.

In particular, the role of protein binding to immobile scaffolds has been extensively explored (Carrero et al. 2003; Sprague et al. 2004; Beaudouin et al. 2006; Mueller et al. 2008). This situation is often encountered in living cells, as it describes, for example, the interaction of proteins with chromatin in the nucleus and the interaction of vesicles or ligands with certain plasma membrane proteins (see Figure 8.2c). The utility of such models is demonstrated in several fields. For example, the diffusion and binding models have been used to analyze FRAP data acquired for various nuclear factors, yielding estimates for residence times on chromatin (Hinow et al. 2006; van Royen et al. 2007; Mueller et al. 2008). As was mentioned earlier (Section 8.2.2), estimating the residence time of the RNA polymerase II molecule while in its elongation phase permits calculation of the *in vivo* elongation rate of the transcript, given also knowledge of the length of the transcribed region (Darzacq et al. 2007). Diffusion and binding measurements have also been applied to the cytoplasm and to membranes. For example, FRAP of fluorescently labeled synaptic receptor ligands has been analyzed to obtain estimates of the ligand residence time at membrane contact sites between neighboring cells, and these estimates have been found to be a hundredfold longer than solution binding measurements (Tolentino 2008).

Besides the models based on simple diffusion and binding, efforts have been made to include complex behaviors such as flow and directed transport in the FRAP analysis (Axelrod et al. 1976; Sullivan et al. 2009), as well as molecular crowding and the distribution of subcellular obstacles (Bancaud et al. 2009). This trend in developing increasingly sophisticated FRAP models should continue, resulting ultimately in robust and accurate procedures to detect and quantify an assortment of complex *in vivo* behaviors.

8.2.4 FLUORESCENCE LOSS IN PHOTOBLEACHING

FRAP is capable of providing quantitative information about the motion of the labeled molecule on a local scale (i.e., within the bleaching spot). However, the standard FRAP protocol cannot provide much information about those dynamics on a larger, more global spatial scale. To supply information about the long-range kinetics of fluorescently tagged molecules, alternative photobleaching techniques have been developed. One of these techniques, called FLIP, is a powerful method to identify if two distinct subcompartments within a specimen are interconnected by determining whether a fluorescent probe can freely diffuse from one compartment to the other. FLIP is based on the periodic photobleaching of a small region of interest (the perturbation region) within the specimen while simultaneously observing of the fluorescence intensity in another region within the specimen (the measurement region). If the fluorescent probe can diffuse from the perturbation region to the measurement region, a loss of fluorescence in the measurement region will be detected, due to the influx of the bleached molecules created in the perturbation region (Figure 8.3a).

FLIP has been used to monitor the exchange of cargos and coating proteins such as coating protein (COPI) between the cytosol and the Golgi (Presley et al. 2002), suggesting that the assembly of vesicle coats is stochastic. More recently, FLIP has also been used to investigate active transport in axons (Iliev and Wouters 2007) and connectivity of the endoplasmic reticulum in protozoa (Teixeira and Huston 2008), as well as to cross validate chromatin-immunoprecipitation experiments on DNA-binding proteins (McNairn and Gerton 2009). Although most FLIP studies have been qualitative, FLIP can be used to obtain quantitative information about diffusion and binding (Luijsterburg et al. 2010).

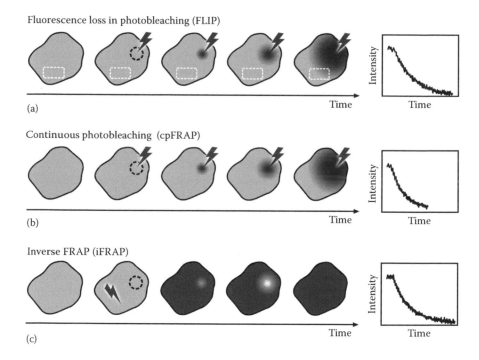

Figure 8.3 Other photobleaching techniques. (a) In FLIP, periodic application of a bleach pulse in the perturbation zone (dotted black circle) can lead to loss of fluorescence in a measurement zone somewhere else in the cell (dotted white circle), if the two zones are in the same diffusional compartment. (b) In cpFRAP, a region is continuously photobleached, and the fluorescence intensity in the same zone is measured over time. (c) In iFRAP, the region of interest is spared from the photobleach, while the rest of the cellular compartment is bleached. Curves in all cases indicate the fluorescence intensity in the measurement zone as a function of time.

8.2.5 CONTINUOUS PHOTOBLEACHING FRAP

Another quantitative approach to nonlocal measurements of cellular dynamics is cpFRAP. In cpFRAP, molecules are continuously bleached in a selected region of the sample, typically with a stationary laser beam (Figure 8.3b). In this situation, fluorescence intensity within the photobleaching region depends on several factors, including (1) the rate of fluorescent decay induced by the bleaching laser, (2) the residence time of the fluorescent molecules within the photobleaching region, and (3) the total pool of fluorescent molecules within the cellular compartment under study. These features can be incorporated into a quantitative model for a cpFRAP. Such models have been used to estimate the size of a cellular compartment in which the fluorescent molecules are free to diffuse (Delon et al. 2006) or to estimate residence times of DNA proteins on chromatin (Wachsmuth et al. 2003).

8.2.6 INVERSE FRAP

In iFRAP, virtually all the fluorescent molecules in the sample are photobleached, except for those in a small region of interest. Following this extensive photobleach, the loss of fluorescence from the remaining unbleached region is then monitored (Figure 8.3c). As with FRAP, it is also possible to construct a quantitative model for iFRAP and obtain estimates of cellular binding rates, as has been done for analysis of polymerase I kinetics (Dundr et al. 2002). It is a common misconception that iFRAP specifically quantifies dissociation rates by measuring the escape rate from the region of interest. In reality, iFRAP does not differ significantly from a conventional FRAP experiment, since an iFRAP experiment can be computationally converted into a FRAP experiment simply by inverting the gray scale of the time-lapse sequence. The direct measurement of dissociation rates by iFRAP is only possible if no significant rebinding of the tagged proteins occurs before the molecules have escaped the selected region. With the aid of quantitative models, iFRAP can be used to estimate both association and dissociation rates. As a complement to FRAP, iFRAP does provide a useful control to test whether the intentional photobleach is causing photodamage at the site of interest, since this region is not bleached in an iFRAP (see also Section 8.2.10).

8.2.7 PHOTOACTIVATON FRAP AND PHOTOCONVERSION FRAP

All of the preceding techniques for making measurements of cellular protein dynamics rely on photobleaching. However, there is continuing debate on whether and to what extent photobleaching induces photodamage in cells (see Section 8.2.10). In the next three sections, we describe alternatives to the photobleaching microscopy methods described earlier that also permit measurement of cellular protein dynamics.

The photoactivatable FPs (such as paGFP) and the photoconvertable FPs (such as EosFP) provide the simplest alternative to photobleaching techniques. As discussed elsewhere in this volume (Chapter 9), these FPs can be converted by illumination with an appropriate wavelength from a dark state to a bright state (photoactivation) or from a bright state emitting at one wavelength to a bright state emitting at another wavelength (photoconversion). In conventional FRAP, the photobleach is used to convert molecules from a bright state to a permanently bleached ("dark" state and "bleached" state are different things) state. The photoactivatable or photoconvertible proteins permit the opposite, namely, selected molecules are converted from a dark state to a bright state by photoactivation (FRAPa) or from one bright state into another bright state by photoconversion (FRAPc) (Figure 8.4a). The advantage over conventional forms of photobleaching microscopy is that the illumination energy required to perform the photoactivation or photoconversion is usually much lower (up to a hundredfold [Calvert et al. 2007]) than that required for photobleaching. Thus, concerns about photodamage are considerably reduced.

FRAPa has been performed by a number of different groups, providing information about nuclear pore structure (Dultz et al. 2009), chromatin protein dynamics (Beaudouin et al. 2006), and the shuttling of GTPases between the plasma membrane and endocytic vesicles (Palamidessi et al. 2008). FRAPa is especially powerful when combined with two-photon microscopy to provide a highly localized photoactivation volume at a specific site in a specimen (Palamidessi et al. 2008). Furthermore, some specific fluorescent probes have been developed to maximize the advantages of FRAPa. One example is Phamret (Matsuda et al. 2008), a tandem FRET-probe composed of CFP and paGFP that shows fast photoconversion, thus minimizing the time to induce the photoactivation.

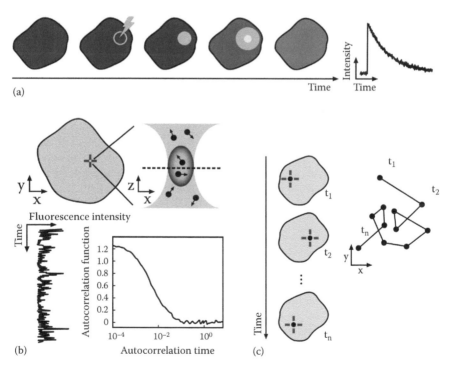

Figure 8.4 Nonphotobleaching techniques. (a) In FRAPa, a photoactivatable or photoconvertible probe is activated in a selected region, and then the loss of fluorescence from that region is monitored. (b) In FCS, fluctuations in fluorescence intensity as a function of time (vertical plot) are measured within a small focal volume (crosshairs, ellipsoid in zoomed-up view). The fluctuation data are autocorrelated at different time lags to yield an autocorrelation curve (decaying curve). (c) In SMT, single molecules are tracked (crosshairs) and their positions as a function of time are determined in the xy plane (connected dots).

8.2.8 FLUORESCENCE CORRELATION SPECTROSCOPY

FCS is another procedure to measure cellular protein dynamics that does not involve intentional photobleaching (Elson 1986; Elson 2004; Kohl and Schwille 2005). It provides information about protein dynamics by measuring the fluctuations of fluorescent intensity produced by fluorescently tagged molecules moving into and out of a precisely defined focal volume. Typically, the focal volume is imaged by a laser beam capable of inducing either one or two-photon excitation.

The fluctuations in fluorescent intensity measured from the focal volume will exhibit a certain degree of correlation, as a molecule located there at time t will have a certain probability of still being there at a later time t + τ. This probability will decay for increasing τ, with a faster decay rate seen for faster diffusing molecules. Therefore, calculating the autocorrelation of the fluorescence fluctuations as function of the autocorrelation time (t) can be used to determine the average time spent by fluorescent molecules in the focal volume (Figure 8.4b).

Analogous to FRAP curves, autocorrelation functions in FCS can be fitted with different quantitative models to estimate diffusion coefficients, anomaly exponents, and binding parameters of the fluorescent molecule. Among others applications, FCS has been widely used to study the dynamics and compartmentalization of macromolecules in the cellular membrane (Kusumi et al. 2010; Machán and Hof 2010). Applications of FCS to the nuclear environment include the study of polyadenylated RNA shuttling in and out of nuclear speckles (Politz et al. 2006), the effect of crowding of the nuclear environment on the diffusivity of inert tracers (Dross et al. 2009), as well as the measurement of diffusion and binding to the DNA of nuclear proteins (Michelman-Ribeiro et al. 2009; Hendrix et al. 2010; Stasevich et al. 2010).

Conventional FCS is typically applied to the measurement of fast dynamics (corresponding to average times in the observation spot shorter than 1 s), since molecules residing for longer times are likely to be photobleached. To address this limitation, various modifications to conventional FCS have been developed to attenuate the light dose delivered to the specimen. These alternatives include scanning FCS, image correlation spectroscopy (ICS), and temporal ICS (TICS). These approaches extend the range of residence times that can be measured by fluorescence correlation, allowing measurement on the seconds time scale of growth factor mobility on cell membranes of intact organisms (Ries et al. 2009) and the interactions between TFs and DNA (Stasevich et al. 2010), plus the construction of spatiotemporal maps of protein velocity (Hebert et al. 2005).

8.2.9 SINGLE MOLECULE TRACKING

SMT permits the analysis of cellular protein dynamics by tracking the motion of an individual protein over an extended field of view, with a localization precision that exceeds the diffraction limit of the microscope. The observation and tracking of individual molecules usually rely on the use of a widefield microscope equipped with highly sensitive electron multiplying charged coupled device (EMCCD) detectors, combined with strategies to maximize the number of photons collected over the background. The result is the ability to measure individual molecular behaviors rather than the population averages produced by all the preceding techniques (Figure 8.4c).

SMT has found its greatest success in the analysis of 2D systems, where total internal reflection microscopy (TIRF) can be used to considerably reduce the out-of-focus

Applications

light. This has enabled tracking of molecules either in reconstituted systems *in vitro* or in plasma membranes *in vivo*. For example, TIRF-SMT has been used to follow the stop-and-go motion of molecular motors on *in-vitro*-reconstituted actin filaments (Yildiz et al. 2003) and to measure the sliding of TFs such as p53 on stretched DNA molecules (Tafvizi et al. 2008). Similarly, by tracking the motion of proteins and lipids in membranes both *in vitro* and *in vivo*, TIRF-SMT has demonstrated that molecules in the membrane rarely move by simple diffusion. These tracking studies in membranes have refined our understanding of plasma membrane architecture, providing evidence for multiple domains, obstacles, and corrals (Ritchie et al. 2005; Kusumi et al. 2010). More recently, SMT has been extended to analysis of thicker specimens. Although challenged by the higher background present in these samples, SMT is starting to provide important information about intranuclear mobility. For example, SMT has been used to quantify the time required by a single mRNA particle to cross a nuclear pore (Kubitscheck et al. 2005; Grünwald and Singer 2010) or measuring the interactions of individual TFs with specific and nonspecific sites on DNA (Elf et al. 2007).

8.2.10 LIMITATIONS OF FRAP

In the previous sections, we have discussed the different techniques now in use to measure cellular protein dynamics. The most widely exploited are the photobleaching approaches, and among these, FRAP accounts for the vast majority of published studies. FRAP however has a number of limitations.

The most obvious limitation is that FRAP relies on high irradiation to produce the intentional photobleach, which can lead to photodamage of the sample. Many different types of photodamage might arise. A few studies have shown that photobleaching can cause a fluorescent molecule to unbind (Akaaboune et al. 2002; Heinze et al. 2009). Photodamage can also arise from scattering of the intense photobleaching irradiation, which has been shown to induce DNA damage within neighboring cells (Dobrucki et al. 2007). These or other deleterious effects of photobleaching could ultimately manifest as a disruption of cellular structures. In fact, some studies have found evidence for structural damage after a photobleach (Vigers et al. 1988; Keith and Farmer 1993), whereas other studies have not (Jacobson et al. 1978; Takeda 1995). In summary, it is clear that under certain conditions, the FRAP photobleach can have undesirable effects, and it is important to characterize the experimental system for potential artifacts that arise from photodamage.

A second limitation of FRAP is that the quantitative analysis of the resultant data might be erroneous due to the approximations that are necessary for the mathematical modeling. These approximations may oversimplify both the experimental details and the complexity of the cellular milieu, so accurately describing these subtleties remains challenging. For example, at the moment, it is not clear how the approximations in quantitative FRAP for the measurement of cellular binding affect either the qualitative and quantitative interpretations derived from the analysis (Mueller et al. 2010). A third limitation of FRAP is that it provides only a description of the average behavior of molecules and can fail to account for heterogeneities of either the diffusing species or the environment. Much can be learned from studying average behaviors, but ultimately a rich biology lies in the exceptions.

8.2.11 CROSS VALIDATION AMONG THE TECHNIQUES FOR MEASURING CELLULAR PROTEIN DYNAMICS

Fortunately, each of the limitations of FRAP noted earlier is well addressed by cross validation using some of the other techniques described in the preceding sections. FCS has been the most common alternative. A few studies show reasonable agreement between the diffusion rates measured by FCS and FRAP (Lellig et al. 2004; Guo et al. 2008). More recently, these cross validations have been extended to comparisons of both diffusion and binding rates estimated by FRAP and FCS, and again, reasonable agreement has been found (Michelman-Ribeiro et al. 2009; Stasevich et al. 2010).

This agreement between quantitative estimates obtained by FRAP and FCS suggests that several of the potential limitations of FRAP are not of concern, at least for the specific cases analyzed. Notably, FCS does not use a photobleach, and so quantitative agreement between FRAP and FCS estimates suggests that the FRAP photobleach does not alter the diffusion or binding rates of the cellular proteins that were studied. This conclusion is also supported by somewhat less direct comparisons between FRAP and iFRAP or FRAPa data. Although the latter two approaches do not subject the region of interest to an intentional photobleach, protein dynamics for H2B (Kimura and Cook 2001; Beaudouin et al. 2006) as well as for freely diffusing FPs (Calvert et al. 2007) appears to be similar regardless of the approach. Thus, at the moment, direct comparison of photobleaching and nonphotobleaching methods argues that the intentional photobleach in FRAP is not detrimental, but more such comparisons are needed in a variety of other systems before a general conclusion can be reached.

The agreement between the FCS and FRAP comparisons cited earlier also provides some evidence that the quantitative models used in FRAP are sufficiently accurate. This is because some of the simplifying assumptions for the FRAP model are not present in the FCS model, and vice versa. For example, most FRAP models presume that the photobleach profile can be well approximated by a cylinder and most FCS models presume that the diffraction-limited volume can be well approximated by a Gaussian. These are both approximations, but agreement between the FRAP and FCS estimates suggests that the approximations are reasonable. However, in general, both the FRAP and the FCS models make use of similar biological assumptions, which are not factored out by the cross validation. Furthermore, both the FRAP and FCS models may presume that diffusion is simple rather than anomalous, so it is possible that each method will yield the same wrong answer.

Thus, further cross validation of FRAP and FCS with other approaches is still needed. Here, SMT offers a significant advantage as it offers a closer and richer view on the actual dynamic behavior of macromolecules. SMT and FCS have yielded similar estimates for the diffusion rates of FPs in solution (Grünwald et al. 2006). Furthermore, SMT, FCS, and FRAP have also yielded similar estimates for the diffusion rates of lipids in bilayers, although here, SMT detected a broader range of diffusion coefficients encompassing the smaller and not completely overlapping ranges detected by FCS and FRAP (Guo et al. 2008). This suggests that the richer data set available with SMT provides a more accurate picture of the full range of diffusing behaviors. Consistent with this, another SMT study on a different model membrane system revealed that an immobile fraction detected by FRAP actually resulted from anomalous restricted diffusion (Schütz et al. 1997).

Applications

8.3 SUITABILITY OF FLUORESCENT PROTEINS FOR FRAP, FCS, AND SMT

The preceding sections have summarized the many capabilities of current methods for measuring protein dynamics. The FPs have played key roles in virtually all of these experiments, providing the genetically encoded fluorescent tag that enabled facile labeling of the cellular protein of interest. Here, we consider the suitability of the common FPs for photobleaching microscopy, as exemplified by FRAP, and for some of the most powerful complementary approaches, namely, FCS and SMT.

8.3.1 FLUORESCENT PROTEINS IN FRAP

The enhanced GFP (eGFP) has become the FP of choice for most FRAP experiments in living cells, although a relatively small fraction of studies (<10%) have used other FPs (such as eCFP, eYFP, or mCherry). In most cases, the selection of a FP has been motivated by convenience (e.g., availability of constructs), rather than by the actual performance of the FP in FRAP. In this section, we discuss the advantages and disadvantages of the most widely used FP for FRAP, namely, eGFP. However, these conclusions for FRAP also apply more generally to the other forms of photobleaching microscopy.

Two of eGFP's key advantages for FRAP are its brightness and photostability. Relatively high brightness is important in FRAP because it permits using a relatively low laser power to collect the recovery curve, thereby minimizing observational photobleaching. Too much observational photobleaching is problematic because the FRAP recovery becomes dominated by this effect, and so it becomes very sensitive to the method used to correct for observational photobleaching (Figure 8.5a). Here, different correction procedures may lead to different FRAP curves. This can be a problem even for qualitative comparisons of FRAP curves, but it is especially troublesome for quantitative analysis of FRAP. The virtue of eGFP is that it is bright enough to be detected at relatively low laser powers. The same is true of eYFP, and although some of the blue- or red-shifted FPs (such as eCFP and mCherry) are somewhat less bright, they still provide a sufficient photon count for FRAP.

A related property of eGFP that is also well suited for FRAP is its photostability. The photostability corresponds to the number of emission cycles that occur before a molecule of eGFP photobleaches. An optimal FP for FRAP should have a photostability that is at an intermediate level. The FPs that photobleach too rapidly have increased observational photobleaching, whereas FPs that photobleach too slowly require many iterations of the intentional photobleach (Figure 8.5b). This can complicate the FRAP analysis since a substantial amount of diffusion will occur during the photobleach, plus the resultant depth of the photobleach may still be small thereby reducing the dynamic range of the recovery data. The advantage of eGFP, as well as eYFP and mCherry, for FRAP is that their photobleaching half time is intermediate between dyes with very short half times such as fluorescein and dyes with very long half times such as the Alexa or Atto dyes. As a result, the FPs strike a good compromise, yielding a tolerable observational photobleaching during measurement of the FRAP recovery and fast photobleaching during the intentional photobleach.

Applications

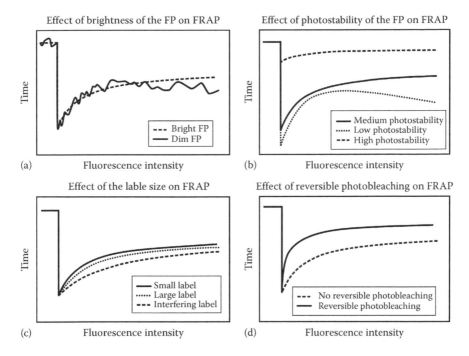

Figure 8.5 The impact of FP deficiencies on FRAP. The black curve in all cases reflects the "true" FRAP behavior. Gray curves reflect the perturbations of the true curve imposed by a specific limitation of the FP. (a) Dim FPs lead to noisy data. (b) Probes with high photostability (light-gray curve) are difficult to bleach, yielding a shallow beach depth, but show no observational photobleaching in the recovery. In contrast, probes with low photostability (dark-gray curve) yield a deep bleach but then do not recover fully due to observational photobleaching. (c) The size of the fusion protein slows down the FRAP recovery only slightly if free diffusion can occur (dark-gray curve), while the retardation is more severe if diffusion occurs in a complex environment with different "pore" sizes that can impede movement of larger molecules (light-gray curve). (d) Reversible photobleaching can produce an artifactual speedup caused by a reversion of bleached molecules, rather than by entry of fluorescent molecules from outside the bleached zone.

A modest disadvantage of eGFP and most other FPs for FRAP is their size. Size affects how fast a molecule diffuses inside a cell. Most of the FPs are ~30 kDa, which when fused to a small cellular protein may in a worst-case scenario approximately double its size. On the one hand, a doubling in size is not so severe, as it will only change the diffusion rate by a factor of $(1/2^{1/3}) = 1.26$ (since the diffusion rate goes as the inverse cube root of size). Thus, even in this worst-case scenario, the measured diffusion rate should not be radically altered by the increase in size following fusion of the FP (Figure 8.5c).

On the other hand, diffusion in the cellular milieu may not always be as simple as summarized by the preceding inverse cubic relationship. There is much evidence to suggest that cellular structures can impede diffusion, and this effect varies depending on the size and shape of the molecule. Thus, making a protein larger with a FP fusion may increase the "pore size" through which the protein can pass thereby significantly decreasing its diffusion rate. Here, the tendency of some FPs to oligomerize becomes

a concern, since larger aggregates will be more impacted by this effect. Hence, it is important to use the "monomeric" forms of the FPs for FRAP-based studies (see Chapter 3). Unfortunately, there is still much to be learned about exactly what the range of "pore sizes" is in different cellular compartments, so at the moment, there is no simple rule to define how much of a size or shape increase will significantly alter a molecule's diffusion rate.

Related to change in size caused by an eGFP fusion is the more generic issue of whether the fusion protein retains normal function. This concern affects not only measurements of cellular dynamics but all forms of microscopy using FP fusions. Thus, controls must always be performed on the fusion protein to assay its function either *in vitro* or *in vivo* or both.

While the size of an eGFP fusion protein remains a potential concern, perhaps the biggest disadvantage of eGFP for FRAP is its reversible photobleaching. This is because the fundamental presumption of FRAP is that the photobleached molecules are bleached permanently. Here, the observed fluorescence recovery will reflect only the entry of unbleached molecules from outside the bleached region. However, if some eGFP molecules are in a reversible dark state and can become fluorescent again, it is impossible to distinguish fluorescence entry from fluorescence reversion. Therefore, reversible photobleaching can completely confound FRAP analysis. For example, a significant reversible fraction could easily be mistaken for a fast mobile fraction when none actually exists.

Various studies have demonstrated that eGFP can reversibly photobleach (Periasamy et al. 1996; Dickson et al. 1997; Weber et al. 1999; Sinnecker et al. 2005). The fraction of eGFP molecules entering the reversible dark state depends strongly on the precise conditions of both the photobleach and the subsequent imaging. Under some circumstances, eGFP reversible fractions are relatively modest (~3%), while under other circumstances, the reversible fractions are substantial (~50%). Thus, to some extent, it is possible to select the photobleach and imaging conditions that minimize the reversible fraction of eGFP, but there are limits since the bleach and imaging conditions also must be tuned to the concentration and the kinetics of the protein under study. The very wide range of eGFP reversible behaviors makes it very difficult to compare different published FRAP studies of the same protein, since bleach and imaging conditions typically change from one laboratory to another. It is highly likely that some of the published differences between mobile fractions observed for the same protein in different studies are due to different amounts of eGFP reversible photobleaching arising under different experimental conditions (Figure 8.5d). The problem of reversible photobleaching is not restricted to GFP, as other FPs such as CFP, mCherry, TagRFP, and mTFP1 also exhibit reversible behavior to varying degrees (Mueller et al. 2012). Of the FPs tested to date, mCherry and YFP appear to have the smallest reversible fractions, and so these provide valuable alternatives to eGFP.

8.3.2 FLUORESCENT PROTEINS IN FCS

As in FRAP, eGFP is also the most commonly used FP for FCS, and similar to FRAP, eGFP's brightness is also an advantage for FCS. Brightness is valuable for FCS because the best autocorrelation data are obtained when only a few molecules are present in the focal volume. Thus, to achieve a good signal under such conditions, the molecules should be bright and easily detected, and eGFP is adequate for this purpose.

FCS is also similar to FRAP in that the increased size of fusion proteins carries the same concerns about altering the protein's dynamics as in FRAP (see Section 8.3.1). FCS differs from FRAP in that reversible photobleaching is not as much a concern. This is because the photobleach in FRAP can trigger a relatively large reversible fraction, whereas the continuous imaging at low intensities in FCS produces a much smaller reversible fraction. Since the transitions between bright and dark states of eGFP occur rapidly, reversible bleaching or "blinking" of eGFP manifests in FCS as a rapid decay in the autocorrelation curve (in the μs to ms range). This so-called triplet-state relaxation is typically much faster than the biological process of interest and so is in general clearly separable. Its magnitude is usually small and dependent on the imaging conditions, such as whether the illumination is continuous or pulsed and whether it is one photon or two photon, so under some conditions, there is virtually no reversible or blinking behavior.

Another difference between FCS and FRAP is that eGFP's intermediate photostability offers no advantage in FCS, since FCS measurements can be seriously impacted by photobleaching. Fortunately, photobleaching in FCS can often be avoided by imaging at low laser power. Further, photobleaching can be identified in FCS by examining the time series of intensity fluctuations, which can be corrected for prior to analyzing the FCS data. In the worst case, however, photobleaching occurs, but the measured intensity does not decay over time. Such "cryptic photobleaching" arises when the photobleaching of molecules within the focal volume is at equilibrium with the influx of unbleached molecules into the volume. The intensity therefore remains constant over time. However, this steady-state condition can artifactually shorten residence times in the focal volume, because the bleaching of molecules mimics their departure. This effect can arise for residence times longer than a few tenths of a second. To avoid the artifacts of cryptic photobleaching, residence time estimates must be made at decreasing laser powers until a plateau value is reached, indicating that there is no further lengthening of the residence time due to reductions in photobleaching.

8.3.3 FLUORESCENT PROTEINS IN SMT

The qualities of eGFP for SMT are in many ways similar to FCS. Brightness is adequate, as it is possible to detect single eGFP molecules with reasonable localization precision. The size of the fusion protein carries the same concerns as for both FRAP and FCS. Reversible photobleaching in SMT is a somewhat more serious concern than for FCS, since blinking of the eGFP tag can lead to a temporary disappearance of the tracked molecule, thereby shortening the lengths of tracks.

However, the most serious deficiency of eGFP for SMT is its photostability. Bleaching makes it impossible to track eGFP fusion proteins for more than a few frames. Similar problems apply to all of the FPs (there are newer FPs that have been selected for photostability). As a result, most SMT experiments are not done with FPs, but rather with organic dyes or quantum dots as labels. Tracking of single molecules for extended periods with eGFP is only possible if multiple eGFPs are attached to the same molecule. Then not all of the eGFPs are bleached at once, and it is possible to follow the molecule for much longer. This strategy has been used, for example, to track mRNA motion in the nucleus. Such multiple labeling however substantially increases concern about artifacts induced by an increased size of the tracked molecule.

Applications

Photoactivatable FPs provide an interesting alternative for SMT. These molecules still suffer from relatively fast bleaching, which again severely limits the lengths of tracks. However, significantly more tracks per cell can be obtained by activating only a few molecules at a time and then iterating that procedure. This is a considerable advantage over conventional probes, whether they are FPs or organic dyes, since in these cases, low concentrations must be used to keep background fluorescence low enough to permit single-molecule detection. As a result, a much smaller number of molecules per cell are available for tracking with quantum dots or organic dyes. With photoactivatable FPs, a large compendium of tracks in the same cell can be accumulated over time, making it possible in principle to generate a spatial map of cellular dynamics (Manley et al. 2008).

8.4 FUTURE PROSPECTS FOR MEASURING CELLULAR DYNAMICS WITH FLUORESCENT PROTEINS

The ability to measure cellular dynamics by light microscopy has led to the explosive growth in all of the methods discussed in this chapter. This burgeoning interest is not likely to diminish anytime soon, since many areas of application are still largely unexplored. A major thrust in recent years has been the use of FRAP and its sister techniques as *in vivo* tools to probe either cellular binding events or cellular architecture. Extracting such information from light microscopy measurements is a relatively new discipline and so we can expect to see significant improvements in the quantitative techniques that are now in use for this purpose. Gold standards will become established making it possible to identify the most accurate quantitative approaches.

In addition to technical advances in microscopy and analysis, it is also likely that the measurement of cellular dynamics will also see significant advances due to the development and use of new FPs. As we have outlined here, eGFP has been largely a default choice for most of these investigations. Although there are advantages of using eGFP as a fluorescent tag, it is not an ideal probe for either the photobleaching techniques or for their nonphotobleaching counterparts, such as FCS and SMT.

An improvement in FPs that would benefit all of the techniques for measuring cellular dynamics would be reduction in the size of the fluorescent tag. While this is not possible for the beta-barrel structure of the FPs from marine organisms, progress is being made with, for example, iLov, a flavin-based peptide of 10 kDa (see Chapter 10). In its current form, this tag is unfortunately not suitable for FRAP given the reversible nature of its photobleaching state (Chapman et al. 2008). Here, a promising recent development is the creation of small peptide tags (13 amino acids) that can be recognized by enzymes that mediate ligation of a membrane-permeable small organic fluorophore (Uttamapinant et al. 2010).

Other than being smaller, the perfect probe for FRAP would also be easy to bleach, would show no reversible photobleaching, and would show no bleaching during measurement of the recovery. Such a dream probe is not as futuristic as it might seem at first glance. Indeed, photoactivatable or photoconvertible proteins could already be close to this ideal. These probes generally require lower activation energies than a photobleach requires, and some are reported to undergo only one round of conversion, meaning that they are not reversible (Lippincott-Schwartz and Patterson 2009). If the photoactivated or photoconverted state is also very photostable, all of the criteria of the "dream probe" would be met. In this regard, further testing of current FPs and development of new ones with these criteria in mind would be very valuable for FRAP approaches.

A perfect probe for both FCS and SMT would also show no photobleaching during the measurement. This is especially critical for SMT where the relatively rapid bleaching of current FPs severely limits their applicability. At the moment, the best hope for SMT comes instead from genetically encoded proteins such as the HaloTag (Schröder et al. 2009) or Snap/Clip tags (Gautier et al. 2008), which bind to cell permeable fluorescent ligands that can be very photostable. However, a photostable FP would be more desirable, as it would obviate both the incubation with a fluorescent ligand and the subsequent washing to remove unbind ligand with the attendant risk of nonspecific interactions and relatively high backgrounds.

Apart from modification of current probes to adapt them to current live-cell techniques, it is also worth considering how new probes might help create new techniques. For example, probes that would change their fluorescence upon binding to an immobile substrate could provide a significant advance for the measurement of cellular binding events. The current approaches require binding events to be deconvolved from the measurement of diffusion, and this requires sophisticated mathematical models. Thus, more straightforward fluorescence assays that specifically detect binding would be very valuable.

In sum, it is likely that the successful partnership between FP development and measurements of cellular dynamics will continue. Given the growing importance of these kinds of measurements, more work on the development of proteins specifically suited to these techniques is merited.

REFERENCES

Akaaboune, M. et al., 2002. Neurotransmitter receptor dynamics studied in vivo by reversible photo-unbinding of fluorescent ligands. *Neuron*, 34(6), 865–876.

Alvarez-Manceñido, F. et al., 2006. Characterization of diffusion of macromolecules in konjac glucomannan solutions and gels by fluorescence recovery after photobleaching technique. *International Journal of Pharmaceutics*, 316(1–2), 37–46.

Axelrod, D. et al., 1976. Mobility measurement by analysis of fluorescence photobleaching recovery kinetics. *Biophysical Journal*, 16(9), 1055–1069.

Bancaud, A. et al., 2009. Molecular crowding affects diffusion and binding of nuclear proteins in heterochromatin and reveals the fractal organization of chromatin. *EMBO Journal*, 28(24), 3785–3798.

Beaudouin, J. et al., 2006. Dissecting the contribution of diffusion and interactions to the mobility of nuclear proteins. *Biophysical Journal*, 90(6), 1878–1894.

Braeckmans, K. et al., 2003. Three-dimensional fluorescence recovery after photobleaching with the confocal scanning laser microscope. *Biophysical Journal*, 85(4), 2240–2252.

Braeckmans, K. et al., 2006. Anomalous photobleaching in fluorescence recovery after photobleaching measurements due to excitation saturation—A case study for fluorescein. *Journal of Biomedical Optics*, 11(4), 044013.

Braga, J., Desterro, J.M.P., and Carmo-Fonseca, M., 2004. Intracellular macromolecular mobility measured by fluorescence recovery after photobleaching with confocal laser scanning microscopes. *Molecular Biology of the Cell*, 15(10), 4749–4760.

Brown, E.B. et al., 1999. Measurement of molecular diffusion in solution by multiphoton fluorescence photobleaching recovery. *Biophysical Journal*, 77(5), 2837–2849.

Calvert, P.D. et al., 2007. Fluorescence relaxation in 3D from diffraction-limited sources of PAGFP or sinks of EGFP created by multiphoton photoconversion. *Journal of Microscopy*, 225(1), 49–71.

Carrero, G. et al., 2003. Using FRAP and mathematical modeling to determine the in vivo kinetics of nuclear proteins. *Methods*, 29(1), 14–28.

Chapman, S. et al., 2008. The photoreversible fluorescent protein iLOV outperforms GFP as a reporter of plant virus infection. *Proceedings of the National Academy of Sciences of the United States of America*, 105(50), 20038–20043.

Chen, Y. et al., 2006. Methods to measure the lateral diffusion of membrane lipids and proteins. *Methods*, 39(2), 147–153.

Dange, T., Joseph, A., and Grünwald, D., 2011. A perspective of the dynamic structure of the nucleus explored at the single-molecule level. *Chromosome Research*, 19(1), 117–129.

Darzacq, X. et al., 2007. In vivo dynamics of RNA polymerase II transcription. *Nature Structural and Molecular Biology*, 14(9), 796–806.

Delon, A. et al., 2006. Continuous photobleaching in vesicles and living cells: A measure of diffusion and compartmentation. *Biophysical Journal*, 90(7), 2548–2562.

De Smedt, S.C. et al., 2005. Studying biophysical barriers to DNA delivery by advanced light microscopy. *Advanced Drug Delivery Reviews*, 57(1), 191–210.

Dickson, R.M. et al., 1997. On/off blinking and switching behaviour of single molecules of green fluorescent protein. *Nature*, 388(6640), 355–358.

Dobrucki, J., Feret, D., and Noatynska, A., 2007. Scattering of exciting light by live cells in fluorescence confocal imaging: Phototoxic effects and relevance for FRAP studies. *Biophysical Journal*, 93(5), 1778–1786.

Dross, N. et al., 2009. Mapping eGFP oligomer mobility in living cell nuclei. *PloS One*, 4(4), e5041.

Dultz, E., Huet, S., and Ellenberg, J., 2009. Formation of the nuclear envelope permeability barrier studied by sequential photoswitching and flux analysis. *Biophysical Journal*, 97(7), 1891–1897.

Dundr, M. et al., 2002. A kinetic framework for a mammalian RNA polymerase in vivo. *Science*, 298(5598), 1623–1626.

Dundr, M. and Misteli, T., 2003. Measuring dynamics of nuclear proteins by photobleaching. *Current Protocols in Cell Biology*, Chapter 13, Unit 13.5, doi: 10.1002/0471143030. cb1305s18. PubMed PMID: 18228420.

Elf, J., Li, G., and Xie, X.S., 2007. Probing transcription factor dynamics at the single-molecule level in a living cell. *Science*, 316(5828), 1191–1194.

Elson, E.L., 1986. Membrane dynamics studied by fluorescence correlation spectroscopy and photobleaching recovery. *Society of General Physiologists Series*, 40, 367–383.

Elson, E.L., 2004. Quick tour of fluorescence correlation spectroscopy from its inception. *Journal of Biomedical Optics*, 9(5), 857–864.

Erdel, F. et al., 2011. Dissecting chromatin interactions in living cells from protein mobility maps. *Chromosome Research*, 19(1), 99–115.

Gautier, A. et al., 2008. An engineered protein tag for multiprotein labeling in living cells. *Chemistry and Biology*, 15(2), 128–136.

Grünwald, D. et al., 2006. Direct observation of single protein molecules in aqueous solution. *Chemphyschem*, 7(4), 812–815.

Grünwald, D. and Singer, R.H., 2010. In vivo imaging of labelled endogenous β-actin mRNA during nucleocytoplasmic transport. *Nature*, 467(7315), 604–607.

Guo, L. et al., 2008. Molecular diffusion measurement in lipid bilayers over wide concentration ranges: A comparative study. *Chemphyschem*, 9(5), 721–728.

Hager, G.L., McNally, J.G., and Misteli, T., 2009. Transcription dynamics. *Molecular Cell*, 35(6), 741–753.

Hebert, B., Costantino, S., and Wiseman, P.W., 2005. Spatiotemporal image correlation spectroscopy (STICS) theory, verification, and application to protein velocity mapping in living CHO cells. *Biophysical Journal*, 88(5), 3601–3614.

Heinze, K.G. et al., 2009. Beyond photobleaching, laser illumination unbinds fluorescent proteins. *Journal of Physical Chemistry B*, 113(15), 5225–5233.

Hemmerich, P. et al., 2008. Dynamics of inner kinetochore assembly and maintenance in living cells. *Journal of Cell Biology*, 180(6), 1101–1114.

Hendrix, J. et al., 2010. The transcriptional co-activator LEDGF/p75 displays a dynamic scan-and-lock mechanism for chromatin tethering. *Nucleic Acids Research*. Available at: http://www.ncbi.nlm.nih.gov/pubmed/20974633. Accessed November 30, 2010.

Hinow, P. et al., 2006. The DNA binding activity of p53 displays reaction-diffusion kinetics. *Biophysical Journal*, 91(1), 330–342.

Hou, L., Lanni, F., and Luby-Phelps, K., 1990. Tracer diffusion in F-actin and Ficoll mixtures. Toward a model for cytoplasm. *Biophysical Journal*, 58(1), 31–43.

Iliev, A.I. and Wouters, F.S., 2007. Application of simple photobleaching microscopy techniques for the determination of the balance between anterograde and retrograde axonal transport. *Journal of Neuroscience Methods*, 161(1), 39–46.

Jacobson, K., Hou, Y., and Wojcieszyn, J., 1978. Evidence for lack of damage during photobleaching measurements of the lateral mobility of cell surface components. *Experimental Cell Research*, 116(1), 179–189.

Kang, M. et al., 2009. A generalization of theory for two-dimensional fluorescence recovery after photobleaching applicable to confocal laser scanning microscopes. *Biophysical Journal*, 97(5), 1501–1511.

Keith, C.H. and Farmer, M.A., 1993. Microtubule behavior in PC12 neurites: Variable results obtained with photobleach technology. *Cell Motility and the Cytoskeleton*, 25(4), 345–357.

Kimura, H. and Cook, P.R., 2001. Kinetics of core histones in living human cells: Little exchange of H3 and H4 and some rapid exchange of H2B. *Journal of Cell Biology*, 153(7), 1341–1353.

Kohl, T. and Schwille, P., 2005. Fluorescence correlation spectroscopy with autofluorescent proteins. *Advances in Biochemical Engineering/Biotechnology*, 95, 107–142.

Kubitscheck, U. et al., 2005. Nuclear transport of single molecules: Dwell times at the nuclear pore complex. *Journal of Cell Biology*, 168(2), 233–243.

Kusumi, A. et al., 2010. Hierarchical organization of the plasma membrane: Investigations by single-molecule tracking vs. fluorescence correlation spectroscopy. *FEBS Letters*, 584(9),1814–1823.

Lellig, C. et al., 2004. Self-diffusion of rodlike and spherical particles in a matrix of charged colloidal spheres: A comparison between fluorescence recovery after photobleaching and fluorescence correlation spectroscopy. *Journal of Chemical Physics*, 121(14), 7022–7029.

Lippincott-Schwartz, J. and Patterson, G.H., 2009. Photoactivatable fluorescent proteins for diffraction-limited and super-resolution imaging. *Trends in Cell Biology*, 19(11), 555–565.

Lorén, N., Nydén, M., and Hermansson, A., 2009. Determination of local diffusion properties in heterogeneous biomaterials. *Advances in Colloid and Interface Science*, 150(1), 5–15.

Lubelski, A. and Klafter, J., 2008. Fluorescence recovery after photobleaching: The case of anomalous diffusion. *Biophysical Journal*, 94(12), 4646–4653.

Luijsterburg, M.S. et al., 2010. Stochastic and reversible assembly of a multiprotein DNA repair complex ensures accurate target site recognition and efficient repair. *Journal of Cell Biology*, 189(3), 445–463.

Machán, R. and Hof, M., 2010. Recent developments in fluorescence correlation spectroscopy for diffusion measurements in planar lipid membranes. *International Journal of Molecular Sciences*, 11(2), 427–457.

Manley, S. et al., 2008. High-density mapping of single-molecule trajectories with photoactivated localization microscopy. *Nature Methods*, 5(2), 155–157.

Matsuda, T., Miyawaki, A., and Nagai, T., 2008. Direct measurement of protein dynamics inside cells using a rationally designed photoconvertible protein. *Nature Methods*, 5(4), 339–345.

Mazza, D. et al., 2008. A new FRAP/FRAPa method for three-dimensional diffusion measurements based on multiphoton excitation microscopy. *Biophysical Journal*, 95(7), 3457–3469.

Applications

McNairn, A.J. and Gerton, J.L., 2009. Intersection of ChIP and FLIP, genomic methods to study the dynamics of the cohesin proteins. *Chromosome Research*, 17(2), 155–163.

Meyvis, T.K. et al., 1999. Fluorescence recovery after photobleaching: A versatile tool for mobility and interaction measurements in pharmaceutical research. *Pharmaceutical Research*, 16(8), 1153–1162.

Michelman-Ribeiro, A. et al., 2009. Direct measurement of association and dissociation rates of DNA binding in live cells by fluorescence correlation spectroscopy. *Biophysical Journal*, 97(1), 337–346.

Mueller, F. et al., 2010a. FRAP and kinetic modeling in the analysis of nuclear protein dynamics: What do we really know? *Current Opinion in Cell Biology*, 22(3), 403–411.

Mueller F., Morisaki, T., Mazza, D., McNally, J.G. 2012. Minimizing the impact of photoswitching of fluorescent proteins on FRAP analysis. *Biophysical Journal*, 102(7), 1656–1665.

Mueller, F., Wach, P., and McNally, J.G., 2008. Evidence for a common mode of transcription factor interaction with chromatin as revealed by improved quantitative fluorescence recovery after photobleaching. *Biophysical Journal*, 94(8), 3323–3339.

Palamidessi, A. et al., 2008. Endocytic trafficking of Rac is required for the spatial restriction of signaling in cell migration. *Cell*, 134(1), 135–147.

Periasamy, N., Bicknese, S., and Verkman, A.S., 1996. Reversible photobleaching of fluorescein conjugates in air-saturated viscous solutions: Singlet and triplet state quenching by tryptophan. *Photochemistry and Photobiology*, 63(3), 265–271.

Politz, J.C.R. et al., 2006. Rapid, diffusional shuttling of poly(A) RNA between nuclear speckles and the nucleoplasm. *Molecular Biology of the Cell*, 17(3), 1239–1249.

Presley, J.F. et al., 2002. Dissection of COPI and Arf1 dynamics in vivo and role in Golgi membrane transport. *Nature*, 417(6885), 187–193.

Reits, E.A. and Neefjes, J.J., 2001. From fixed to FRAP: Measuring protein mobility and activity in living cells. *Nature Cell Biology*, 3(6), E145–E147.

Ries, J. et al., 2009. Modular scanning FCS quantifies receptor-ligand interactions in living multicellular organisms. *Nature Methods*, 6(9), 643–645.

Ritchie, K. et al., 2005. Detection of non-Brownian diffusion in the cell membrane in single molecule tracking. *Biophysical Journal*, 88(3), 2266–2277.

Sauer, M. et al., 2008. C-terminal diversity within the p53 family accounts for differences in DNA binding and transcriptional activity. *Nucleic Acids Research*, 36(6), 1900–1912.

Saxton, M.J., 2001. Anomalous subdiffusion in fluorescence photobleaching recovery: A Monte Carlo study. *Biophysical Journal*, 81(4), 2226–2240.

Schram, V. and Thompson, T., 1997. Influence of the intrinsic membrane protein bacteriorhodopsin on gel-phase domain topology in two-component phase-separated bilayers. *Biophysical Journal*, 72(5), 2217–2225.

Schröder, J. et al., 2009. In vivo labeling method using a genetic construct for nanoscale resolution microscopy. *Biophysical Journal*, 96(1), L01–L03.

Schütz, G.J., Schindler, H., and Schmidt, T., 1997. Single-molecule microscopy on model membranes reveals anomalous diffusion. *Biophysical Journal*, 73(2), 1073–1080.

Sinnecker, D. et al., 2005. Reversible photobleaching of enhanced green fluorescent proteins. *Biochemistry*, 44(18), 7085–7094.

Snapp, E.L., Altan, N., and Lippincott-Schwartz, J., 2003. Measuring protein mobility by photobleaching GFP chimeras in living cells. *Current Protocols in Cell Biology*, Chapter 21, Unit 21.1, doi: 10.1002/0471143030.cb2101s19. PubMed PMID: 18228432.

Soumpasis, D.M., 1983. Theoretical analysis of fluorescence photobleaching recovery experiments. *Biophysical Journal*, 41(1), 95–97.

Sprague, B.L. et al., 2004. Analysis of binding reactions by fluorescence recovery after photobleaching. *Biophysical Journal*, 86(6), 3473–3495.

Stasevich, T.J. et al., 2010. Cross-validating FRAP and FCS to quantify the impact of photobleaching on in vivo binding estimates. *Biophysical Journal*, 99(9), 3093–3101.

Applications

Sullivan, K.D. et al., 2009. Improved model of fluorescence recovery expands the application of multiphoton fluorescence recovery after photobleaching in vivo. *Biophysical Journal*, 96(12), 5082–5094.

Tafvizi, A. et al., 2008. Tumor suppressor p53 slides on DNA with low friction and high stability. *Biophysical Journal*, 95(1), L01–L03.

Takeda, S., 1995. Tubulin dynamics in neuronal axons of living zebrafish embryos. *Neuron*, 14(6), 1257–1264.

Teixeira, J.E. and Huston, C.D., 2008. Evidence of a continuous endoplasmic reticulum in the protozoan parasite *Entamoeba histolytica*. *Eukaryotic Cell*, 7(7), 1222–1226.

Tolentino, T., 2008. Measuring diffusion and binding kinetics by contact area FRAP. *Biophysical Journal*, 95(2), 920–930.

Uttamapinant, C. et al., 2010. A fluorophore ligase for site-specific protein labeling inside living cells. *Proceedings of the National Academy of Sciences*, 107(24), 10914–10919.

van Royen, M.E. et al., 2007. Compartmentalization of androgen receptor protein-protein interactions in living cells. *Journal of Cell Biology*, 177(1), 63–72.

van Royen, M.E. et al., 2009. Fluorescence recovery after photobleaching (FRAP) to study nuclear protein dynamics in living cells. *Methods in Molecular Biology*, 464, 363–385.

Vigers, G.P., Coue, M., and McIntosh, J.R., 1988. Fluorescent microtubules break up under illumination. *Journal of Cell Biology*, 107(3), 1011–1024.

Wachsmuth, M. et al., 2003. Analyzing intracellular binding and diffusion with continuous fluorescence photobleaching. *Biophysical Journal*, 84(5), 3353–3363.

Weber, W. et al., 1999. Shedding light on the dark and weakly fluorescent states of green fluorescent proteins. *Proceedings of the National Academy of Sciences of the United States of America*, 96(11), 6177–6182.

Yguerabide, J., Schmidt, J.A., and Yguerabide, E.E., 1982. Lateral mobility in membranes as detected by fluorescence recovery after photobleaching. *Biophysical Journal*, 40(1), 69–75.

Yildiz, A. et al., 2003. Myosin V walks hand-over-hand: Single fluorophore imaging with 1.5-nm localization. *Science*, 300(5628), 2061–2065.

Optical highlighters: Applications to cell biology

Malte Renz and *Jennifer Lippincott-Schwartz*
National Institutes of Health

Contents

9.1 INTRODUCTION

Optical highlighters are fluorescent proteins that shift their spectral properties in response to irradiation with light of specific wavelength and intensity. That is, they either become bright or change color upon being irradiated. In combination with spatially and temporally confined irradiation, optical highlighters permit the marking of a subset of molecules and then the tracking of the highlighted molecules over time. Such highlighted molecules can serve to mark structures of different sizes: these can be entire cells, subcellular organelles, groups of proteins, or even single proteins, and they can be followed on different time scales ranging from milliseconds to days. Here, we provide a survey of the approaches, the biological systems they have been used in, and the questions they have helped to address. The presented inventory may thereby inspire new applications of this class of fluorescent proteins.

Following the route of optically highlighted species helps to answer "if" they are moving, "where" they are going, and "how" they move. In so doing, optical highlighting permits study of the dynamics of a system. The approach is not affected by synthesis of new protein or protein folding during an experiment because only molecules that have been irradiated are fluorescent. This makes optical highlighting applications especially suitable for analyzing protein dynamics, including the mobility, binding properties, and turnover rates of a protein. These dynamics can be measured for proteins within the cytoplasm, within a compartment lumen, or on a membrane surface, enabling novel properties of the proteins in their natural environment to be addressed. For example, protein-binding properties and, in particular, dissociation constants within protein complexes can be assessed. Assuming that an optical highlighter is degraded along with the protein of interest, protein degradation can be quantified. Finally, using a special group of optical highlighters that change their color upon irradiation, protein synthesis is accessible.

Early applications of green fluorescent protein (GFP) photoactivation and photoconversion were reported in the mid-1990s. Using the phenomenon that wild-type GFP green fluorescence can be enhanced by UV light, Yokoe and Meyer were able to determine K-ras mobility and its membrane dissociation constant.[1] Elowitz et al. described that GFP green fluorescence converts into red fluorescence upon irradiation with 488 nm light under conditions of low oxygen. They exploited this to characterize protein diffusion in bacteria.[2] Limitations in both approaches, however, prevented them from having broad applicability: wtGFP photoactivation triggered only a threefold increase in fluorescence, whereas green-to-red conversion of GFP required low-oxygen conditions.

Only in 2002 were the first broadly applicable photoactivatable and photoconvertible fluorescent proteins described: PAGFP (photoactivatable GFP)[3] and Kaede (Japanese for maple leaf).[4] Ever since, continued developments of these proteins and discovery of new proteins have enriched the available palette of optical highlighters. The main categories of optical highlighters are the following, assuming this classification must be temporary given future developments:

- Irreversible dark to bright: "photoactivatable" proteins of different colors (e.g., PAGFP, PA-mCherry[5])
- Irreversible photoconverters: green to red (e.g., Kaede, EosFP,[6] Dendra,[7] KikGR[8]) and blue to green (e.g., PSCFP[9])
- Reversible highlighters: green (e.g., Dronpa[10]) and red (e.g., rsCherry[11])
- Mixed irreversible/reversible highlighters (e.g., KFP,[12] irisFP[13])

These diverse groups of optical highlighters have permitted the broad range of biological applications that we will itemize in this chapter. The presented inventory consists of two parts: a narrative and a commented reference section. Two main categories of optical highlighter applications will be addressed. The first comprises all applications in which at least a few proteins have been highlighted simultaneously, that is, optical highlighters for ensemble studies. Applications in this category, italicized for easy identification, encompass those for entire cells, subcellular organelles, and specific proteins of interest. The second category addresses the use of optical highlighters for single-molecule studies.

9.2 OPTICAL HIGHLIGHTERS FOR ENSEMBLE STUDIES

9.2.1 TRACKING CELLS

Optical highlighters offer great potential as tools for tracking cells in diverse physiological contexts. Two general strategies are available for making cells of interest express optical highlighters in a live organism. The genetic information for a particular optical highlighter can be inserted into the cells by, for example, microinjection or electroporation resulting in a transient expression. Alternatively, transgenic organisms can be generated that permanently express optical highlighters either widespread in all cells or by means of specific promoters—defined in a certain cell type. Optical highlighters expressed in this way have been used to generate transgenic mice, fish, flies, worms, sea squirts, and plants. Spatially confined irradiation, especially two-photon irradiation, is very suitable for "switching on" only one cell out of larger group of cells expressing the optical highlighters. The single, fluorescently marked cell can then be traced through a living organism.

In the field of embryogenesis, following "embryonic cells" as they migrate and differentiate is necessary for addressing many questions. However, it has been challenging to selectively mark, trace, and analyze migratory cells in living embryos. Optical highlighters are providing promising tools for addressing this challenge.[14–24] For example, Murray and Saint used PAGFP-tagged tubulin to study cell fate and mesoderm migration in *Drosophila* embryos. They observed distinct directions of cell movement and showed these were dependent upon the fibroblast growth factor (FGF) "Heartless."[14] Microinjection of purified recombinant Kaede into "zebrafish" embryos allowed Brown et al. to isolate cells from different embryonic regions and stages so that patterns of genes expression could be analyzed. Flow cytometry and subsequent transcriptome analysis enabled identification of novel genes decisive for early development.[15] The generation of embryonic stem cell lines expressing, for example, KikGR and tdKaede has made it possible to explore other aspects of cell movement and behavior during early development.[16,17]

The "developing nervous system" is a specific field of embryogenesis where cell tracking is proving useful.[25–32] Horie et al. established stable expression of the green-to-red photoconvertible protein Kaede in the larva brain of a sea squirt (*Ciona intestinalis*). Using different promoters, the authors guided expression of Kaede in specific cell types. Photoconversion of cells in early development enabled tracing of the fate of photoconverted red cells and distinguished them from newly formed green cells. With these tools, the authors showed that most larval neurons disappeared during development and that only a subset of cholinergic and glutamergic neurons were retained. Unexpectedly, ependymal cells, which line brain ventricles and the central canal of the spinal cord, were found to contribute significantly to the construction of the adult central nervous system. Some of the ependymal cells even differentiated into neurons exhibiting stem-cell-like characteristics.[25] In order to dissect neural crest cell (NCC) migration patterns, Kulesa and coworkers microinjected KikGR into the neural tube of avian embryos. They found that leading and trailing NCCs displayed similar average speed and directionality. Interestingly, leading cells proliferated along the migratory path and eventually outnumbered trailing cells almost threefold. Cell division resulted in red daughter cells having only 5% loss in the detected red-to-green intensity ratio.[26]

In the "developed neuronal system," cell communication and neural circuitry need to be deciphered.[33–42] Here, the challenge is not to follow a moving cell but to trace all the neural extensions, untangling the complex network of dendrites and axons. Photoactivating or photoconverting a freely diffusing optical highlighter in the soma of a neuron and watching as it gradually fills the entire neuron with a specific color is one way to visualize all the extensions of a particular neuron. This approach has been used in several studies. For example, the *Drosophila* male pheromone cVA (11-*cis* vaccenyl acetate), which elicits sexually dimorphic behavior through the same sensory neurons (SNs), was analyzed. In male *drosophila*, cVA triggers male–male aggression and suppresses courtship, while it promotes receptivity in females. Combining circumscribed photoactivation of PAGFP and electrophysiological methods, Axel and coworkers dissected and mapped the distinct neural circuits in male and female flies that are responsible for the observed dimorphic behavior.[33,34] In a different study, Neumann et al. examined axonal regeneration of SNs in *Caenorhabditis elegans*. Using a transgenic strain expressing Kaede in mechanosensory neurons, they analyzed how axonal fusion reestablished anterograde and retrograde diffusion.[35]

"Immune cells"[43–46] circulate through blood and lymph stream and migrate into tissues to reach sites of inflammation and immune response. Victora et al. used a combination of optical highlighting and flow cytometry to follow such movement, focusing on immune cell behavior within lymph nodes. They generated transgenic mice in which all hematopoietic cells express PAGFP. Using two-photon irradiation, PAGFP could be activated in lymph nodes with 10 μm precision, approximating the diameter of one cell. The half-life of PAGFP in naïve B cells was estimated to be 30 h, long enough to reveal B cell dynamics in the germinal center of lymph nodes. The study revealed that B cell division is restricted to the dark zone of a germinal center. The naïve B cells tracked with PAGFP moved to the light zone of the germinal center, where T cells help control the return of B cells to the dark zone for clonal expansion.[43]

Similarly, Kaede was used to track the dynamic of T cells between skin and the draining lymph nodes. When the skin of a Kaede transgenic mouse is exposed to UV light, the skin turns entirely red within minutes. Tomura et al. used this effect to label T cells that are present in the skin. Immediately after irradiation, none of the draining auxiliary lymph node cells or blood cells were photoconverted. Following the traffic of nonregulatory and regulatory T cells between skin and draining lymph nodes, these authors found that upon immune activation, more regulatory, in particular inhibitory, T cells migrate to the draining lymph nodes. These cells may thus contribute to the downregulation of the cutaneous immune response.[44]

In the field of "tumor biology," the imaging of optical highlighters in transplanted tumors, known as orthotopic imaging, has been used to trace individual mammary cancer cells in mice. Kedrin et al. injected a metastatic breast cancer cell line that stably expressed Dendra2 into the mammary fat of a mouse. Controlled photoconversion of one to several hundred cancer cells *in situ* was possible. Twenty-four hour after photoconversion, limited migration was detected in regions containing no detectable vessels, whereas in regions with vessels, the tumor cells were lined up along the vessels and single red metastatic cells appeared in the lungs.[47]

9.2.2 TRACKING ORGANELLES

Optical highlighting approaches have provided new information about overall dynamics of organelles. This includes organelle movements between locations, their fission or fusion, and their sites of formation. Among the organelles studied in this manner are "mitochondria" and organelles involved in "secretory,"[48–52] "endocytic,"[3,9,53] and "degradative"[54–57] pathways.

To study exocytosis, Baltrusch et al. labeled insulin secretory granules with PAGFP- and Dendra-tagged cargo protein neuropeptide Y (NPY) and the granule membrane phosphatase phogrin. They traced the labeled secretory granules after highlighting them in insulin-secreting cells. The track speed and displacement of insulin secretory vesicles were reduced after starvation. Correspondingly, glucose-induced insulin secretion was decreased.[48]

Studying the dynamics of early endosomes, Antignani and Youle labeled the small GTPase Rab5 with PAGFP. After performing localized photoactivation to highlight Rab5 molecules associated with endosomes, the authors then followed the lifetime of Rab5 on early endosomes. Interestingly, the fluorescence intensity of highlighted endosomes decayed more slowly in diphtheria toxin-treated cells than in control cells. This suggested that diphtheria toxin treatment reduces Rab5 exchange activity on endosomes.[53]

A similar fluorescence pulse–chase approach was used to track the dynamics of autophagosomes. By photoactivating the entire population of PAGFP-Atg8- and PAGFP-LC3-labeled autophagosomes in serum-starved cells, Hailey et al. showed that there was fast autophagosome turnover with a half time of about 30 min. Here, the loss of PAGFP signal was not due to photobleaching or release of labeled protein from autophagosomes, since signal loss was prevented by chloroquine treatment, which inactivates lysosomal proteases and blocks fusion with lysosomes.[54,55] This same approach was used to follow PAGFP-tagged Pex16, an early peroxisomal protein, in the endoplasmic reticulum (ER). When the tag labeling Pex 16 was selectively photoactivated and followed over time, Kim et al. found that the protein began to appear in the peroxisomes. This provided critical evidence indicating that peroxisomes originate from the ER.[56] These authors then photoactivated all the peroxisomes within the cells, cultured the cells further, and then photoactivated for a second time. Comparison of the cells before and after the second photoactivation showed that the new peroxisomes did not contain any photoactivated material. Thus, peroxisomes under these conditions proliferate by outgrow from the ER rather than by division.[56]

These approaches have also been applied to the mitochondria. Mitochondria are motile, multifunctional organelles that are highly dynamic in shape and form. This dynamism is caused by the mitochondria's ability to undergo fission and fusion. To relate mitochondrial shape changes to the variety of cellular functions that these organelles accomplish, it is important to be able to image these dynamic events inside living cells. The visualization of fusion events, however, can be complicated when the mitochondria are only in close apposition or are partially fused. Furthermore, fusions of inner and outer mitochondrial membranes are independent events. To assess mitochondrial connectivity, several groups have employed photoactivation or photoconversion of a matrix-targeted optical highlighter.[58–66] Adjacent and intertwined structures were found to often represent distinct networks, and fission frequently occurred without any gross morphological changes. For further characterization, photoactivation can be coupled with real-time monitoring of mitochondrial membrane potential.[58] Using photoactivation of mitochondrial targeted PAGFP, Molina et al. showed that mitochondria in β-cells under noxious conditions fragment. Shifting the dynamic balance toward fusion, however, protected β-cells from nutrient-induced apoptosis.[59]

9.2.3 TRACKING PROTEINS

Optical highlighters have been used as "inert markers to probe compartmentalization" and permeability of compartment borders such as the nuclear envelope[67–69] and at cell-to-cell boundaries.[70–72] Dultz et al. characterized the reestablishment of the nuclear envelope as a permeability barrier after mitosis using a photoswitchable optical highlighter called Dronpa (see Chapter 6). Dronpa was the first reversible highlighter—that is,

Applications

it can be selectively turned on and off.[10] Dultz et al. used Dronpa to measure nuclear shuttling at multiple time points after onset of anaphase in the same cell. The nuclear envelope remained relatively permeable for passive diffusion for up to 2 h, indicating only a slow recovery of the diffusion barrier constituted by nuclear pores.[67] In a different study, O'Brien et al. reported that PAGFP diffusion into the cell nucleus increases after vasopressin, angiotensin, and phenylephrine stimulation resulting in nuclear accumulation of PAGFP indicating increased permeability of the nuclear envelope.[68]

Another example of the use of optical highlighters to track protein movement between compartments are the studies in the tobacco plant leaf. Glandular trichomes are specialized organs that function in metabolite production in the leaves of higher plants. The basal trichome cell in the tobacco leaf is connected to the epidermis by numerous plasmodesmata. When PAGFP was photoactivated in an epidermal cell, it moved apically across the epidermal/trichome boundary. In contrast, when PAGFP was activated in the trichome, it did not cross the boundary. Rather, it only diffused apically into the distal trichome indicating unidirectional flow through trichome plasmodesmata.[70]

"Fused to proteins of interest," the optical highlighters allow the functions of proteins to be studied in their various cellular contexts, dissecting input and outflow pathways from the steady-state system as a whole. Most applications of optical highlighters, by far, fall into this category.

One important area where optical highlighters have revealed new insights is "histone/chromatin mobility within the nucleus,"[73–78] "nucleocytoplasmic shuttling" of transcription factors,[10,79–88] and shuttling between nuclear compartments.[89–94] To address the relationship between histone mobility and DNA repair *in vivo*, Kruhlak et al. induced double-strand breaks in cells expressing PAGFP-tagged histone 2b. The damaged chromatin showed limited mobility but underwent an ATP-dependent local expansion, probably to establish an accessible subnuclear environment facilitating DNA repair.[73] Schmierer and Hill analyzed the nucleocytoplasmic shuttling of the Smad proteins and intracellular signal transducers of the transforming growth factor (TGF) receptor and determined import and export rates. They reported that TGF-beta induced nuclear Smad2 accumulation by decreasing export and intranuclear mobility.[79]

Using the photoswitchable Dronpa, Miyawaki and coworkers monitored the nuclear import and export of extracellular signal-regulated kinase 1 (ERK1) in the same cell before and after stimulation with epidermal growth factor (EGF). Under EGF stimulation, nuclear ERK1 import and export rates are markedly increased, such that ERK1 does not accumulate in cell nucleus. In contrast, bidirectional flow rates of importin-beta, which imports proteins containing a nuclear localization signal, remained constant after EGF stimulation.[10] Finally, several groups studied protein shuttling between different nuclear compartments, including Cajal bodies (CBs), nucleoli, and speckles.[89–94] Deryuhseva et al., for example, analyzed the dynamics of PAGFP-labeled coilin in CBs of isolated *Xenopus* oocytes. Its slow diffusional behavior was not influenced by the transcriptional state or nucleocytoplasmic exchange.[89]

Various facets of the "cytoskeleton" have been studied applying optical highlighting methods.[7,95–108] Kiuchi et al. used Dronpa-labeled actin to study stimulus-induced actin filament assembly and lamellipodium extension, by providing an abundant pool of monomeric actin. They showed that the rate of fluorescence decay of Dronpa-actin activated in the cytoplasm reflects the distribution of actin monomers. It decreased after the addition of the actin-polymerizing Jasplakinolide B and increased in response to the depolymerizing Latrunculin A. To address how the actin-binding protein cofilin contributes to these processes, they showed that cofilin inactivation and knockdown decreased the cytoplasmic actin monomer pool. Furthermore, cofilin was required for EGF-induced filament assembly in the cell periphery (Figure 9.1).[95]

Neurofilaments may be transported along axons in a stop-and-go-like fashion switching between mobile and stationary phases. In cultured neurons, pulse–escape experiments using optical highlighting revealed biphasic decay kinetics, consistent with a mobile state with intermittent short pauses of 30 s and a stationary state with long pauses of about 60 min.[96] In a subsequent study, the same group of researchers reported that myosin Va functions as a short-range motor decreasing long-term pauses and enhancing efficiency of neurofilament transport.[97]

Photoconversion of a subpopulation of actin–tdEOS, a green-to-red photoconvertible protein (see Chapter 6), at the leading edge of migrating cells enabled Burnette et al. to track the fate and turnover of these actin filaments. Their findings revealed that during edge protrusion, the actin filaments completely turn over within 1–2 min. During edge retraction, however, a subset of filaments is long-lived. These compact into an arc-shaped bundle that helps produce the stack of contractile actin filaments in the lamella. The

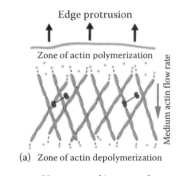

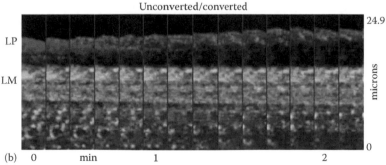

Figure 9.1 Actin filament dynamics during cell edge protrusion observed by photoconversion of actin–tdEOS. (a) Diagram of actin organization in the lamellipodia (LP) at the leading edge of migrating cells. A steady-state network of actin driving edge protrusion is maintained by a zone of actin polymerization near the cell edge and a zone of actin depolymerization at the rear. (b) Photoconversion of the LP actin pool reveals converted actin molecules (in red) being replaced by fresh unconverted actin molecules (in green) over time. No converted red molecules reach the lamella (LM) region in the rear. The data reveal that during edge protrusion, the LP pool of actin filaments undergoes complete depolymerization without reaching the LM zone. (Adapted from Burnette, D.T. et al., *Nat. Cell Biol.*, 13(4), 371, 2011.)

results suggested that actin filaments in the leading edge and in the lamella are spatially and temporally connected.[98]

Optical highlighters have also been used to study protein dynamics in the "plasma membrane" and within "endocytic and secretory pathways."[109–132] In one example, Raab-Graham et al. used the photoconvertible Kaede as reporter for local expression of the voltage-gated potassium channel Kv1.1 in dendrites. They photoconverted Kaede-Kv1.1 from green to red at time point zero and monitored the green fluorescence over time. The increase in green fluorescence indicated newly synthesized protein in the absence of Kaede-Kv1.1 movement. Treatment with Rapamycin, PI3K inhibitors, or NMDA antagonists resulted in a stronger increase in green fluorescence in dendrites as compared to control cells. Thus, mTOR and synaptic excitation may cause local suppression of dendritic Kv1.1 channels by reducing local expression.[118]

The endocytic pathways in plants have not been well characterized. Using EosFP-tagged PIN auxin efflux carriers, Dhonukshe et al. were able to demonstrate constitutive clathrin-mediated endocytosis in plants. They converted Eos at the plasma membrane and followed its accumulation in intracellular vesicles; inversely they converted Eos in vesicles and followed it back to the plasma membrane.[109]

9.3 OPTICAL HIGHLIGHTERS FOR SINGLE-MOLECULE STUDIES

Ultimately, optical highlighting can be used to switch on just a single fluorescent protein and thereby mark the protein attached to it. The emitted light from such a single source of about 2.5 nm in size will be blurred and spread by the imaging optics. The lateral size of the blurred spot is wavelength dependent and about 250–300 nm. This prevents details in the spot to be resolved. However, the prior knowledge that just a single molecule is being viewed allows the mathematical localization of the intensity center of the spot with a much greater precision. Optical highlighters permit controlled activation of only single fluorophores at a given time and inspired recent developments in fluorescence microscopy. Single-molecule-based superresolution microscopy using photoactivatable fluorescent proteins has been realized in photoactivated localization microscopy (PALM[133]) and fluorescence photoactivation localization microscopy (FPALM[134]). Similarly exploiting the on–off blinking behavior of organic dyes is the technique of stochastic optical reconstruction microscopy (STORM[135]) and direct STORM (dSTORM[136]) (see Chapter 12). In all these techniques, only a few sparsely distributed fluorophores are stochastically photoactivated to fluoresce. Neighboring molecules remain dark. The individual fluorophores are imaged, localized, and bleached. Then, in the next cycle, adjacent fluorophores can be switched on. Merging all obtained single-molecule localizations yields the final superresolution image. Thus, repeated activation and sampling permit densely expressed fluorescent proteins to be resolved in time, even though they are spatially inseparable when viewed all together at once (Figure 9.2).

We will assign cell biological applications of superresolution microscopy to two sections. "Imaging single molecules" refers to the use of superresolution microscopy to reveal steady-state distributions of molecules in a fixed cell.[133,137–144] "Tracking single molecules," on the other hand, lists applications that gained insight into the dynamic distribution of single molecules.[145–151] To keep to the chapter focused on optical highlighters, only applications employing photoactivatable or photoconvertible fluorescent proteins will be discussed.

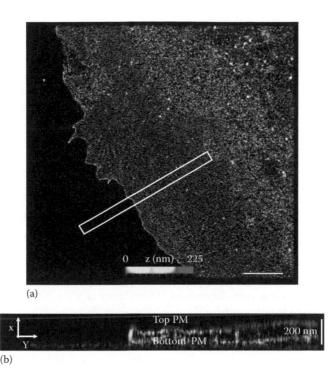

(a)

(b)

Figure 9.2 Interferometric photoactivated localization microscopy (iPALM) showing the 3D distribution of VSVG molecules tagged with tdEOS in the plasma membrane of a COS-7 cell. Single molecules are color-coded based on the z position (see the color scale in [a]) with red molecules closest to the coverslip, followed by yellow, blue, and purple at further distances. Both top (a) and side views (b) of the plasma membrane are shown. (Adapted from Shtengel, G. et al., *Proc. Natl. Acad. Sci. USA*, 106(9), 3125, 2009.)

9.3.1 IMAGING SINGLE MOLECULES

Initially, Betzig et al. demonstrated that PALM images of the Kaede-tagged lysosomal transmembrane protein CD63 in cryoprepared thin cell sections resolved lysosomal membranes down to 10 nm.[133] Later, Greenfield et al. used PALM to map three proteins central to bacterial chemotaxis and analyzed *Escherichia coli* sensory clusters. Detected clusters were shown to form through stochastic self-assembly without any cytoskeletal involvement or active transport.[137] Dual-color superresolution imaging on focal adhesions visualized distinct interlocking aggregates of α-actinin and vinculin that could not be resolved in conventional fluorescence microscopy.[138,139]

In order to combine single-molecule-based imaging of PALM with the exquisite structural information provided by transmission electron microscopy (TEM), correlative PALM/TEM has been performed. Superposition of PALM and TEM images obtained from cryoprepared cells revealed that tdEos-tagged mitochondrial matrix reporter molecules extend up to, but not into, the ~20 nm outer mitochondrial membrane.[133] Recently, Watanabe et al. introduced new embedding solutions for correlative PALM/EM that include optimized fixatives and plastics. This approach preserves fluorescence during fixation and embedding while maintaining enhanced tissue contrast through polymer embedding and paves the way for an improved PALM/EM correlation.[140]

Applications

Hess and coworkers combined interferometry with PALM to achieve 3D nanoscale resolution of cellular structures.[141] Using this microscopy approach, the precise architecture of integrin-based focal adhesions could be successfully resolved: integrins and actin were shown to be vertically separated by a ~40 nm focal adhesion core. This core consists of a membrane-adjacent signaling layer, an intermediate force-transducing layer comprising talin and vinculin, and an upper actin-regulatory layer with zyxin, vasp, and α-actinin.[142]

9.3.2 TRACKING SINGLE MOLECULES

To gain insight into diffusional behavior of single molecules, Manley et al. implemented single-particle tracking into PALM imaging strategies. Thereby, hundreds of single-molecule tracks could be assessed in a single cell. The mobility of vesicular stomatitis virus glycoprotein (VSVG) and the HIV-1 structural protein Gag at the plasma membrane were analyzed, revealing differences in diffusion coefficients and clustering characteristics of the proteins.[145,146]

In a similar PALM approach, Frost et al. mapped discrete velocities of actin molecules along the actin filaments of dendritic spines, which were previously not accessible because of a spine's submicron dimensions. The velocity of actin–tdEos molecules was elevated in discrete foci throughout the dendritic spine. In general, however, velocities were enhanced at the synapse and decreased at the endocytotic zone.[147]

In E. coli, the mobility of a major component of the bacterial cytoskeleton, FtsZ, was analyzed on the single-molecule level. Two populations of molecules were detected: the FtsZ molecules forming the Z-ring near the center of the bacterium were stationary, while the rest of FtsZ underwent Brownian motion throughout the entire cell.[148]

Hess et al. analyzed PAGFP-labeled hemagglutinin of influenza A virus and found that hemagglutinin forms irregular clusters with length scales of 40 nm to many micrometers at the plasma membrane of mammalian cell. Since it can be difficult to identify a particular molecule in an earlier frame as the same in a later frame and thus potentially complicate single-particle tracking approaches, the authors analyzed all distances between each localized molecule from one frame to the next. From this, they determined an effective diffusion coefficient for PAGFP-tagged hemagglutinin, which was found to be similar to that determined from bulk diffusion analyses.[149]

9.4 CONCLUSION

The advent of optical highlighters has permitted unparalleled insight into the dynamics, synthesis, and degradation of entire cells, cellular organelles, and proteins. Furthermore, optical highlighters have inspired and enabled recent developments in fluorescence microscopy. In particular, their on–off switching behavior has been a decisive component of single-molecule-based superresolution microscopy, which by building up images one molecule at a time can provide unprecedented insight into the molecular architecture of cell structures. As made obvious in this review, the approaches now being employed and questions being addressed using this new class of fluorescent proteins are enormous in scope and variety. We can expect numerous new possibilities in the future, where optical highlighters will play key roles in unraveling biological problems.

REFERENCES

1. Yokoe, H. and T. Meyer, Spatial dynamics of GFP-tagged proteins investigated by local fluorescence enhancement. *Nat Biotechnol*, 1996. **14**(10): 1252–1256.
2. Elowitz, M.B. et al., Photoactivation turns green fluorescent protein red. *Curr Biol*, 1997. **7**(10): 809–812.
3. Patterson, G.H. and J. Lippincott-Schwartz, A photoactivatable GFP for selective photolabeling of proteins and cells. *Science*, 2002. **297**(5588): 1873–1877.
4. Ando, R. et al., An optical marker based on the UV-induced green-to-red photoconversion of a fluorescent protein. *Proc Natl Acad Sci USA*, 2002. **99**(20): 12651–12656.
5. Subach, F.V. et al., Photoactivatable mCherry for high-resolution two-color fluorescence microscopy. *Nat Methods*, 2009. **6**(2): 153–159.
6. Wiedenmann, J. et al., EosFP, a fluorescent marker protein with UV-inducible green-to-red fluorescence conversion. *Proc Natl Acad Sci USA*, 2004. **101**(45): 15905–15910.
7. Gurskaya, N.G. et al., Engineering of a monomeric green-to-red photoactivatable fluorescent protein induced by blue light. *Nat Biotechnol*, 2006. **24**(4): 461–465.
8. Tsutsui, H. et al., Semi-rational engineering of a coral fluorescent protein into an efficient highlighter. *EMBO Rep*, 2005. **6**(3): 233–238.
9. Chudakov, D.M. et al., Photoswitchable cyan fluorescent protein for protein tracking. *Nat Biotechnol*, 2004. **22**(11): 1435–1439.
10. Ando, R., H. Mizuno, and A. Miyawaki, Regulated fast nucleocytoplasmic shuttling observed by reversible protein highlighting. *Science*, 2004. **306**(5700): 1370–1373.
11. Stiel, A.C. et al., Generation of monomeric reversibly switchable red fluorescent proteins for far-field fluorescence nanoscopy. *Biophys J*, 2008. **95**(6): 2989–2997.
12. Chudakov, D.M. et al., Kindling fluorescent proteins for precise in vivo photolabeling. *Nat Biotechnol*, 2003. **21**(2): 191–194.
13. Adam, V. et al., Structural characterization of IrisFP, an optical highlighter undergoing multiple photo-induced transformations. *Proc Natl Acad Sci USA*, 2008. **105**(47): 18343–18348.

A: OPTICAL HIGHLIGHTERS FOR ENSEMBLE STUDIES

TRACKING CELLS

EMBRYONIC CELLS

14. Murray, M.J. and R. Saint, Photoactivatable GFP resolves *Drosophila* mesoderm migration behaviour. *Development*, 2007. **134**(22): 3975–3983.

PAGFP-labeled tubulin was used to study cell fate during gastrulation in *Drosophila* embryogenesis. The direction of movement of cells was dependent upon FGF receptor "Heartless."

15. Brown, J.L. et al., Transcriptional profiling of endogenous germ layer precursor cells identifies dusp4 as an essential gene in zebrafish endoderm specification. *Proc Natl Acad Sci USA*, 2008. **105**(34): 12337–12342.

The authors isolated "zebrafish" cells containing stably expressed Kaede from different embryonic regions and stages. They then combined FACS and transcriptome analysis to identify germ-layer-specific genes. Loss of function studies subsequently revealed MAPK is essential for early development.

Applications

16. Nowotschin, S. and A.K. Hadjantonakis, Use of KikGR a photoconvertible green-to-red fluorescent protein for cell labeling and lineage analysis in ES cells and mouse embryos. *BMC Dev Biol*, 2009. **9**: 49.

The authors generated CAG::KikGR (chicken β-actin promoter) transgenic embryonic stem cells and mice, demonstrating their widespread expression.

17. Shigematsu, Y. et al., Novel embryonic stem cells expressing tdKaede protein photoconvertible from green to red fluorescence. *Int J Mol Med*, 2007. **20**(4): 439–444.

The study established a tdKaede embryonic stem cell line derived from transgenic mice blastocysts.

18. Wacker, S.A. et al., A green to red photoconvertible protein as an analyzing tool for early vertebrate development. *Dev Dyn*, 2007. **236**(2): 473–480.

mRNA or purified tdEosFP was injected for cell lineage tracing in frog embryos. The labeled cells were stable for up to 2 weeks. Injection at the two-cell stage resulted in green frog embryos.

19. Stark, D.A. and P.M. Kulesa, Photoactivatable green fluorescent protein as a single-cell marker in living embryos. *Dev Dyn*, 2005. **233**(3): 983–992.
20. Stark, D.A. and P.M. Kulesa, An in vivo comparison of photoactivatable fluorescent proteins in an avian embryo model. *Dev Dyn*, 2007. **236**(6): 1583–1594.
21. Stark, D.A., J.C. Kasemeier-Kulesa, and P.M. Kulesa, Photoactivation cell labeling for cell tracing in avian development. CSH Protoc, 2008. **2008**: pdb prot4975.
22. Kulesa, P.M. et al., Watching the assembly of an organ a single cell at a time using confocal multi-position photoactivation and multi-time acquisition. *Organogenesis*, 2009. **5**(4): 238–247.

PA-FPs were used to study migratory cells in the avian embryo. The photoactivated cells maintained normal migratory behavior. The use of PS-CFP2 allowed cell monitoring for up to 48 h, revealing that initially neighboring NCCs can populate different branchial arches.

23. Pisharath, H. et al., Targeted ablation of beta cells in the embryonic zebrafish pancreas using *E. coli* nitroreductase. *Mech Dev*, 2007. **124**(3): 218–229.

Using a transgenic ins::Kaede "zebrafish," the researchers examined beta-cell regeneration. After photoconversion to red, newly formed beta-cells containing only green signal could be analyzed.

24. Griswold, S.L. et al., Generation and characterization of iUBC-KikGR photoconvertible transgenic mice for live time lapse imaging during development. *Genesis*, 2011. 49(7): 591–598.

A iUBC::KikGR transgenic mouse with widespread expression was established. *Ex vivo* organ imaging of the photoconverted red was possible for up to 24 h.

NEURONAL CELLS

25. Horie, T. et al., Ependymal cells of chordate larvae are stem-like cells that form the adult nervous system. *Nature*, 2011. **469**(7331): 525–528.

The authors stably expressed Kaede in *C. intestinalis* larva brain. Most larval neurons disappeared during development except a subset of cholinergic and glutamergic neurons.

Unexpectedly, ependymal cells were found to contribute to the construction of the adult CNS and differentiate into neurons.

26. Hamada, M. et al., Expression of neuropeptide- and hormone-encoding genes in the *Ciona intestinalis* larval brain. *Dev Biol*, 2011. **352**(2): 202–214.

A follow-up transcriptome analysis of Kaede stably expressing *C. intestinalis* identified 565 genes preferentially expressed in larva brain.

27. Kulesa, P.M. et al., Neural crest invasion is a spatially-ordered progression into the head with higher cell proliferation at the migratory front as revealed by the photoactivatable protein, KikGR. *Dev Biol*, 2008. **316**(2): 275–287.

KikGR DNA was injected into the neural tube of chick embryos followed by electroporation. This allowed the migratory pattern of NCC invasion to be followed. Leading and trailing NCCs had similar speeds and directionalities. However, leading cells proliferated along the migratory path, eventually outnumbering the trailing cells threefold.

28. Imai, J.H., X. Wang, and S.H. Shi, Kaede-centrin1 labeling of mother and daughter centrosomes in mammalian neocortical neural progenitors. *Curr Protoc Stem Cell Biol*, 2010. **Chapter 5**: Unit 5A 5.

In utero electroporation of Kaede-centrin1 was used to study the inheritance of mother and daughter centrosomes during asymmetric cell division of the developing mice cortex.

29. Sato, T., M. Takahoko, and H. Okamoto, HuC:Kaede, a useful tool to label neural morphologies in networks in vivo. *Genesis*, 2006. **44**(3): 136–142.

Huc::Kaede transgenic "zebrafish" line with neurons expressing Kaede.

30. Caron, S.J. et al., *In vivo* birthdating by BAPTISM reveals that trigeminal sensory neuron diversity depends on early neurogenesis. *Development*, 2008. **135**(19): 3259–3269.

Using Kaede, the temporal pattern of neurogenesis in trigeminal SNs in "zebrafish" was described. Photoconversion of Kaede in trigeminal neurons made it possible to distinguish old and new neurons and to follow their behavior. Whereas early born neurons (which had both red and green colors) gave rise to multiple neuron classes, late born (containing only green color) were restricted in their fate.

31. Mutoh, T. et al., Dynamic behavior of individual cells in developing organotypic brain slices revealed by the photoconvertible protein Kaede. *Exp Neurol*, 2006. **200**(2): 430–437.

Kaede DNA was injected into ventricles and then electroporated. Shape changes and behavior of progenitor cells in living mouse brain slices could then be studied.

32. Curran, K. et al., Interplay between Foxd3 and Mitf regulates cell fate plasticity in the zebrafish neural crest. *Dev Biol*, 2010. **344**(1): 107–118.

Cell lineage tracing of "zebrafish" NCCs was performed using EosFP tagged to a nuclear localization signal. After photoconversion, nuclei stayed red for 48 h. Cells were scored based on whether they differentiated into melanophores (accumulate melanin) or iridophores (iridescent). The data revealed that a Foxd3/mitfa transcriptional switch determines cell fate of bipotent precursor.

NEURAL CIRCUITS

33. Datta, S.R. et al., The *Drosophila* pheromone cVA activates a sexually dimorphic neural circuit. *Nature*, 2008. **452**(7186): 473–477.
34. Ruta, V. et al., A dimorphic pheromone circuit in *Drosophila* from sensory input to descending output. *Nature*, 2010. **468**(7324): 686–690.

Drosophila male pheromone cVA (11-*cis*vaccenyl acetate) elicits sexually distinct behavior through the same SNs. A combination of photoactivation and electrophysiological methods helped to dissect distinct neural circuits in male and female flies. Flies were studied in which fru[GAL4] directed expression of PA-FPs.

35. Neumann, B. et al., Axonal regeneration proceeds through specific axonal fusion in transected *C. elegans* neurons. *Dev Dyn*, 2011. 240(6): 1365–1372.

Axonal regeneration of SNs was studied in *C. elegans* using a transgenic strain that expresses a Pmec-4::Kaede construct in mechanosensory neurons. Regeneration occurred through axonal fusion reestablishing anterograde and retrograde diffusion.

36. Aramaki, S. and K. Hatta, Visualizing neurons one-by-one in vivo: Optical dissection and reconstruction of neural networks with reversible fluorescent proteins. *Dev Dyn*, 2006. **235**(8): 2192–2199.

UAS–Dronpa DNA was injected into transgenic Tg(deltaD:Gal4) "zebrafish." The Dronpa was then preferentially expressed in neurons. After one- and two-photon activation, neuronal networks were optically dissected and reconstituted.

37. Hatta, K., H. Tsujii, and T. Omura, Cell tracking using a photoconvertible fluorescent protein. *Nat Protoc*, 2006. **1**(2): 960–967.

The study describes protocols for mRNA and DNA injection and creation of permanent transgenic "zebrafish" (UAS-Gal4) for neural tracing.

38. Scott, E.K. et al., Targeting neural circuitry in zebrafish using GAL4 enhancer trapping. *Nat Methods*, 2007. **4**(4): 323–326.
39. Arrenberg, A.B., F. Del Bene, and H. Baier, Optical control of zebrafish behavior with halorhodopsin. *Proc Natl Acad Sci USA*, 2009. **106**(42): 17968–17973.

The study reveals how "zebrafish" expression of a microbial light-sensitive chloride pump (halorhodopsin) can be used to silence neurons. With the light on, the fish stopped swimming, whereas with the light off, there was a rebound forward swimming. Kaede constructs were used to identify brain areas that trigger the rebound effect.

40. Kimura, Y., Y. Okamura, and S. Higashijima, alx, a zebrafish homolog of Chx10, marks ipsilateral descending excitatory interneurons that participate in the regulation of spinal locomotor circuits. *J Neurosci*, 2006. **26**(21): 5684–5697.

Transgenic "zebrafish" expressing Kaede in alx cells showed early born cells in dorsal region and later-born cells without dorsal movement. The alx cells are ipsilateral descending neurons involved in the regulation of motoneuron activity, such as escape and swimming.

41. McLean, D.L. and J.R. Fetcho, Spinal interneurons differentiate sequentially from those driving the fastest swimming movements in larval zebrafish to those driving the slowest ones. *J Neurosci*, 2009. **29**(43): 13566–13577.

Using alx::Kaede and Huc::Kaede constructs, the authors assessed the emergence of spinal circuits according to swim speed. First, dorsal-most neurons driving high-frequency movement appeared, then ventral for low frequency.

42. Davison, J.M. et al., Transactivation from Gal4-VP16 transgenic insertions for tissue-specific cell labeling and ablation in zebrafish. *Dev Biol*, 2007. **304**(2): 811–824.

Increased expression by hybrid transcription vector Gal4-VP16 in zebrafish; used to drive UAS::Kaede.

IMMUNE CELLS

43. Victora, G.D. et al., Germinal center dynamics revealed by multiphoton microscopy with a photoactivatable fluorescent reporter. *Cell*, 2010. **143**(4): 592–605.

The authors studied cell dynamics in germinal centers using a transgenic PAGFP mouse in which all hematopoietic cells express PAGFP. PAGFP was activated in lymph nodes with 10 μm precision using two-photon laser-scanning microscopy. The estimated half-life of PAGFP in naïve B cells was 30 h. Analysis of B cell dynamics in germinal center of lymph nodes revealed that B cell division is restricted to dark zone.

44. Tomura, M. et al., Activated regulatory T cells are the major T cell type emigrating from the skin during a cutaneous immune response in mice. *J Clin Invest*, 2010. **120**(3): 883–893.

The skin of a Kaede mouse was exposed to UV, labeling T cells in skin. Following nonregulatory and regulatory T cells, the data revealed these cell types traffic between skin and draining lymph nodes. Upon immune activation, more inhibitory regulatory T cells migrate to lymph nodes, contributing to the downregulation of the cutaneous immune response.

45. Tomura, M. et al., Monitoring cellular movement in vivo with photoconvertible fluorescence protein "Kaede" transgenic mice. *Proc Natl Acad Sci USA*, 2008. **105**(31): 10871–10876.

A transgenic CAG::Kaede mouse was generated with widespread Kaede expression. Photoconverted Kaede was stable for 7 day in the inguinal lymph node. Green cells (representing newly formed) in the inguinal nodes immigrated after conversion, whereas red cells (representing old cells) stayed.

46. Tomura, M., K. Itoh, and O. Kanagawa, Naive CD4+ T lymphocytes circulate through lymphoid organs to interact with endogenous antigens and upregulate their function. *J Immunol*, 2010. **184**(9): 4646–4653.

This study showed that naïve CD4+ T cells to interact with endogenous antigens after migration to lymphoid organs. The interaction induces CD69 expression, which prolongs the stay of the cells in the lymphoid organ and increases cytokine production (IL-2, TNF-α).

TUMOR CELLS

47. Kedrin, D. et al., Intravital imaging of metastatic behavior through a mammary imaging window. *Nat Methods*, 2008. **5**(12): 1019–1021.

Orthotopic imaging of metastatic breast cancer cells stably expressing Dendra2. Twenty-four hours after photoswitching in tumor regions with detectable vessels, red cancer cells lined up at vessels and single red cells had redistributed to the lung.

TRACKING ORGANELLES

SECRETORY PATHWAY

48. Baltrusch, S. and S. Lenzen, Monitoring of glucose-regulated single insulin secretory granule movement by selective photoactivation. *Diabetologia*, 2008. **51**(6): 989–996.

Cargo protein NPY and granule membrane phosphatase phogrin were labeled with PAGFP/Dendra and insulin secretory granule tracked: track speed, displacement, and glucose-induced insulin secretion were lower after starvation.

49. Jones, V.C. et al., A lentivirally delivered photoactivatable GFP to assess continuity in the endoplasmic reticulum of neurones and glia. *Pflugers Arch*, 2009. **458**(4): 809–818.
50. Brown, S.C. et al., Exploring plant endomembrane dynamics using the photoconvertible protein Kaede. *Plant J*, 2010. **63**(4): 696–711.
51. Aubert, A. et al., Sphingolipids involvement in plant endomembrane differentiation: The BY2 case. *Plant J*, 2011. **65**(6): 958–971.

The authors studied the effect of ceramide synthase inhibitor (FB1) on plant cell Golgi. SIT-Kaede was photoconverted to red, newly synthesized appeared green. After some time, green–red overlap was assessed. The control showed a perfect overlap, meaning all newly synthesized reached preexisting compartments. After 4 h FB1 treatment, there was some overlap and some green accumulated in non-Golgi structures; after 24 h FB1, however, newly synthesized green SIT-Kaede was found only in non-Golgi compartments.

52. Salvarezza, S.B. et al., LIM kinase 1 and cofilin regulate actin filament population required for dynamin-dependent apical carrier fission from the trans-Golgi network. *Mol Biol Cell*, 2009. **20**(1): 438–451.

Post-Golgi actin trafficking was addressed using actin-PAGFP. Actin diffusion away from Golgi in the presence or absence of Latrunculin, Jasplakinolide, and dominant negative cofilin was determined.

ENDOSOMES/LYSOSOMES

53. Antignani, A. and R.J. Youle, Endosome fusion induced by diphtheria toxin translocation domain. *Proc Natl Acad Sci USA*, 2008. **105**(23): 8020–8025.

Antignani et al. studied early endosome dynamics and showed that diphtheria toxin (DT) reduces PAGFP-Rab5 exchange on endosomes: fluorescence intensity of endosomes decayed more slowly in DT-treated cells than in control cells.

9. Chudakov, D.M. et al., Photoswitchable cyan fluorescent protein for protein tracking. *Nat Biotechnol*, 2004. **22**(11): 1435–1439.

The authors addressed trafficking of human dopamine transporter (hDAT) labeled with PS-CFP.

3. Patterson, G.H. and J. Lippincott-Schwartz, A photoactivatable GFP for selective photolabeling of proteins and cells. *Science*, 2002. **297**(5588): 1873–1877.

Twenty minutes after photoactivating a subset of PAGFP-marked lysosomes, the green signal was found in all lysosomes indicating that lysosomes communicate with other lysosomes/endosomes in a microtubule-dependent manner.

AUTOPHAGOSOMES/PEROXISOMES

54. Hailey, D.W. and J. Lippincott-Schwartz, Using photoactivatable proteins to monitor autophagosome lifetime. *Methods Enzymol*, 2009. **452**: 25–45.
55. Hailey, D.W. et al., Mitochondria supply membranes for autophagosome biogenesis during starvation. *Cell*, 2010. **141**(4): 656–667.

In a pulse-labeling experiment, all autophagosomes were photoactivated in serum-starved cells and their degradation followed over time showing a high turnover with a half time of 30 min.

56. Kim, P.K. et al., The origin and maintenance of mammalian peroxisomes involves a de novo PEX16-dependent pathway from the ER. *J Cell Biol*, 2006. **173**(4): 521–532.

Kim et al. selectively photoactivated PAGFP-labeled Pex16, an early event peroxisomal protein, in the ER, followed it over time, and eventually found it in peroxisomes. They showed that peroxisomes proliferate by outgrow from the ER rather than by division. The authors photoactivated all peroxisomes, cultured cells further, and then photoactivated them for a second time. New peroxisomes did not contain already photoactivated material.

57. Sinclair, A.M. et al., Peroxule extension over ER-defined paths constitutes a rapid subcellular response to hydroxyl stress. *Plant J*, 2009. **59**(2): 231–242.

The use of EosFP-SKL in plants suggested that exposure to reactive oxygen species results in peroxisome formation by rapid expansion of a single peroxisomal domain. The authors show evidence for ER–peroxisome connectivity.

MITOCHONDRIA

58. Twig, G. et al., Tagging and tracking individual networks within a complex mitochondrial web with photoactivatable GFP. *Am J Physiol Cell Physiol*, 2006. **291**(1): C176–C184.

Photoactivation of matrix-targeted PAGFP was coupled with real-time monitoring of mitochondrial membrane potential to assess mitochondrial connectivity.

59. Molina, A.J. et al., Mitochondrial networking protects beta-cells from nutrient-induced apoptosis. *Diabetes*, 2009. **58**(10): 2303–2315.

Molina et al. used mtPAGFP to assess mitochondrial connectivity. Under noxious conditions, mitochondria were shown to fragment. Shifting balance toward fusion protected beta-cells from nutrient-induced apoptosis.

60. Arimura, S. et al., Frequent fusion and fission of plant mitochondria with unequal nucleoid distribution. *Proc Natl Acad Sci USA*, 2004. **101**(20): 7805–7808.

Mito-Kaede showed multiple mitochondria fusion events in plants.

61. Young, K.W. et al., Mitochondrial Ca2+ signalling in hippocampal neurons. *Cell Calcium*, 2008. **43**(3): 296–306.

Mitochondrial Ca-fluxes were shown to be dependent upon synaptic activity but independent of cross talk with ER or presence of a mitochondrial network.

62. Molina, A.J. and O.S. Shirihai, Monitoring mitochondrial dynamics with photoactivatable [corrected] green fluorescent protein. *Methods Enzymol*, 2009. **457**: 289–304.

Applications

63. Koutsopoulos, O.S. et al., Human Miltons associate with mitochondria and induce microtubule-dependent remodeling of mitochondrial networks. *Biochim Biophys Acta*, 2010. **1803**(5): 564–574.

The ectopic expression of Milton proteins induced formation of extended mitochondrial tubules and bulbous that are connected to the adjacent mitochondrial network as shown by matrix-targeted Dendra. The remodeling was microtubule-dependent.

64. Wakamatsu, K. et al., Fusion of mitochondria in tobacco suspension cultured cells is dependent on the cellular ATP level but not on actin polymerization. *Plant Cell Rep*, 2010. **29**(10): 1139–1145.

Mitochondrial fusion in plants was dependent upon ATP but not actin, shown by the use of the inhibitors DNP, CCCP, oligomycin, and Lat B.

65. Mathur, J. et al., mEosFP-based green-to-red photoconvertible subcellular probes for plants. *Plant Physiol*, 2010. **154**(4): 1573–1587.
66. McKinney, S.A. et al., A bright and photostable photoconvertible fluorescent protein. *Nat Methods*, 2009. **6**(2): 131–133.

TRACKING PROTEINS

OPTICAL HIGHLIGHTERS ONLY

NUCLEAR ENVELOPE

67. Dultz, E., S. Huet, and J. Ellenberg, Formation of the nuclear envelope permeability barrier studied by sequential photoswitching and flux analysis. *Biophys J*, 2009. **97**(7): 1891–1897.

Dultz et al. studied the reestablishment of the nuclear envelope as a permeability barrier after mitosis by monitoring Dronpa flux from cytoplasm into the nucleus. The reversible photoswitcher Dronpa allowed many measurements at different time points in the same cell, indicating only a slow recovery of the diffusion barrier by nuclear pores.

68. O'Brien, E.M. et al., Hormonal regulation of nuclear permeability. *J Biol Chem*, 2007. **282**(6): 4210–4217.

PAGFP diffusion into nucleus was shown to increase after vasopressin, angiotensin, and phenylephrine stimulation, resulting in nuclear PAGFP accumulation.

69. Shimozono, S., H. Tsutsui, and A. Miyawaki, Diffusion of large molecules into assembling nuclei revealed using an optical highlighting technique. *Biophys J*, 2009. **97**(5): 1288–1394.

Tetrameric KikGR was used to monitor reestablishment of nuclear envelope (NE) as a diffusion barrier for molecules >60 kDa. Twenty minutes after cytokinesis, NE was still permeable for KikGR (103 kDa), but not for mECFP-KikGR (210 kDa).

CELL-TO-CELL BOUNDARY

70. Christensen, N.M., C. Faulkner, and K. Oparka, Evidence for unidirectional flow through plasmodesmata. *Plant Physiol*, 2009. **150**(1): 96–104.

The basal trichome is connected to the epidermis by plasmodesmata. PAGFP activated in epidermal cell moved apically across epidermal/trichome boundary; PAGFP activated in trichome, however, did not cross it, indicating unidirectional flow through plasmodesmata.

71. Mavrakis, M., R. Rikhy, and J. Lippincott-Schwartz, Plasma membrane polarity and compartmentalization are established before cellularization in the fly embryo. *Dev Cell*, 2009. **16**(1): 93–104.

A highlighted EosFP-labeled plasma membrane marker stayed confined and did not spread across plasma membrane of syncytial blastoderm embryo demonstrating compartmentalization in fly embryos.

72. Calvert, P.D., W.E. Schiesser, and E.N. Pugh, Jr., Diffusion of a soluble protein, photoactivatable GFP, through a sensory cilium. *J Gen Physiol*, 2010. **135**(3): 173–196.

The connecting cilium of *Xenopus* retinal rod photoreceptors does not pose a major diffusion barrier. However, there is differential axial diffusion in different rod cell compartments.

OPTICAL HIGHLIGHTERS IN FUSION PROTEINS

CELL NUCLEUS

- Histones/chromatin

73. Kruhlak, M.J. et al., Changes in chromatin structure and mobility in living cells at sites of DNA double-strand breaks. *J Cell Biol*, 2006. **172**(6): 823–834.

The authors used PAGFP-H2B to determine the mobility of chromatin containing double-strand breaks. There was limited mobility, but ATP-dependent local expansion after DNA damage, which may establish an accessible environment facilitating DNA repair.

74. Wiesmeijer, K. et al., Chromatin movement visualized with photoactivable GFP-labeled histone H4. *Differentiation*, 2008. **76**(1): 83–90.

PAGFP-H4 movement was analyzed by two-photon irradiation showing constrained chromatin movement not affected by transcriptional inhibition.

75. Post, J.N. et al., One- and two-photon photoactivation of a paGFP-fusion protein in live *Drosophila* embryos. *FEBS Lett*, 2005. **579**(2): 325–330.

This is a comparison of one- and two-photon irradiation of H2A-PAGFP in *Drosophila* embryos. After one-photon activation cells did not divide, two-photon irradiation, however, maintained cell viability.

76. Pedersen, D.S. et al., Nucleocytoplasmic distribution of the *Arabidopsis* chromatin-associated HMGB2/3 and HMGB4 proteins. *Plant Physiol*, 2010. **154**(4): 1831–1841.

The analysis of nucleocytoplasmic shuttling of *Arabidopsis* chromatin-associated high mobility group proteins (HMGBs) showed that HMGB2 and four shuttled between cytoplasm and nucleus, while HMGB1 stayed in the nucleus.

77. Eichinger, C.S. et al., Aberrant DNA polymerase alpha is excluded from the nucleus by defective import and degradation in the nucleus. *J Biol Chem*, 2009. **284**(44): 30604–30614.

Eichinger et al. showed nuclear export of point mutated PAGFP-tagged DNA polymerase alpha subunit p180.

78. Cvackova, Z. et al., Chromatin position in human HepG2 cells: Although being non-random, significantly changed in daughter cells. *J Struct Biol*, 2009. **165**(2): 107–117.

Histone H4-Dendra2 was used to address the maintenance of chromatin territory after cell division. Chromatin distribution differed in mother and daughter cells.

- Transcription factors

79. Schmierer, B. and C.S. Hill, Kinetic analysis of Smad nucleocytoplasmic shuttling reveals a mechanism for transforming growth factor beta-dependent nuclear accumulation of Smads. *Mol Cell Biol*, 2005. **25**(22): 9845–9858.

The authors determined nuclear import and export rates of Smads. TGF-beta induced nuclear accumulation by decreasing export and nuclear mobility of PAGFP-Smad2.

10. Ando, R., H. Mizuno, and A. Miyawaki, Regulated fast nucleocytoplasmic shuttling observed by reversible protein highlighting. *Science*, 2004. **306**(5700): 1370–1373.

Extracellular receptor kinases, ERK1-Dronpa, were used to monitor nuclear import and export in same cell in the presence or absence of EGF. Under EGF stimulation, ERK import and export rates increased. In contrast, bidirectional flow rates of importin-beta remained unchanged after EGF stimulation.

80. Kwon, O.Y. et al., Real-time imaging of NF-AT nucleocytoplasmic shuttling with a photoswitchable fluorescence protein in live cells. *Biochim Biophys Acta*, 2008. **1780**(12): 1403–1477.

In this study, Dronpa-tagged nuclear factor of activated T cells (NF-AT) was used. The addition of ionomycin increased nuclear NF-AT import. The addition of cyclosporine increased nuclear export. The overexpression of glycogen synthase kinase-3 opposed Ca^{2+}/calcineurin-induced nuclear import.

81. Kfoury, Y. et al., Tax ubiquitylation and SUMOylation control the dynamic shuttling of Tax and NEMO between Ubc9 nuclear bodies and the centrosome. *Blood*, 2011. **117**(1): 190–199.

Dendra-tagged T-lymphotrophic virus type I oncoprotein (Tax) shuttled between nuclear bodies and centrosome, targeted by ubiquitylation and SUMOylation.

82. Nakrieko, K.A., I.A. Ivanova, and L. Dagnino, Analysis of nuclear export using photoactivatable GFP fusion proteins and interspecies heterokaryons. *Methods Mol Biol*, 2010. **647**: 161–170.
83. Anobile, J.M. et al., Nuclear localization and dynamic properties of the Marek's disease virus oncogene products Meq and Meq/vIL8. *J Virol*, 2006. **80**(3): 1160–1166.

Anobile et al. studied PAGFP-tagged Meq protein and the splice variant Meq/vIL8 of Marek's disease virus (MDV). Meq/vIL8 lacks transcription regulation domain. After photoactivation, Meq equilibrated in nucleoplasm/nucleoli with a $t_{1/2}$ of 35 s and Meq/vIL8 with a $t_{1/2}$ of 20 min.

84. Rey, O. et al., The nuclear import of protein kinase D3 requires its catalytic activity. *J Biol Chem*, 2006. **281**(8): 5149–5157.

In contrast to wtPKD (protein kinase D3), PKD mutants (point mutations rendering catalytic domain inactive) did not enter cell nucleus despite interaction with importins. PKD variants were tagged with PAGFP.

85. Lummer, M. et al., Reversible photoswitchable DRONPA-s monitors nucleocytoplasmic transport of an RNA-binding protein in transgenic plants. *Traffic*, 2011. 12(6): 693–702.

The authors generated transgenic AtGRP7–AtGRP-Dronpa-s plants. They developed a codon-optimized Dronpa for use in plants tagged to *Arabidopsis thaliana* glycine-rich

RNA-binding protein 7 (AtGRP7) and analyzed its nucleocytoplasmic shuttling. Transport was faster when transcript levels were maximal.

86. Jones, D.M. et al., Mobility analysis of an NS5A-GFP fusion protein in cells actively replicating hepatitis C virus subgenomic RNA. *J Gen Virol*, 2007. **88**(Pt 2): 470–475.
87. Rossman, J.S. et al., POLKADOTS are foci of functional interactions in T-Cell receptor-mediated signaling to NF-kappaB. *Mol Biol Cell*, 2006. **17**(5): 2166–2176.

T cell receptor induce cytoplasmic "punctate and oligomeric killing or activating domains transducing signals" (Polkadots). These domains incorporated stably and exchanged rapidly Bcl10-PAGFP.

88. Kam, Y. and V. Quaranta, Cadherin-bound beta-catenin feeds into the Wnt pathway upon adherens junctions dissociation: Evidence for an intersection between beta-catenin pools. *PLoS One*, 2009. **4**(2): e4580.

Upon treatment with lysophosphatidic acid (LPA), β-catenin-PAGFP initially incorporated in adherens junctions redistributed into cell nucleus.

NUCLEAR COMPARTMENTS

89. Deryusheva, S. and J.G. Gall, Dynamics of coilin in Cajal bodies of the *Xenopus* germinal vesicle. *Proc Natl Acad Sci USA*, 2004. **101**(14): 4810–4814.

PAGFP-tagged coilin in CBs of isolated *Xenopus* oocytes showed slow diffusion of coilin within and out of CBs, which was not influenced by transcriptional state or nucleocytoplasmic exchange.

90. Stanek, D. et al., Spliceosomal small nuclear ribonucleoprotein particles repeatedly cycle through Cajal bodies. *Mol Biol Cell*, 2008. **19**(6): 2534–2543.

Mature spliceosomal snRNP accumulated in CBs and shuttled between different CBs. Only a small fraction was imported from cytoplasm.

91. Sleeman, J., A regulatory role for CRM1 in the multi-directional trafficking of splicing snRNPs in the mammalian nucleus. *J Cell Sci*, 2007. **120**(Pt 9): 1540–1550.

Newly imported snRNPs exchanged between CBs and accessed cytoplasm; there was no accumulation in speckles. At steady-state, however, snRNPs exchanged between CBs and speckles and did not access the cytoplasm. A role of leptomycin B (LMB) was shown.

92. Muro, E. et al., In nucleoli, the steady state of nucleolar proteins is leptomycin B-sensitive. *Biol Cell*, 2008. **100**(5): 303–313.

The dynamics of nucleolar proteins PAGFP-B23, Nop52, and fibrillarin within nucleolus and between nucleoli was LMB sensitive. Fibrillarin trafficking between nucleoli and CB, however, was LMB insensitive.

93. Wang, M., C.M. Trim, and W.J. Gullick, Localisation of Neuregulin 1-beta3 to different sub-nuclear structures alters gene expression. *Exp Cell Res*, 2011. **317**(4): 423–432.

Wang et al. showed PAGFP-tagged Neuregulin 1 redistribution between nuclear compartments that had an effect on phosphorylation and/or expression of HER4 and HER2.

94. Onischenko, E. et al., Role of the Ndc1 interaction network in yeast nuclear pore complex assembly and maintenance. *J Cell Biol*, 2009. **185**(3): 475–491.

Applications

The authors used Nup82-2xDendra to separately track old and newly synthesized nucleoporins. In nup53Δ nup59Δ cells, proper Nup82 localization was depended on POM34: when POM34 was absent, Nup82 accumulated in cytoplasmic foci.

CYTOSKELETON

95. Kiuchi, T. et al., Cofilin promotes stimulus-induced lamellipodium formation by generating an abundant supply of actin monomers. *J Cell Biol*, 2007. **177**(3): 465–476.

The fluorescence decay of Dronpa-actin in the cytoplasm reflects the distribution of actin monomer pool: it decreased by actin polymerization with Jasplakinolide B and increased by depolymerizing actin with Latrunculin A. Cofilin inactivation/knockdown decreased actin monomer pool. Cofilin was required for EGF-induced filament assembly in the cell periphery.

96. Trivedi, N., P. Jung, and A. Brown, Neurofilaments switch between distinct mobile and stationary states during their transport along axons. *J Neurosci*, 2007. **27**(3): 507–516.

Stop-and-go hypothesis for neurofilament transport along axons: pulse–escape experiments showed switch between mobile with intermittent short pauses of 30 s and stationary states with long pauses of 60 min.

97. Alami, N.H., P. Jung, and A. Brown, Myosin Va increases the efficiency of neurofilament transport by decreasing the duration of long-term pauses. *J Neurosci*, 2009. **29**(20): 6625–6634.

In a follow-up study, myosin Va, a short-range motor, was shown to decrease long-term pauses and enhance the efficiency of neurofilament transport.

98. Burnette, D.T. et al., A role for actin arcs in the leading-edge advance of migrating cells. *Nat Cell Biol*, 2011. 13(4): 371–381.

Actin–tdEos was photoconverted during retraction of leading edge. A subset condensed into actin arcs moving rearward into the lamella showing the lamellipodium evolved into the lamella. Role of myosin II.

99. Rusan, N.M. and P. Wadsworth, Centrosome fragments and microtubules are transported asymmetrically away from division plane in anaphase. *J Cell Biol*, 2005. **168**(1): 21–28.

PAGFP–tubulin showed sevenfold higher microtubule (MT) release in anaphase than in metaphase in motion away from equatorial region.

100. Ferenz, N.P. and P. Wadsworth, Prophase microtubule arrays undergo flux-like behavior in mammalian cells. *Mol Biol Cell*, 2007. **18**(10): 3993–4002.

Ferenz et al. showed poleward flux in prophase nuclei. MT motion toward and away from centrosomes exhibited a wide range of rates in prophase nuclei: rapid motion was dynein dependent, while slow motion was not. The slow motion rates were similar to poleward flux and sensitive to Eg5 and Kif2a perturbation.

101. Ferenz, N.P. et al., Imaging protein dynamics in live mitotic cells. *Methods*, 2010. **51**(2): 193–196.
102. Tulu, U.S., N.P. Ferenz, and P. Wadsworth, Photoactivatable green fluorescent protein-tubulin. *Methods Cell Biol*, 2010. **97**: 81–90.

103. Goodson, H.V., J.S. Dzurisin, and P. Wadsworth, Generation of stable cell lines expressing GFP-tubulin and photoactivatable-GFP-tubulin and characterization of clones. *Cold Spring Harb Protoc*, 2010. **2010**(9): pdb prot5480.
104. Colakoglu, G. and A. Brown, Intermediate filaments exchange subunits along their length and elongate by end-to-end annealing. *J Cell Biol*, 2009. **185**(5): 769–777.

PAGFP-tagged neurofilaments and vimentin filaments elongated end to end and incorporated intercalating subunits along their length.

105. Dovas, A. et al., Visualization of actin polymerization in invasive structures of macrophages and carcinoma cells using photoconvertible beta-actin-Dendra2 fusion proteins. *PLoS One*, 2011. **6**(2): e16485.
106. Flynn, K.C. et al., Growth cone-like waves transport actin and promote axonogenesis and neurite branching. *Dev Neurobiol*, 2009. **69**(12): 761–779.

Dendra- and PAGFP-tagged actin was used to analyze actin waves in neurons. In nonwave regions, actin was diffusive. In wave structures, actin was immobilized.

107. Hong, S., R.B. Troyanovsky, and S.M. Troyanovsky, Spontaneous assembly and active disassembly balance adherens junction homeostasis. *Proc Natl Acad Sci USA*, 2010. **107**(8): 3528–3533.

Dendra-tagged β-catenin turnover in adherens junctions was ATP-dependent. Its removal was unaffected by mutation of endocytosis motives.

108. Schenkel, M. et al., Visualizing the actin cytoskeleton in living plant cells using a photo-convertible mEos::FABD-mTn fluorescent fusion protein. *Plant Methods*, 2008. **4**: 21.
7. Gurskaya, N.G. et al., Engineering of a monomeric green-to-red photoactivatable fluorescent protein induced by blue light. *Nat Biotechnol*, 2006. **24**(4): 461–465.

Mobility of fibrillarin and vimentin was studied.

ENDOCYTIC AND SECRETORY PATHWAY

109. Dhonukshe, P. et al., Clathrin-mediated constitutive endocytosis of PIN auxin efflux carriers in *Arabidopsis*. *Curr Biol*, 2007. **17**(6): 520–527.

EosFP-tagged PIN auxin efflux carriers were used to show constitutive clathrin-mediated endocytosis in plants: Eos was converted at the plasma membrane and accumulation in intracellular vesicles followed, and vice versa.

110. Fang, Z. et al., The membrane-associated protein, supervillin, accelerates F-actin-dependent rapid integrin recycling and cell motility. *Traffic*, 2010. **11**(6): 782–799.

tdEos–supervillin shuttled to endosomes. Vesicles were photoconverted to monitor movement of red and reappearance of green in converted vesicles.

111. Chisari, M. et al., Shuttling of G protein subunits between the plasma membrane and intracellular membranes. *J Biol Chem*, 2007. **282**(33): 24092–24098.

Dronpa-tagged α- and γ-subunits of heterotrimeric G-proteins shuttled between Golgi and plasma membrane indicating that G-proteins may not permanently reside at plasma membrane but continuously test plasma membrane and endomembranes.

112. Toyooka, K. and K. Matsuoka, Exo- and endocytotic trafficking of SCAMP2. *Plant Signal Behav*, 2009. **4**(12): 1196–1198.

Dronpa-tagged secretory carrier membrane proteins (SCAMP2) were activated at the plasma membrane of plant cells. They showed no lateral spread but recycled into intracellular vesicles. When activated in cytokinesis at the cell plate, they appeared intracellularly in daughter cells.

113. Schmidt, A. et al., Use of Kaede fusions to visualize recycling of G protein-coupled receptors. *Traffic*, 2009. **10**(1): 2–15.

Schmidt et al. used Kaede-tagged G protein-coupled receptors. After agonist-induced internalization, the authors photoswitched fluorescence in endosomes. Corticotropin-releasing factor receptor 1 and vasopressin receptor 1a recycled back to the plasma membrane, while vasopressin receptor 2 did not.

114. Bergeland, T. et al., Cell-cycle-dependent binding kinetics for the early endosomal tethering factor EEA1. *EMBO Rep*, 2008. **9**(2): 171–178.

PAGFP-labeled early endosomal antigen (EEA1) was used to determine off rates from endosomes. The on rate was faster than off rate, the residency time determined by the off rate. Rab5 overexpression reduced EEA1 off rate. Binding kinetics were cell-cycle dependent, for example, in mitosis, the off rate increased.

115. Sutter, J.U. et al., Selective mobility and sensitivity to SNAREs is exhibited by the *Arabidopsis* KAT1 K+ channel at the plasma membrane. *Plant Cell*, 2006. **18**(4): 935–954.

PAGFP-tagged KAT1 K^+ channel of *Arabidopsis* was immobile and in clusters at the plasma membrane. When SNARE function was disrupted, fewer KAT1 were found at the plasma membrane; these were highly mobile.

116. Baltrusch, S. and S. Lenzen, Novel insights into the regulation of the bound and diffusible glucokinase in MIN6 beta-cells. *Diabetes*, 2007. **56**(5): 1305–1315.

PSCFP–glucokinase in beta-cell clones showed increased mobility at high glucose concentrations.

117. Gerbin, C.S. and R. Landgraf, Geldanamycin selectively targets the nascent form of ERBB3 for degradation. *Cell Stress Chaperones*, 2010. **15**(5): 529–544.

The authors studied the effect of Geldanamycin (GA) on ErbB3-Dendra2. After cycloheximide addition, the entire cell was photoconverted: there was still an increase in green fluorescence due to delay between synthesis, folding, and chromophore maturation. After GA addition, however, almost no increase in green was detected indicating GA sensitivity late in protein maturation.

NEURAL PROTEINS

118. Raab-Graham, K.F. et al., Activity- and mTOR-dependent suppression of Kv1.1 channel mRNA translation in dendrites. *Science*, 2006. **314**(5796): 144–148.

The authors photoconverted the Kaede-labeled voltage-gated potassium channel Kv1.1 in dendrites and monitored green fluorescence. An increase in green fluorescence indicated newly synthesized protein. Rapamycin, PI3K inhibitors, and NMDA antagonists resulted in a stronger increase than controls.

119. Tsuriel, S. et al., Exchange and redistribution dynamics of the cytoskeleton of the active zone molecule bassoon. *J Neurosci*, 2009. **29**(2): 351–358.

The PAGFP-tagged Bassoon, a scaffold protein of cytoskeletal matrix associated with active zone of presynaptic membrane, showed restricted mobility with an exchange rate of several hours, was only slightly increased after stimulation.

120. Sturgill, J.F. et al., Distinct domains within PSD-95 mediate synaptic incorporation, stabilization, and activity-dependent trafficking. *J Neurosci*, 2009. **29**(41): 12845–12854.

The immobile PAGFP-tagged PSD-95 is important in stabilizing postsynaptic density (PSD). The authors identified domains that stabilize PSD-95. NMDAR activation resulted in PSD-95 loss from dendritic spines indicating destabilization may be decisive for synaptic plasticity.

121. Leung, K.M. et al., Asymmetrical beta-actin mRNA translation in growth cones mediates attractive turning to netrin-1. *Nat Neurosci*, 2006. **9**(10): 1247–1256.

Kaede was linked to beta-actin 3′ UTR and injected in blastomeres for monitoring local translation. It was converted entirely to red. The recurrence of green served as measure of new translation. Translation increased by netrin-1. Axon severing showed monitored translation as local.

122. Banerjee, S., P. Neveu, and K.S. Kosik, A coordinated local translational control point at the synapse involving relief from silencing and MOV10 degradation. *Neuron*, 2009. **64**(6): 871–884.

PAGFP and Kaede were used to monitor synthesis and degradation of RISC protein MOV10 and depalmitoylating enzyme Lypla1 (local translational control point during synaptic plasticity).

123. Gray, N.W. et al., Rapid redistribution of synaptic PSD-95 in the neocortex in vivo. *PLoS Biol*, 2006. **4**(11): e370.

In utero electroporated PAGFP-tagged postsynaptic scaffolding protein (PSD-95) for measuring its redistribution in the developing mouse brain; rapid turnover at postnatal day 10–21, decreasing with developmental age and increasing after sensory deprivation.

124. Weissmann, C. et al., Microtubule binding and trapping at the tip of neurites regulate tau motion in living neurons. *Traffic*, 2009. **10**(11): 1655–1668.

Weissmann et al. studied the mobility of PAGFP-tagged tau protein. Taxol, disease-relevant mutations, and preaggregated amyloid β increased diffusion coefficients. Tau enrichment in neurite tip was preserved after taxol but suppressed in disease-related states.

125. Wiegert, J.S., C.P. Bengtson, and H. Bading, Diffusion and not active transport underlies and limits ERK1/2 synapse-to-nucleus signaling in hippocampal neurons. *J Biol Chem*, 2007. **282**(40): 29621–28633.

Dronpa-labeled extracellular signal-regulated kinase 1/2 (ERK1/2) activated on the soma of hippocampal neurons reached nucleus by facilitated diffusion, while ERK1/2 diffused passively in dendrites.

126. Vogelaar, C.F. et al., Axonal mRNAs: characterisation and role in the growth and regeneration of dorsal root ganglion axons and growth cones. *Mol Cell Neurosci*, 2009. **42**(2): 102–115.

Local translation of Kaede-tagged beta-actin in axons was increased by axotomy.

Applications

127. Lai, K.O. et al., Importin-mediated retrograde transport of CREB2 from distal processes to the nucleus in neurons. *Proc Natl Acad Sci USA*, 2008. **105**(44): 17175–17180.

In the *Aplysia* SN–motor neuron (MN) culture system, the authors used CREB2-dendra to determine retrograde transport from dendrite into the nucleus. The transport increased in phenylalanine-methionine-arginine-phenylalanine-amide (FMRFamide)-induced long-term depression.

128. Wang, D.O. et al., Synapse- and stimulus-specific local translation during long-term neuronal plasticity. *Science*, 2009. **324**(5934): 1536–1540.

The 5′ and 3′ UTRs of sensorin was fused to Dendra2 and the *Aplysia* SN–MN culture system used. The increase in green after photoactivation indicated translation. There was a translation of reporter protein in SN only after multiple applications of serotonin, only in stimulated synapses, and only during long-term facilitation, and it required a synapse to MN and Ca-signaling in MN.

OTHERS

129. Sastalla, I. et al., Codon-optimized fluorescent proteins designed for expression in low-GC gram-positive bacteria. *Appl Environ Microbiol*, 2009. **75**(7): 2099–2110.
130. Vorvis, C., S.M. Markus, and W.L. Lee, Photoactivatable GFP tagging cassettes for protein-tracking studies in the budding yeast *Saccharomyces cerevisiae*. *Yeast*, 2008. **25**(9): 651–659.
131. Hamer, G., O. Matilainen, and C.I. Holmberg, A photoconvertible reporter of the ubiquitin-proteasome system in vivo. *Nat Methods*, 2010. **7**(6): 473–478.
132. Ivanchenko, S. et al., Dynamics of HIV-1 assembly and release. *PLoS Pathog*, 2009. **5**(11): e1000652.

Gag-mEos was preferentially activated at the plasma membrane using TIRF illumination. An increase in green fluorescence indicated Gag recruitment to the budding site from cytoplasm.

B: OPTICAL HIGHLIGHTERS FOR SINGLE-MOLECULE STUDIES

IMAGING SINGLE MOLECULES

133. Betzig, E. et al., Imaging intracellular fluorescent proteins at nanometer resolution. *Science*, 2006. **313**(5793): 1642–1645.
134. Hess, S.T., T.P. Girirajan, and M.D. Mason, Ultra-high resolution imaging by fluorescence photoactivation localization microscopy. *Biophys J*, 2006. **91**(11): 4258–4272.
135. Rust, M.J., M. Bates, and X. Zhuang, Sub-diffraction-limit imaging by stochastic optical reconstruction microscopy (STORM). *Nat Methods*, 2006. **3**(10): 793–795.
136. Heilemann, M. et al., Subdiffraction-resolution fluorescence imaging with conventional fluorescent probes. *Angew Chem Int Ed Engl*, 2008. **47**(33): 6172–6176.
137. Greenfield, D. et al., Self-organization of the *Escherichia coli* chemotaxis network imaged with super-resolution light microscopy. *PLoS Biol*, 2009. **7**(6): e1000137.

tdEos-tagged CheW and CheY and mEos-tagged Tar (central to bacterial chemotaxis) were mapped to analyze *E. coli* sensory clusters. Clusters form through stochastic self-assembly without cytoskeletal involvement or active transport.

138. Shroff, H. et al., Dual-color superresolution imaging of genetically expressed probes within individual adhesion complexes. *Proc Natl Acad Sci USA*, 2007. **104**(51): 20308–20313.

Applications

Focal adhesion proteins were mapped. Dronpa-tagged α-actinin and tdEos-tagged vinculin in distinct interlocking aggregates.

139. Shroff, H., H. White, and E. Betzig, Photoactivated localization microscopy (PALM) of adhesion complexes. *Curr Protoc Cell Biol*, 2008. Chapter 4: Unit 4 21.
140. Watanabe, S. et al., Protein localization in electron micrographs using fluorescence nanoscopy. *Nat Methods*, 2011. **8**(1): 80–84.
141. Shtengel, G. et al., Interferometric fluorescent super-resolution microscopy resolves 3D cellular ultrastructure. *Proc Natl Acad Sci USA*, 2009. **106**(9): 3125–3130.
142. Kanchanawong, P. et al., Nanoscale architecture of integrin-based cell adhesions. *Nature*, 2010. **468**(7323): 580–584.

Imaging of the 3D architecture of integrin-based cell adhesions showed integrins and actin to be separated by a ~40 nm focal adhesion core consisting of membrane-adjacent signaling layer, intermediate force-transduction layer with talin and vinculin, and upper actin-regulatory layer with zyxin, vasp, and α-actinin.

143. Lillemeier, B.F. et al., TCR and Lat are expressed on separate protein islands on T cell membranes and concatenate during activation. *Nat Immunol*, 2010. **11**(1): 90–96.

CD3ζ-PSCFP2, Lat-PSCFP2, and anti-TCRβ-scFv-PSCFP2 were used to analyze plasma membrane distribution and dynamics of T cell receptor. In quiescent T cells, TCR and Lat were distributed in separate membrane domains, while upon activation, TCR and Lat concatenated.

144. Owen, D.M. et al., PALM imaging and cluster analysis of protein heterogeneity at the cell surface. *J Biophotonics*, 2010. **3**(7): 446–454.

PALM imaging of lymphocyte-specific tyrosine kinase (Lck$_{N10}$-tdEos) and sarcoma nonreceptor tyrosine kinase (Src$_{N15}$-PSCFP2).

TRACKING SINGLE MOLECULES

145. Manley, S. et al., High-density mapping of single-molecule trajectories with photoactivated localization microscopy. *Nat Methods*, 2008. **5**(2): 155–157.

Determination of VSVG-dEos and Gag-tdEos diffusion coefficients.

146. Manley, S., J.M. Gillette, and J. Lippincott-Schwartz, Single-particle tracking photoactivated localization microscopy for mapping single-molecule dynamics. *Methods Enzymol*, 2010. **475**: 109–120.
147. Frost, N.A. et al., Single-molecule discrimination of discrete perisynaptic and distributed sites of actin filament assembly within dendritic spines. *Neuron*, 2010. **67**(1): 86–99.

Discrete velocities of actin–tdEos along actin filaments of dendritic spines were shown. Molecular velocity was elevated in discrete foci throughout the spines, generally enhanced at synapse, and decreased at endocytotic zone.

148. Niu, L. and J. Yu, Investigating intracellular dynamics of FtsZ cytoskeleton with photoactivation single-molecule tracking. *Biophys J*, 2008. **95**(4): 2009–2016.

Niu et al. used Dendra2-tagged prokaryotic homologues of tubulin (FtsZ). They describe immobile FtsZ molecules near the center forming the Z-ring of *E. coli* cytokinesis apparatus. The rest of FtsZ molecules underwent anomalous diffusion through the entire bacterium which was apparently spatially restricted to helical-shaped regions.

Applications

149. Hess, S.T. et al., Dynamic clustered distribution of hemagglutinin resolved at 40 nm in living cell membranes discriminates between raft theories. *Proc Natl Acad Sci USA*, 2007. **104**(44): 17370–17375.

PAGFP-labeled hemagglutinin (HA) of influenza virus at the plasma membrane of fixed mammalian cell exhibited irregular clusters with length scales of 40 nm to many micrometers. In live cells, all distances between each localized molecule from one frame to the next were analyzed to determine the effective diffusion coefficient.

150. Shroff, H. et al., Live-cell photoactivated localization microscopy of nanoscale adhesion dynamics. *Nat Methods*, 2008. **5**(5): 417–423.

tdEosFP-Paxillin was used for the characterization of how focal adhesions evolve over time in stationary fibroblasts.

151. Burghardt, T.P., J. Li, and K. Ajtai, Single myosin lever arm orientation in a muscle fiber detected with photoactivatable GFP. *Biochemistry*, 2009. **48**(4): 754–765.

Human cardiac myosin regulatory light chain (HCRLC) was tagged to PAGFP and exchanged with native light chains in permeabilized skeletal muscle fibers. The authors used polarized fluorescence to determine single myosin lever arm orientations (bent and straight).

10 Optogenetic tools derived from plant photoreceptors

John M. Christie
University of Glasgow

Contents

10.1 INTRODUCTION

Light is ubiquitously recognized throughout nature as a source of energy. For many organisms including plants and algae, light is an important environmental stimulus that directs their development, morphogenesis, and physiology. This is achieved by specialized photoreceptors that detect and respond to changes in light intensity, quality, direction, and duration. These light-responsive proteins typically contain an organic cofactor or chromophore that enables them to interact with light. In animals, photoreceptors are predominantly associated with visual perception. Given the importance of light in shaping their growth, plants have evolved a versatile collection of photoreceptors that enable them to sample a large portion of the electromagnetic spectrum. Over the past two decades, much progress has been made in uncovering the molecular identity of the plant photoreceptor network, as well as their mechanism of action. This rapid increase in knowledge has recently instigated the design of novel fluorescent reporters and synthetic photoswitches with desirable applications in cell biology and biotechnology. These advances in synthetic biology have given rise to the field of optogenetics, which will be discussed, as will the biochemical and photochemical properties of the photoreceptor proteins involved.

10.2 PLANT PHOTORECEPTORS

To date, five different types of photoreceptors have been identified in higher vascular plants [1], namely, the phytochromes (PHYs), cryptochromes, phototropins, members of the Zeitlupe (ZTL) family, and the recently characterized Per-ARNT-Sim (PAS)/light, oxygen, voltage (LOV) or LOV/LOV protein (PLP/LLP) [2]. PHYs are photoreversible red/far-red (FR) photoreceptors, whereas cryptochromes, phototropins, PLP/LLP, and ZTL family members specifically absorb UV-A/blue wavelengths. Photoreceptor composites of both PHY and phototropin are known to exist in lower vascular plants such as ferns where they function as dual red/blue photoreceptors [3]. The presence of such a hybrid photoreceptor is proposed to enhance light sensitivity and aid the prevalence of species such as ferns in low light conditions typically found under the canopy of dense forests [4]. Plants also respond to UV-B [5] and green light [6]; although the photosensor responsible for these green light responses remains elusive, recent studies have uncovered the molecular identity of a plant UV-B photoreceptor known as UV-Resistance locus 8 (UVR8) [7].

Nonvascular plants including algae possess an additional array of photoresponsive proteins since their aquatic environment has unique properties of light transmission. For instance, the unicellular green alga *Chlamydomonas reinhardtii* contains two archaeal types of rhodopsin that are activated by blue and green light, respectively [8–10]. Other unicellular photosynthetic organisms such as euglenids possess a novel class of photoreceptor identified as a blue-light-activated adenylyl cyclase [11]. This wide selection of photoreceptors from both plant and algal sources offers diverse starting templates for synthetic engineering. Successful strategies based on the plant photoreceptors so far have been largely based on phototropin-related photoreceptors and the red-light-responding PHYs. Other notable examples have exploited the light-sensing mechanism of algal channelrhodopsin. The following sections aim to provide a brief summary regarding the structure and function of these photoreceptors before discussing their utility as tools to photomodulate and monitor cellular activities.

10.3 PHOTOTROPIN AND RELATED PHOTORECEPTORS

Like many aspects of plant biology, much of our understanding of photoreceptor function has come from genetic analysis of the model plant *Arabidopsis thaliana*. *Arabidopsis* contains two phototropins (phot1 and phot2) that have both overlapping and distinct functions [12]. Phototropins regulate a range of processes that collectively serve to optimize photosynthetic efficiency and promote plant growth especially under weak light conditions [13]. These include phototropism [14], stomatal opening [15], light-induced chloroplast movements [16], and leaf expansion and movement [17,18]. By contrast, phototropin in *Chlamydomonas* regulates the algal sexual life cycle in response to blue light [19]. However, despite their functional differences, the mode action of higher plant and algal phototropins appears to be highly conserved [20].

The primary amino acid structure of phototropins can be separated into two parts: an N-terminal photosensory input region coupled to a C-terminal effector region containing a canonical serine/threonine kinase motif (Figure 10.1a). The N-terminal region comprises two so-called LOV domains, each of which bind the vitamin-B derived cofactor flavin

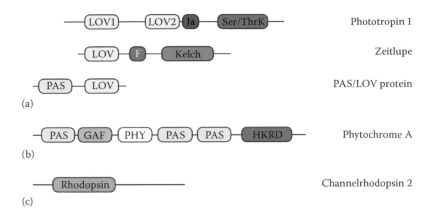

Figure 10.1 Domain structures of plant and algal photoreceptors. (a) LOV-domain-containing blue-light receptors from *Arabidopsis*. Domain abbreviations are as follows: LOV (light, oxygen, voltage), Jα (Jα helix), Ser/ThrK (serine/threonine kinase), F (F-box), kelch (kelch repeats), and PAS (Per-ARNT-Sim). (b) *Arabidopsis* PHY A. Domain abbreviations are as follows: PAS (Per-ARNT-Sim), GAF (GAF domain), PHY (phytochrome), HKRD (histidine kinase-related domain). (c) *Chlamydomonas* channelrhodopsin 2.

mononucleotide (FMN) noncovalently as a UV-A/blue-light-absorbing chromophore [21,22]. LOV domains form a subset of the versatile PAS superfamily [23] that exhibit sequence homology to motifs found in a diverse range of eukaryotic and prokaryotic proteins involved in sensing light, oxygen, or voltage, hence the acronym LOV [24]. X-ray crystallography has shown that the LOV domain consists primarily of five antiparallel β-sheets and two α-helices [25], binding the FMN tightly inside an enclosed structure (Figure 10.2a). Upon UV-A/blue-light excitation, LOV domains undergo a reversible photocycle involving the formation of a covalent bond between the FMN chromophore and a conserved cysteine residue within the protein [26]. This light-activated state thermally decays to the dark state within tens to thousands of seconds depending on the LOV domain (Figure 10.3a). These differences confer LOV domains with different photocycle properties and thus offer interesting new possibilities for fine-tuning the photoreactions of LOV photosensory modules for specific optogenetic applications.

The LOV2 domain of phot1 (Figure 10.1a) is the predominant light sensor controlling receptor kinase activity [27–29]. A common feature of LOV domain signaling is that their photoactivation leads to structural changes in N- or C-terminal sequences outside the LOV core. For example, photoexcitation of LOV2 leads to displacement of a conserved α-helix from the β-scaffold surface of the domain [30]. Artificial unfolding of this helix, designated Jα (Figure 10.1a), results in activation of the C-terminal kinase domain in the absence of light [31]. In particular, protein rearrangements within the central β-sheet scaffold play a key role in propagating the photochemical signal generated within the LOV2 domain out to the Jα-helix [32–34].

In addition to phototropins, *Arabidopsis* contains a second LOV-containing photoreceptor family consisting of three members: ZTL, flavin binding, kelch repeat, F-box 1 (FKF1), and LOV kelch protein 2 (LKP2). These photoresponsive proteins play important roles in regulating targeted degradation of components associated with circadian clock function and flowering in a light-dependent manner [35,36]. ZTL, FKF1, and LKP2 share three characteristic domains: an LOV domain at the N-terminus followed by an F-box

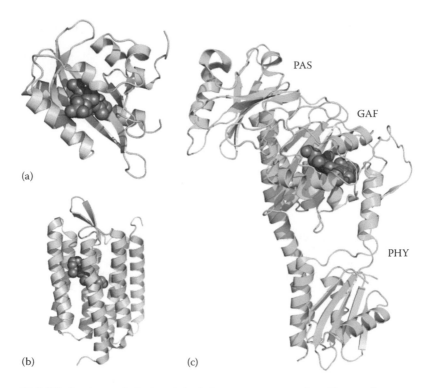

Figure 10.2 3D structures of plant and algal photoreceptor proteins. Chromophores are shown in blue as space-filling models. (a) LOV2 domain of *A. sativa* phot1 (PDB entry 2V0U) [95]. (b) *Halobacterium salinarum* bacteriorhodopsin (PDB entry 1C3W) [97]. (c) Single subunit of the N-terminal photosensory region from *Pseudomonas aeruginosa* bacteriophytochrome (PDB entry 3C2W) [96]. PAS, GAF, and PHY domains are indicated.

motif and six kelch repeats at the C-terminus (Figure 10.1a). The F-box motif is typically found in E3 ubiquitin ligases that target proteins for degradation *via* the ubiquitin–proteasome system [37], whereas the kelch repeats form a β-propeller structure thought to be involved in mediating protein–protein interactions. The LOV domains of ZTL, FKF1, and LKP2 exhibit photochemical properties analogous to those of phototropin LOV domains but fail to revert to their dark state in the absence of light [38,39].

More recently, a third LOV-containing photoreceptor has been identified in higher plants. PLP or LLP, as it is known (Figure 10.1a), contains two PAS/LOV domains but lacks an effector or output region [2]. However, the biological significance of LLP photoreceptors in plants and mosses has yet to be determined.

10.4 PHYTOCHROMES

PHYs control many aspects of plant development and growth. *Arabidopsis* contains five PHYs (phyA-E) that mainly absorb red and FR wavelengths of light *via* a covalently attached linear tetrapyrrole (bilin) chromophore that is synthesized in the chloroplast from heme [40,41]. PHYs interconvert between red and FR absorbing forms called Pr and Pfr, where the Pfr form is considered to be the active form because many physiological responses are promoted by red light [41,42]. Moreover, PHYs sense the ratio of red and FR light and use

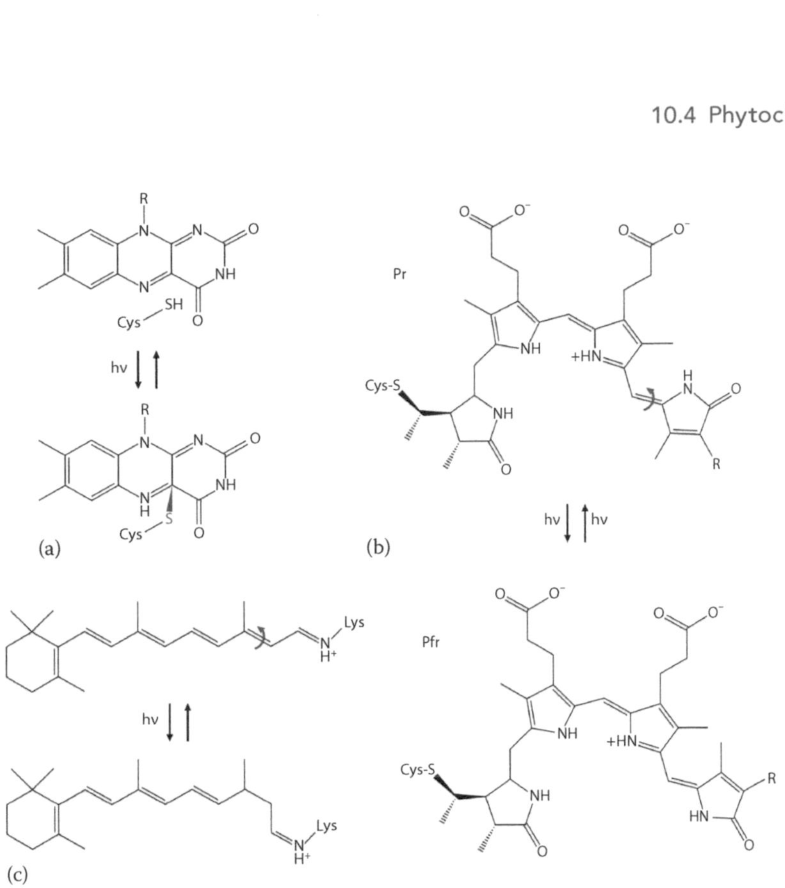

Figure 10.3 Chromophore composition and photochemistry of plant and algal photoreceptors. (a) LOV (light, oxygen, voltage). (b) PHY. (c) Rhodopsin.

this information to monitor spectral qualities such as sunset and sunrise or as an indicator of shading [43]. PHYs also occur in fungi and bacteria (known as bacteriophytochromes). Fungal and bacterial PHYs bind biliverdin whereas plant PHYs bind phycocyanobilin (PCB) or phytochromobilin (PφB) as chromophores [41,42]. Cyanobacterial PHYs also bind PCB, PφB, as well as other linear tetrapyrrole chromophores.

Like phototropin, the PHY protein can be divided into two segments: an N-terminal photosensory input domain and a C-terminal histidine kinase-related domain that is traditionally regarded as the regulatory, dimerization, and signal output domain [43]. The photosensory region of plant and bacterial PHYs comprises three domains denoted PAS, GAF (cGMP phosphodiesterase/adenylyl cyclase/FhlA) and PHY, respectively (Figure 10.1b). PHY and GAF domains are closely similar to each other and resemble PAS domains in terms of architecture [41,42]. X-ray crystallography has established that the PAS and GAF domains of the photosensory core are knotted together [44]. The bilin chromophore is bound within a cleft of the GAF domain (Figure 10.2b) and is covalently attached to a conserved cysteine residue *via* the C3 side chain of the bilin A-ring (Figure 10.3b). While the bilin cofactor predominantly interacts with the GAF domain, it also forms contacts with the PHY and PAS domains, which are required for maximal photochemical reactivity [41]. This is not the case for some cyanobacterial PHYs that lack the PAS or PHY domain, or both [42].

The primary step involved in Pr–Pfr photoconversion involves isomerization of the bilin chromophore around the C15 = C16 double bond between the C and D rings, resulting in flipping of the D ring (Figure 10.3b). Isomerization is also proposed to occur *via* linkage between the A and B rings in some PHYs [45]. However, the protein structural changes required for signal propagation following PHY photoexcitation are only just beginning to be elucidated [42].

10.5 CHANNELRHODOPSIN

Although higher plant counterparts of rhodopsin have not been identified, rhodopsin-type proteins serve as photoreceptors in *Chlamydomonas* and other green flagellate algae. The channelrhodopsins (Figure 10.1c) are a subfamily of opsin proteins that function as light-gated cation channels controlling phototaxis [8,10]. *Chlamydomonas* contains two channelrhodopsins ChR1 and ChR2. Both appear to be nonselective cation channels that conduct H^+, Na^+, K^+, and Ca^{2+} [8–10]. Like rhodopsin, ChRs are seven-transmembrane proteins (Figure 10.2c) that contain an all-*trans*-retinal chromophore, an aldehyde derivative of vitamin A. The retinal chromophore is covalently linked to the protein *via* a protonated Schiff base (Figure 10.3c). ChR2 absorbs blue light maximally at 480 nm [46]. Photoexcitation induces a conformational change from all-*trans* to 13-*cis*-retinal (Figure 10.3c) instigating conformational changes in the transmembrane protein that ultimately lead to opening of the channel pore to at least 6 Å (Figure 10.4a).

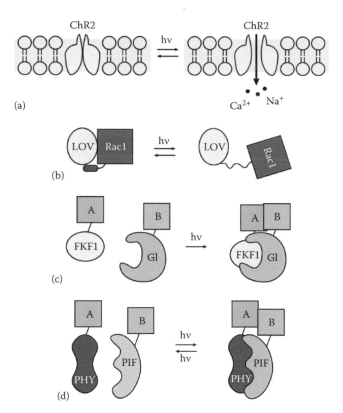

Figure 10.4 Optogenetic applications for plant and algal photoreceptors. (a) Channelrhodopsin expression to regulate light-driven membrane depolarization. (b) LOV2-Jα photoregulation of Rac1 activity. (c) Optogenetic dimerization of target proteins A and B by fusion to FKF1 and GIGANTEA (GI) interaction sites. (d) Reversible optogenetic dimerization of target proteins A and B by fusion to PHY and PIF interaction sites.

Within a few milliseconds following photoexcitation, retinal relaxes back to the all-*trans* form, closing the pore and thereby stopping the flow of ions. While most seven-transmembrane proteins are G-protein-coupled receptors that function to open ion channels indirectly *via* intracellular second messengers, channelrhodopsins directly mediate rapid changes in cellular depolarization [8]. This property of ChRs makes them extremely useful for bioengineering and neuroscience applications.

10.6 CHROMOPHORE-BINDING DOMAINS AS NOVEL FLUORESCENT PROBES

Fluorescent proteins (FPs) have revolutionized the imaging of protein dynamics within living cells [47]. The best example is green fluorescent protein (GFP) from the jellyfish *Aequorea victoria* [48]. The impact of GFP on molecular biology is now unquestionable since its pioneers were awarded the Nobel Prize in Chemistry in 2008. However, despite its wide application, the use of GFP and its derivatives has several limitations. Their use is restricted to aerobic systems because formation of the chromophore is dependent on oxygen. For certain applications the use of GFP is limited owing to its relatively large size (~25 kDa). Generation of smaller derivatives is unlikely as the 11-stranded β-barrel structure of the protein is intrinsic to its function [49]. Furthermore, the GFP fluorescence is also unstable at pH <5, which prompted the development of pH-resistant variants. As a result, there is considerable interest in the development of alternative protein-based fluorescent probes.

LOV domains have considerable potential as novel fluorescent probes with advantages over the current GFP technology [50]. The flavin chromophore gives the LOV domain a weak intrinsic green fluorescence when excited with UV-A/blue light [22]. Replacement of the active-site cysteine residue (Figure 10.3a) by site directed mutagenesis abolishes LOV-domain photochemistry and increases its fluorescence [26,51]. Further mutagenesis has been used to generate a monomeric variant with improved fluorescent properties. This variant designated iLOV is relatively small in size (~10 kDa) and has been shown to outperform GFP as a reporter for plant viruses, as well as conferring improved functionality when fused to viral proteins essential for movement [52]. Moreover, the dependence of iLOV on a cellular cofactor for fluorescence does not limit its use in subcellular targeting, nor impact its expression in human cells [52].

In addition to their smaller size, LOV-based FPs have other favorable attributes as fluorescent reporters [52]. LOV-mediated fluorescence is stable over a wide pH range [51] and may circumvent the pH sensitivity commonly associated with GFP-related FPs [49]. Similarly, LOV-mediated fluorescence does not depend on molecular oxygen [50,53], which is of particular interest with respect to the study of biological processes under conditions of hypoxia. Consequently, LOV-based FPs hold promise over GFP derivatives as real-time reporters for probing cell biomass and product formation in micro- and macrobioreactors [54] or for evaluating the differentiation of embryonic stem cells. Recently, a rationally designed LOV-based FP that efficiently generates singlet oxygen upon blue-light illumination can facilitate correlative light and electron microscopy of intact tissues [55].

Impairment of PHY photochemistry through mutagenesis can also render these proteins fluorescent [56,57]. The utility of bilin-binding proteins as fluorescent probes has the advantage that they can be used in the near-infrared region of the spectrum [56,58]. This property makes them especially suited for whole-body imaging since their

Applications

fluorescence can penetrate tissue, unlike that of GFP. *In vivo* optical imaging of underlying tissues in animals is more feasible between 650 and 900 nm because these wavelengths minimize the absorbance by hemoglobin, water, and lipids, as well as light scattering.

The efficacy of PHY–chromophore-binding domains as infrared FPs (IFPs) was demonstrated in a recent important development. The truncated version of a bacteriophytochrome from *Deinococcus radiodurans* containing the PAS and GAF domains was engineered for use as a fluorescent marker in mammalian tissues [57]. The biliverdin cofactor of PHY IFPs is endogenously produced in animal cells as a breakdown product of heme and can be further supplemented by direct administration. PHY-based IFPs were found to outperform the existing FR coral FP called mKate when delivered into the intact mouse liver by adenovirus [57]. Rational mutagenesis has already enhanced the fluorescent properties of IFPs [59]. Thus, their application offers viable alternatives as reporters of deep tissue imaging. Given their different genetic sources, PHY- and LOV-based FPs should also prove useful for double- and triple-labeling studies where expression of multiple GFP derivatives can result in gene silencing.

10.7 OPTOGENETICS AND SYNTHETIC PHOTOSWITCHES

Further application of photofunctional proteins has developed into an exciting new field coined optogenetics [60,61]. As the name suggests, optogenetics combines optical and genetic techniques to create molecular tools that enable light-mediated control or monitoring of cellular processes. Light is particularly attractive in this regard as it provides a rapid, noninvasive means to control enzyme activity. The hallmark of optogenetics has come from the application of light-activated cation channels to probe neural circuits in animals. This field is rapidly expanding to include construction of artificial gene products from natural photoreceptor sources that can be exploited to elegantly photomodulate target cell activities. Unlike FPs, these synthetic photoswitches are designed to noninvasively interfere with cellular processes with exquisite spatiotemporal control. The remainder of this chapter is devoted to summarizing some of these recent advances.

10.7.1 CHANNELRHODOPSIN APPLICATIONS IN NEUROBIOLOGY

The advent of optogenetics was firmly established through the discovery and application of natural light-sensitive cation channels such as ChR. The key to its success is that ChR can be genetically encoded and spatially expressed in any cell type. Another advantage is that ChR, like other algal and microbial opsins, is fully functional in mammals where its chromophore retinal is in plentiful supply. Since neural activity is dependent upon depolarization of the neuron cell membrane, electrodes can be used to elicit artificial stimulation. Optogenetics has taken this mode of neural intervention one step further. ChR and related proteins have been used as noninvasive substitutes to genetically and optically depolarize transfected neurons [62]. These proteins therefore provide valuable tools for controlling the activity of selected neural populations in living animals.

ChR2 from *Chlamydomonas* and the halorhodopsin NpHR from the archaeon *Natronomonas pharaonis* are two of the most commonly used tools in neurobiology [63].

ChR2 allows the influx of Ca^{2+} and Na^+ when illuminated by blue light (~470 nm) resulting in membrane depolarization and neuron activation (Figure 10.4a). ChR2 has been used successfully to artificially stimulate neural networks in a number of transgenic animals including the nematode *Caenorhabditis elegans*, *Drosophila*, zebra fish, and mice [62]. In contrast to ChR2, NpHR is a Cl^- channel that is activated by yellow light (~590 nm) [64]. The photoactivation of NpHR causes hyperpolarization of the cell membrane and can be used to inhibit or silence neural activity [63]. Fortunately, the wavelengths used to excite ChR2 and NpHR do not overlap. As a result, both proteins can be expressed simultaneously in the same cell to activate and silence neural activity upon exposure to blue and yellow light, respectively [63,65]. An important issue for optogenetics is how to precisely target the expression of photofunctional proteins to a specific type of cell. However, recent developments in viral delivery approaches are beginning to increase the ability to target cell types not based just on their genetic identity but also on their morphology and tissue topology [66].

Since their initial use as neurobiological tools, much effort has been expended to enhance the biophysical properties of ChR2 and NpHR. Through mutagenesis of existing channel candidates and from the identification of other light-driven channels, optical regulation at FR/infrared wavelengths and extension of this control across the entire visible spectrum are now possible [66]. This second wave of neurobiological tools is now being used to investigate various aspects of neuronal circuits and behavior [62], as well as cardiac function [67]. By fusing opsins to specific G-protein-coupled receptors, chimeric photosensors have been engineered that allow noninvasive manipulation of intracellular messengers such as cGMP, cAMP, and IP3 in individual cells [68] and animals [69]. The emerging repertoire of optogenetic probes now allows cell-type-specific and temporally precise control of various signaling processes.

10.7.2 LOV-BASED PHOTOSWITCHES

The photoswitchable property of LOV-containing proteins has gathered increasing attention as contributors to the optogenetic toolbox. The underlying strategy has been to fuse the chromophore-binding domains of these receptors to enzymatic or DNA-binding motifs to generate synthetic light-activated hybrid proteins [61]. Although introduced relatively recently, the number of optogenetic applications based on this approach is already impressive.

The seminal discovery that blue light induces unfolding of the C-terminal Jα helix from the LOV2 domain of *Avena sativa* phot1 [30] has provided a molecular framework for the design of several engineered photoreceptors. This small flavin-based photoswitch has been used to sterically regulate the DNA-binding activity of the *Escherichia coli* tryptophan repressor protein (TrpR) [70]. An initial variant designated LOV-TAP was found to target DNA sequences poorly in the dark and showed ~fivefold higher affinity following irradiation. However, rational mutagenesis of the LOV-TAP protein has improved the regulatory effect of light to ~64-fold [71]. A similar design approach was employed to create light-regulated variants of dihydrofolate reductase [72] and to control the activity of *Bacillus subtilis* lipase A [73]. Similarly, the LOV domain has been used to reprogram the signal specificity of a bacterial PAS-histidine kinase [74]. By replacing the PAS domains of *Bradyrhizobium japonicum* FixL, protein activity was reprogrammed from an oxygen sensor to the one that is light activated. Photoexcitation led to more than 1000-fold decrease in FixL kinase activity *in vitro*. These studies have recently been extended to demonstrate that it is possible to engineer a dual-sensing FixL variant to integrate both light and oxygen signals [75].

The demonstration that engineered LOV proteins can photomodulate cellular processes *in vivo* has recently been reported. Fusion of LOV2-Jα to the small GTPase Rac1 rendered the protein light-regulated [76]. The activity of the resulting protein, denoted photoactivatable Rac1 or PA-Rac1, is sterically inhibited by the LOV domain that obstructs the active site in darkness as indicated by its crystallographic structure. PA-Rac1 activity is then increased by blue-light irradiation (Figure 10.4b). Since Rac1 is a key protein-regulating actin cytoskeletal dynamics, blue-light irradiation can be used to remotely control the motility of fibroblasts expressing PA-Rac1. This elegant system has now been used to control movements in other cell types including neutrophils of developing zebra fish embryos [77] and *Drosophila* ovary cells [78].

In addition to using the LOV domain to engineer artificial light-activated enzymes, optogeneticists have also exploited the light-dependent nature of protein–protein interactions. In *Arabidopsis*, FKF1 interacts with the protein GIGANTEA (GI) in response to blue light to control the initiation of flowering [36]. The interaction domains of FKF1 and GI have therefore been used to render protein interactions light inducible by fusing these regions to targets of interest (Figure 10.4c). By using this approach, Rac1 can be recruited effectively to the membrane to induce actin polymerization and the formation of cell protrusions in response to blue light [79]. This methodology is likely to be generally applicable since FKF1-GI interactions have also been manipulated to mediate light-induced transcription when coupled, respectively, to the VP16 transactivation and GAL4 DNA-binding domains [79]. Additional so-called blue-light-induced dimerizers have now been developed from cryptochrome-signaling components [80].

10.7.3 EXPLOITING PHYTOCHROME PHOTOREVERSIBILITY

Optogenetic dimerization applications based on the photoreversible properties of PHY and its selective interaction with the basic helix–loop–helix protein known as PHY-interacting factor 3 (PIF3) have also been performed. The interaction between phyB and PIF3 is promoted by red light and negated by FR light [81]. Fusion of the protein regions responsible for this interaction to the DNA-binding domain and transactivation domain of GAL4 has created a red-light-inducible promoter system in yeast [82]. The utility of phyB–PIF interactions for such applications is particularly advantageous over LOV-based photoswitches as this process can be simply reversed by illumination with FR light.

The reversible interaction between phyB and PIFs has been adapted to photoregulate actin polymerization [83] and protein splicing [84] *in vitro*. Furthermore, phyB–PIF interactions have been employed to modulate the activity of target proteins *in vivo* (Figure 10.4d). PhyB and PIF6 can promote plasma membrane recruitment of the nucleotide exchange factors Tiam and intersectin upon exposure of fibroblasts to red light [85]. When localized to the membrane, Tiam and intersectin activate their GTPase effectors Rac1 and Cdc42, respectively, promoting the formation of cell protrusions. Thus, the motility of the fibroblasts expressing PhyB and PIF6 can be controlled by light in a spatiotemporal manner, similar to the observations with PA-Rac1 [76]. One disadvantage compared to the PA-Rac1 system, however, is that cofactors of higher plant PHY are typically required to be administered or synthesized *in vivo* through genetic manipulation.

Work with bacteriophytochromes has demonstrated that the PHY sensor can be adapted to create artificial red-light-sensitive photoswitches. When the sensor region from the cyanobacterial PHY Cph1 is fused to the histidine kinase portion of *E. coli* EnvZ, the resulting engineered variant is placed under red-light control to photoregulate target gene expression instead of being osmotically sensitive [86]. The growing repertoire of different optically sensitive forms of cyanobacterial PHYs [87,88] is now beginning to provide greater versatility with regard to the spectral qualities that can be used to activate and silence the activity of such reversible photoswitches [89].

10.8 CONCLUSIONS AND FUTURE PERSPECTIVES

The increasing number of new optogenetic tools created from plant and algal photoreceptors clearly shows their value in the development of new engineering strategies for biotechnological and biomedical applications. Such photoreceptor sources are not only limited to plants and algae but are also prevalent in bacteria and fungi, thereby enhancing the diversity that can be used for these design principles. For instance, the LOV sensor motif is ubiquitously found in bacteria where it is coupled to a wide variety of effector domains [90], which by themselves should provide useful scaffolds for the design of light-regulated enzymatic activities. Other photosensory motifs, in addition to those outlined here, are now being harnessed for optogenetic applications. Notable examples include the blue-light-utilizing FAD (BLUF) domain and the bacterial PAS-prototype protein photoactive yellow protein (PYP). Recent studies involving PYP, which belongs to the xanthopsin class of photoreceptors [1], already demonstrates its potential for regulating the activity of DNA-binding proteins [91]. BLUF domains bind flavin adenine dinucleotide and regulate the activity of effectors including adenylate cyclases [11]. Photoactivatable adenylate cyclases (PACs) offer considerable utility as general tools for controlling intracellular cAMP levels in animal cells in response to blue-light irradiation [92]. Small bacterial PACs show great promise in this regard [93] with the added benefit that they can be modified to control intracellular cGMP levels [94]. With the field of optogenetics now firmly established, a bright future lies ahead in both generating and maximizing the utility of fluorescent biomarkers and synthetic photoswitches derived from natural photoreceptors.

REFERENCES

1. Moglich, A., X. Yang, R. A. Ayers, and K. Moffat. 2010. Structure and function of plant photoreceptors. *Annu Rev Plant Biol* 61:21–47.
2. Kasahara, M., M. Torii, A. Fujita, and K. Tainaka. 2010. FMN binding and photochemical properties of plant putative photoreceptors containing two LOV domains, LOV/LOV proteins. *J Biol Chem* 285:34765–34772.
3. Nozue, K., T. Kanegae, T. Imaizumi, S. Fukuda, H. Okamoto, K. C. Yeh, J. C. Lagarias, and M. Wada. 1998. A phytochrome from the fern *Adiantum* with features of the putative photoreceptor NPH1. *Proc Natl Acad Sci USA* 95:15826–15830.
4. Kanegae, T., E. Hayashida, C. Kuramoto, and M. Wada. 2006. A single chromoprotein with triple chromophores acts as both a phytochrome and a phototropin. *Proc Natl Acad Sci USA* 103:17997–18001.
5. Jenkins, G. I. 2009. Signal transduction in responses to UV-B radiation. *Annu Rev Plant Biol* 60:407–431.

6. Folta, K. M. and S. A. Maruhnich. 2007. Green light: A signal to slow down or stop. *J Exp Bot* 58:3099–3111.

7. Rizzini, L., Favory, J.-J., Cloix, C., Faggionato, D., O'Hara, A., Kaiserli, E., Baumeister, R. et al. 2011. Perception of UV-B by the *Arabidopsis* UVR8 protein. *Science* 332:103–106.

8. Nagel, G., D. Ollig, M. Fuhrmann, S. Kateriya, A. M. Musti, E. Bamberg, and P. Hegemann. 2002. Channelrhodopsin-1: A light-gated proton channel in green algae. *Science* 296:2395–2398.

9. Nagel, G., T. Szellas, W. Huhn, S. Kateriya, N. Adeishvili, P. Berthold, D. Ollig, P. Hegemann, and E. Bamberg. 2003. Channelrhodopsin-2, a directly light-gated cation-selective membrane channel. *Proc Natl Acad Sci USA* 100:13940–13945.

10. Berthold, P., S. P. Tsunoda, O. P. Ernst, W. Mages, D. Gradmann, and P. Hegemann. 2008. Channelrhodopsin-1 initiates phototaxis and photophobic responses in *Chlamydomonas* by immediate light-induced depolarization. *Plant Cell* 20:1665–1677.

11. Iseki, M., S. Matsunaga, A. Murakami, K. Ohno, K. Shiga, K. Yoshida, M. Sugai, T. Takahashi, T. Hori, and M. Watanabe. 2002. A blue-light-activated adenylyl cyclase mediates photoavoidance in *Euglena gracilis*. *Nature* 415:1047–1051.

12. Christie, J. M. 2007. Phototropin blue-light receptors. *Annu Rev Plant Biol* 58:21–45.

13. Takemiya, A., S. Inoue, M. Doi, T. Kinoshita, and K. Shimazaki. 2005. Phototropins promote plant growth in response to blue light in low light environments. *Plant Cell* 17:1120–1127.

14. Sakai, T., T. Kagawa, M. Kasahara, T. E. Swartz, J. M. Christie, W. R. Briggs, M. Wada, and K. Okada. 2001. *Arabidopsis* nph1 and npl1: Blue light receptors that mediate both phototropism and chloroplast relocation. *Proc Natl Acad Sci USA* 98:6969–6974.

15. Kinoshita, T., M. Doi, N. Suetsugu, T. Kagawa, M. Wada, and K. Shimazaki. 2001. Phot1 and phot2 mediate blue light regulation of stomatal opening. *Nature* 414:656–660.

16. Kagawa, T., T. Sakai, N. Suetsugu, K. Oikawa, S. Ishiguro, T. Kato, S. Tabata, K. Okada, and M. Wada. 2001. *Arabidopsis* NPL1: A phototropin homolog controlling the chloroplast high-light avoidance response. *Science* 291:2138–2141.

17. Sakamoto, K. and W. R. Briggs. 2002. Cellular and subcellular localization of phototropin 1. *Plant Cell* 14:1723–1735.

18. Inoue, S., T. Kinoshita, A. Takemiya, M. Doi, and K. Shimazaki. 2008. Leaf positioning of *Arabidopsis* in response to blue light. *Mol Plant* 1:15–26.

19. Huang, K. and C. F. Beck. 2003. Phototropin is the blue-light receptor that controls multiple steps in the sexual life cycle of the green alga *Chlamydomonas reinhardtii*. *Proc Natl Acad Sci USA* 100:6269–6274.

20. Onodera, A., S. G. Kong, M. Doi, K. Shimazaki, J. Christie, N. Mochizuki, and A. Nagatani. 2005. Phototropin from *Chlamydomonas reinhardtii* is functional in *Arabidopsis thaliana*. *Plant Cell Physiol* 46:367–374.

21. Christie, J. M., P. Reymond, G. K. Powell, P. Bernasconi, A. A. Raibekas, E. Liscum, and W. R. Briggs. 1998. *Arabidopsis* NPH1: A flavoprotein with the properties of a photoreceptor for phototropism. *Science* 282:1698–1701.

22. Christie, J. M., M. Salomon, K. Nozue, M. Wada, and W. R. Briggs. 1999. LOV (light, oxygen, or voltage) domains of the blue-light photoreceptor phototropin (nph1): Binding sites for the chromophore flavin mononucleotide. *Proc Natl Acad Sci USA* 96:8779–8783.

23. Taylor, B. L. and I. B. Zhulin. 1999. PAS domains: Internal sensors of oxygen, redox potential, and light. *Microbiol Mol Biol Rev* 63:479–506.

24. Huala, E., P. W. Oeller, E. Liscum, I. S. Han, E. Larsen, and W. R. Briggs. 1997. *Arabidopsis* NPH1: A protein kinase with a putative redox-sensing domain. *Science* 278:2120–2123.

25. Crosson, S. and K. Moffat. 2001. Structure of a flavin-binding plant photoreceptor domain: Insights into light-mediated signal transduction. *Proc Natl Acad Sci USA* 98:2995–3000.

26. Salomon, M., J. M. Christie, E. Knieb, U. Lempert, and W. R. Briggs. 2000. Photochemical and mutational analysis of the FMN-binding domains of the plant blue light receptor, phototropin. *Biochemistry* 39:9401–9410.

27. Cho, H. Y., T. S. Tseng, E. Kaiserli, S. Sullivan, J. M. Christie, and W. R. Briggs. 2007. Physiological roles of the light, oxygen, or voltage domains of phototropin 1 and phototropin 2 in *Arabidopsis*. *Plant Physiol* 143:517–529.

28. Christie, J. M., T. E. Swartz, R. A. Bogomolni, and W. R. Briggs. 2002. Phototropin LOV domains exhibit distinct roles in regulating photoreceptor function. *Plant J* 32:205–219.

29. Kaiserli, E., S. Sullivan, M. A. Jones, K. A. Feeney, and J. M. Christie. 2009. Domain swapping to assess the mechanistic basis of *Arabidopsis* phototropin 1 receptor kinase activation and endocytosis by blue light. *Plant Cell* 21:3226–3244.

30. Harper, S. M., L. C. Neil, and K. H. Gardner. 2003. Structural basis of a phototropin light switch. *Science* 301:1541–1544.

31. Harper, S. M., J. M. Christie, and K. H. Gardner. 2004. Disruption of the LOV-Jalpha helix interaction activates phototropin kinase activity. *Biochemistry* 43:16184–16192.

32. Iwata, T., D. Nozaki, S. Tokutomi, T. Kagawa, M. Wada, and H. Kandori. 2003. Light-induced structural changes in the LOV2 domain of *Adiantum* phytochrome3 studied by low-temperature FTIR and UV-visible spectroscopy. *Biochemistry* 42:8183–8191.

33. Jones, M. A., K. A. Feeney, S. M. Kelly, and J. M. Christie. 2007. Mutational analysis of phototropin 1 provides insights into the mechanism underlying LOV2 signal transmission. *J Biol Chem* 282:6405–6414.

34. Nozaki, D., T. Iwata, T. Ishikawa, T. Todo, S. Tokutomi, and H. Kandori. 2004. Role of Gln1029 in the photoactivation processes of the LOV2 domain in *Adiantum* phytochrome3. *Biochemistry* 43:8373–8379.

35. Kim, W. Y., S. Fujiwara, S. S. Suh, J. Kim, Y. Kim, L. Han, K. David, J. Putterill, H. G. Nam, and D. E. Somers. 2007. ZEITLUPE is a circadian photoreceptor stabilized by GIGANTEA in blue light. *Nature* 449:356–360.

36. Sawa, M., D. A. Nusinow, S. A. Kay, and T. Imaizumi. 2007. FKF1 and GIGANTEA complex formation is required for day-length measurement in *Arabidopsis*. *Science* 318:261–265.

37. Smalle, J. and R. D. Vierstra. 2004. The ubiquitin 26S proteasome proteolytic pathway. *Annu Rev Plant Biol* 55:555–590.

38. Imaizumi, T., T. F. Schultz, F. G. Harmon, L. A. Ho, and S. A. Kay. 2005. FKF1 F-box protein mediates cyclic degradation of a repressor of CONSTANS in *Arabidopsis*. *Science* 309:293–297.

39. Nakasako, M., D. Matsuoka, K. Zikihara, and S. Tokutomi. 2005. Quaternary structure of LOV-domain containing polypeptide of *Arabidopsis* FKF1 protein. *FEBS Lett* 579:1067–1071.

40. Chen, M., J. Chory, and C. Fankhauser. 2004. Light signal transduction in higher plants. *Annu Rev Genet* 38:87–117.

41. Rockwell, N. C., Y. S. Su, and J. C. Lagarias. 2006. Phytochrome structure and signaling mechanisms. *Annu Rev Plant Biol* 57:837–858.

42. Rockwell, N. C. and J. C. Lagarias. 2010. A brief history of phytochromes. *ChemPhysChem* 11:1172–1180.

43. Franklin, K. A. and P. H. Quail. 2010. Phytochrome functions in *Arabidopsis* development. *J Exp Bot* 61:11–24.

44. Wagner, J. R., J. S. Brunzelle, K. T. Forest, and R. D. Vierstra. 2005. A light-sensing knot revealed by the structure of the chromophore-binding domain of phytochrome. *Nature* 438:325–331.

45. Ulijasz, A. T., G. Cornilescu, C. C. Cornilescu, J. Zhang, M. Rivera, J. L. Markley, and R. D. Vierstra. 2010. Structural basis for the photoconversion of a phytochrome to the activated Pfr form. *Nature* 463:250–254.

46. Bamann, C., T. Kirsch, G. Nagel, and E. Bamberg. 2008. Spectral characteristics of the photocycle of channelrhodopsin-2 and its implication for channel function. *J Mol Biol* 375:686–694.

47. Tsien, R. Y. 2009. Constructing and exploiting the fluorescent protein paintbox (Nobel Lecture). *Angew Chem Int Ed Engl* 48:5612–5626.
48. Shaner, N. C., G. H. Patterson, and M. W. Davidson. 2007. Advances in fluorescent protein technology. *J Cell Sci* 120:4247–4260.
49. Tsien, R. Y. 1998. The green fluorescent protein. *Annu Rev Biochem* 67:509–544.
50. Drepper, T., T. Eggert, F. Circolone, A. Heck, U. Krauss, J. K. Guterl, M. Wendorff, A. Losi, W. Gartner, and K. E. Jaeger. 2007. Reporter proteins for in vivo fluorescence without oxygen. *Nat Biotechnol* 25:443–445.
51. Swartz, T. E., S. B. Corchnoy, J. M. Christie, J. W. Lewis, I. Szundi, W. R. Briggs, and R. A. Bogomolni. 2001. The photocycle of a flavin-binding domain of the blue light photoreceptor phototropin. *J Biol Chem* 276:36493–36500.
52. Chapman, S., C. Faulkner, E. Kaiserli, C. Garcia-Mata, E. I. Savenkov, A. G. Roberts, K. J. Oparka, and J. M. Christie. 2008. The photoreversible fluorescent protein iLOV outperforms GFP as a reporter of plant virus infection. *Proc Natl Acad Sci USA* 105:20038–20043.
53. Tielker, D., I. Eichhof, K. E. Jaeger, and J. F. Ernst. 2009. Flavin mononucleotide-based fluorescent protein as an oxygen-independent reporter in *Candida albicans* and *Saccharomyces cerevisiae*. *Eukaryot Cell* 8:913–915.
54. Drepper, T., R. Huber, A. Heck, F. Circolone, A. K. Hillmer, J. Buchs, and K. E. Jaeger. 2010. Flavin mononucleotide-based fluorescent reporter proteins outperform green fluorescent protein-like proteins as quantitative in vivo real-time reporters. *Appl Environ Microbiol* 76:5990–5994.
55. Shu, X., Lev-Ram, V., Deerinck, T. J., Qi, Y., Ramko, E. B., Davidson, M. W., Jin, Y., Ellisman, M. H., and Tsien, R. Y. 2011. A genetically encoded tag for correlated light and electron microscopy of intact cells, tissues, and organisms. *PLOS Biol.* 9:e1001041.
56. Fischer, A. J. and J. C. Lagarias. 2004. Harnessing phytochrome's glowing potential. *Proc Natl Acad Sci USA* 101:17334–17339.
57. Shu, X., A. Royant, M. Z. Lin, T. A. Aguilera, V. Lev-Ram, P. A. Steinbach, and R. Y. Tsien. 2009. Mammalian expression of infrared fluorescent proteins engineered from a bacterial phytochrome. *Science* 324:804–807.
58. Wagner, J. R., J. Zhang, D. von Stetten, M. Gunther, D. H. Murgida, M. A. Mroginski, J. M. Walker, K. T. Forest, P. Hildebrandt, and R. D. Vierstra. 2008. Mutational analysis of *Deinococcus radiodurans* bacteriophytochrome reveals key amino acids necessary for the photochromicity and proton exchange cycle of phytochromes. *J Biol Chem* 283:12212–12226.
59. Toh, K. C., E. A. Stojkovic, I. H. van Stokkum, K. Moffat, and J. T. Kennis. 2010. Proton-transfer and hydrogen-bond interactions determine fluorescence quantum yield and photochemical efficiency of bacteriophytochrome. *Proc Natl Acad Sci USA* 107:9170–9175.
60. Deisseroth, K., G. Feng, A. K. Majewska, G. Miesenbock, A. Ting, and M. J. Schnitzer. 2006. Next-generation optical technologies for illuminating genetically targeted brain circuits. *J Neurosci* 26:10380–10386.
61. Moglich, A. and K. Moffat. 2010. Engineered photoreceptors as novel optogenetic tools. *Photochem Photobiol Sci* 9:1286–1300.
62. Fiala, A., A. Suska, and O. M. Schluter. 2010. Optogenetic approaches in neuroscience. *Curr Biol* 20:R897–903.
63. Zhang, W., W. Ge, and Z. Wang. 2007. A toolbox for light control of *Drosophila* behaviors through channelrhodopsin 2-mediated photoactivation of targeted neurons. *Eur J Neurosci* 26:2405–2416.
64. Duschl, A., J. K. Lanyi, and L. Zimanyi. 1990. Properties and photochemistry of a halorhodopsin from the haloalkalophile, *Natronobacterium pharaonis*. *J Biol Chem* 265:1261–1267.
65. Han, X. and E. S. Boyden. 2007. Multiple-color optical activation, silencing, and desynchronization of neural activity, with single-spike temporal resolution. *PLoS ONE* 2:e299.

Applications

66. Gradinaru, V., F. Zhang, C. Ramakrishnan, J. Mattis, R. Prakash, I. Diester, I. Goshen, K. R. Thompson, and K. Deisseroth. 2010. Molecular and cellular approaches for diversifying and extending optogenetics. *Cell* 141:154–165.

67. Arrenberg, A. B., D. Y. Stainier, H. Baier, and J. Huisken. 2010. Optogenetic control of cardiac function. *Science* 330:971–974.

68. Kim, J. M., J. Hwa, P. Garriga, P. J. Reeves, U. L. RajBhandary, and H. G. Khorana. 2005. Light-driven activation of beta 2-adrenergic receptor signaling by a chimeric rhodopsin containing the beta 2-adrenergic receptor cytoplasmic loops. *Biochemistry* 44:2284–2292.

69. Airan, R. D., K. R. Thompson, L. E. Fenno, H. Bernstein, and K. Deisseroth. 2009. Temporally precise in vivo control of intracellular signalling. *Nature* 458:1025–1029.

70. Strickland, D., K. Moffat, and T. R. Sosnick. 2008. Light-activated DNA binding in a designed allosteric protein. *Proc Natl Acad Sci USA* 105:10709–10714.

71. Strickland, D., X. Yao, G. Gawlak, M. K. Rosen, K. H. Gardner, and T. R. Sosnick. 2010. Rationally improving LOV domain-based photoswitches. *Nat Methods* 7:623–626.

72. Lee, J., M. Natarajan, V. C. Nashine, M. Socolich, T. Vo, W. P. Russ, S. J. Benkovic, and R. Ranganathan. 2008. Surface sites for engineering allosteric control in proteins. *Science* 322:438–442.

73. Krauss, U., J. Lee, S. J. Benkovic, and K.-E. Jaeger. 2010. LOVely enzymes: Towards engineering light-controllable biocatalysts. *Microb Biotechnol* 3:15–23.

74. Moglich, A., R. A. Ayers, and K. Moffat. 2009. Design and signaling mechanism of light-regulated histidine kinases. *J Mol Biol* 385:1433–1444.

75. Moglich, A., R. A. Ayers, and K. Moffat. 2010. Addition at the molecular level: Signal integration in designed Per-ARNT-Sim receptor proteins. *J Mol Biol* 400:477–486.

76. Wu, Y. I., D. Frey, O. I. Lungu, A. Jaehrig, I. Schlichting, B. Kuhlman, and K. M. Hahn. 2009. A genetically encoded photoactivatable Rac controls the motility of living cells. *Nature* 461:104–108.

77. Yoo, S. K., Q. Deng, P. J. Cavnar, Y. I. Wu, K. M. Hahn, and A. Huttenlocher. 2010. Differential regulation of protrusion and polarity by PI3K during neutrophil motility in live zebrafish. *Dev Cell* 18:226–236.

78. Wang, X., L. He, Y. I. Wu, K. M. Hahn, and D. J. Montell. 2010. Light-mediated activation reveals a key role for Rac in collective guidance of cell movement in vivo. *Nat Cell Biol* 12:591–597.

79. Yazawa, M., A. M. Sadaghiani, B. Hsueh, and R. E. Dolmetsch. 2009. Induction of protein-protein interactions in live cells using light. *Nat Biotechnol* 27:941–945.

80. Kennedy, M. J., R. M. Hughes, L. A. Peteya, J. W. Schwartz, M. D. Ehlers, and C. L. Tucker. 2010. Rapid blue-light-mediated induction of protein interactions in living cells. *Nat Methods* 7:973–975.

81. Ni, M., J. M. Tepperman, and P. H. Quail. 1999. Binding of phytochrome B to its nuclear signalling partner PIF3 is reversibly induced by light. *Nature* 400:781–784.

82. Shimizu-Sato, S., E. Huq, J. M. Tepperman, and P. H. Quail. 2002. A light-switchable gene promoter system. *Nat Biotechnol* 20:1041–1044.

83. Leung, D. W., C. Otomo, J. Chory, and M. K. Rosen. 2008. Genetically encoded photoswitching of actin assembly through the Cdc42-WASP-Arp2/3 complex pathway. *Proc Natl Acad Sci USA* 105:12797–12802.

84. Tyszkiewicz, A. B. and T. W. Muir. 2008. Activation of protein splicing with light in yeast. *Nat Methods* 5:303–305.

85. Levskaya, A., O. D. Weiner, W. A. Lim, and C. A. Voigt. 2009. Spatiotemporal control of cell signalling using a light-switchable protein interaction. *Nature* 461:997–1001.

86. Levskaya, A., A. A. Chevalier, J. J. Tabor, Z. B. Simpson, L. A. Lavery, M. Levy, E. A. Davidson, A. Scouras, A. D. Ellington, E. M. Marcotte, and C. A. Voigt. 2005. Synthetic biology: Engineering *Escherichia coli* to see light. *Nature* 438:441–442.

Applications

87. Rockwell, N. C., S. L. Njuguna, L. Roberts, E. Castillo, V. L. Parson, S. Dwojak, J. C. Lagarias, and S. C. Spiller. 2008. A second conserved GAF domain cysteine is required for the blue/green photoreversibility of cyanobacteriochrome Tlr0924 from *Thermosynechococcus elongatus*. *Biochemistry* 47:7304–7316.

88. Rockwell, N. C., L. Shang, S. S. Martin, and J. C. Lagarias. 2009. Distinct classes of red/far-red photochemistry within the phytochrome superfamily. *Proc Natl Acad Sci USA* 106:6123–6127.

89. Zhang, J., X. J. Wu, Z. B. Wang, Y. Chen, X. Wang, M. Zhou, H. Scheer, and K. H. Zhao. 2010. Fused-gene approach to photoswitchable and fluorescent biliproteins. *Angew Chem Int Ed Engl* 49:5456–5458.

90. Pathak, G. P., A. Ehrenreich, A. Losi, W. R. Streit, and W. Gartner. 2009. Novel blue light-sensitive proteins from a metagenomic approach. *Environ Microbiol* 11:2388–2399.

91. Morgan, S. A. and G. A. Woolley. 2010. A photoswitchable DNA-binding protein based on a truncated GCN4-photoactive yellow protein chimera. *Photochem Photobiol Sci* 9:1320–1326.

92. Schroder-Lang, S., M. Schwarzel, R. Seifert, T. Strunker, S. Kateriya, J. Looser, M. Watanabe, U. B. Kaupp, P. Hegemann, and G. Nagel. 2007. Fast manipulation of cellular cAMP level by light in vivo. *Nat Methods* 4:39–42.

93. Stierl, M., P. Stumpf, D. Udwari, R. Gueta, R. Hagedorn, A. Losi, W. Gartner, L. Petereit, M. Efetova, M. Schwarzel, T. G. Oertner, G. Nagel, and P. Hegemann. 2010. Light-modulation of cellular cAMP by a small bacterial photoactivated adenylyl cyclase, bPAC, of the soil bacterium beggiatoa. *J Biol Chem*. 282:1181–1188.

94. Ryu, M. H., O. V. Moskvin, J. Siltberg-Liberles, and M. Gomelsky. 2010. Natural and engineered photoactivated nucleotidyl cyclases for optogenetic applications. *J Biol Chem* 285(53):41501–41508.

95. Halavaty, A. S. and K. Moffat. 2007. N- and C-terminal flanking regions modulate light-induced signal transduction in the LOV2 domain of the blue light sensor phototropin 1 from *Avena sativa*. *Biochemistry* 46:14001–14009.

96. Yang, X., J. Kuk, and K. Moffat. 2008. Crystal structure of *Pseudomonas aeruginosa* bacteriophytochrome: Photoconversion and signal transduction. *Proc Natl Acad Sci USA* 105:14715–14720.

97. Luecke, H., B. Schobert, H. T. Richter, J. P. Cartailler, and J. K. Lanyi. 1999. Structural changes in bacteriorhodopsin during ion transport at 2 angstrom resolution. *Science* 286:255–261.

Fluorescent proteins for FRET: Monitoring protein interactions in living cells

Richard N. Day
Indiana University School of Medicine

Contents

11.1 INTRODUCTION

The cloning of the green fluorescent protein (GFP) from the jellyfish *Aequorea victoria* and its subsequent production in other organisms sparked a revolution in the study of cellular processes (Prasher et al. 1992; Chalfie et al. 1994; Inouye and Tsuji 1994). For the first time, the genetically encoded fluorescent protein (FP) could be used to directly label any protein inside the living cell. When combined with the extraordinary advances in microscope and camera technologies over the past decade, the FPs opened entirely new avenues of study in cell biology, medicine, and physiology. In the years since its cloning, the sequence encoding the *Aequorea* GFP has been engineered to yield new FPs emitting light from the blue to yellowish-green range of the visible spectrum (Tsien 1998; Cubitt et al. 1999; Nagai et al. 2002; Rizzo et al. 2004; Ai et al. 2007; Day and Davidson, 2009). Furthermore, many marine organisms produce FPs that are homologous to the *Aequorea* GFP (Labas et al. 2002; Matz et al. 2002; Shagin et al. 2004), and some of these GFP-like proteins have extended the fluorescence palette into the deep red spectrum (Matz et al. 1999; Karasawa et al. 2004; Shcherbo et al. 2007; reviewed in Day and Davidson 2009).

These new FPs expand the repertoire of applications from multicolor imaging of protein colocalization and behavior inside living cells to the detection of changes in intracellular activities. However, it is their use for Förster (fluorescence) resonance energy transfer (FRET) microscopy of living cells that has generated the most interest in these probes (Tsien 1998; Lippincott-Schwartz et al. 2001; Zhang et al. 2002; Giepmans et al. 2006; Shaner et al. 2007). This chapter provides an overview of the genetically encoded FPs currently used for FRET microscopy. The objective is to relate the important features and photophysical properties of the different FPs that make them useful as probes for FRET measurements in biological systems. The intent is not to provide a comprehensive listing of all the FPs that are currently available for FRET applications, but rather to highlight a few exceptional probes and discuss their application for FRET studies.

11.2 WHY USE FRET IMAGING?

The optical resolution of the conventional light microscope is limited by the wavelength of the illuminating light source (Inoué 2006). When using blue light illumination, objects labeled with fluorescent probes inside a specimen must be separated by 200 nm or more to be resolved as distinct objects. Consequently, despite the appearance of colocalization in multicolor fluorescent imaging, considerable distances may actually separate the labeled objects. Even with the recent advances in optical techniques that offer severalfold improvement in optical resolution (see Chapter 12), higher resolution is still necessary to detect the interactions of proteins inside living cells. FRET microscopy is one method that can achieve the ångstrom (Å)-scale resolution that is necessary to monitor the interactions of proteins inside living cells. What is more, changes in FRET signals can also report subtle conformation changes in probes that are designed to monitor intracellular activities—probes that are called biosensors (described in Section 11.10).

FRET is the process through which energy absorbed by one fluorophore (the "donor") is transferred directly to another nearby molecule (the "acceptor") via a nonradiative pathway. This transfer of energy depletes the donor's excited-state energy, quenching its fluorescence emission, while causing increased emission from a fluorescent acceptor (Förster 1965; Stryer 1978; Lakowicz 2006). There are three basic requirements for the efficient transfer of energy between the donor and the acceptor fluorophores. First, because energy transfer results from electromagnetic dipolar interactions, the efficiency of energy transfer (E_{FRET}) varies as the inverse of the sixth power of the distance (r) that separates the fluorophores. This sixth-power dependence is described by the following equation:

$$E_{FRET} = \frac{R_0^6}{R_0^6 + r^6}$$

(11.1)

where R_0 is the Förster distance at which the efficiency of energy transfer is 50%. The relationship of E_{FRET} to the distance separating the fluorophores is illustrated in

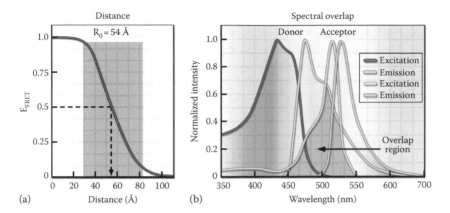

Figure 11.1 The distance dependence and spectral overlap requirements for efficient energy transfer. (a) The efficiency of energy transfer, E_{FRET}, was determined using Equation 11.1 and is plotted as a function of the separation distance (r). (b) The excitation and emission spectra for typical donor and acceptor FPs used for FRET microscopy is shown, with the shaded region indicating the spectral overlap (J_λ) between the donor emission and acceptor excitation.

Figure 11.1a. Because E_{FRET} varies as the inverse of the sixth power of the separation distance between the fluorophores, the efficiency of energy transfer falls off sharply over the range of 0.5–1.5 R_0 (shaded area, Figure 11.1a). This is why energy transfer between the FPs is limited to distances of less than about 80 Å.

The second requirement for efficient transfer of energy is that the electromagnetic dipoles of the donor and acceptor are in a favorable alignment, which is described by the orientation factor, κ^2. The orientation factor κ^2 relates the angular dependence of the dipolar interaction and, depending on the relative orientation of the donor and acceptor, can range in value from 0 to 4 (Gryczynski et al. 2005). However, for many biological applications, where proteins labeled with the donor and acceptor fluorophores freely diffuse within cellular compartments and adopt a variety of conformations, the orientations of the FP tags can be expected to randomize. Under these conditions, κ^2 is often assumed to be 0.667, which reflects the random orientations of the probes. However, it is important to note that because of their large size, the FPs rotate slowly relative to their fluorescence lifetime (typically 3–5 ns), so dynamic isotropic conditions do not apply. Since there will be a negligible change in the dipole orientation during the excited state lifetime, little averaging will occur. The Vogel laboratory used Monte Carlo simulations to demonstrate that the slow rotation of the FPs has the potential to give rise to a bimodal distribution of efficiencies in FRET experiments (Vogel et al. 2012). The third requirement for the efficient transfer of energy is that the donor and acceptor fluorophores share a strong spectral overlap. The efficient transfer of energy requires that the donor emission spectrum strongly overlap with the absorption spectrum of the acceptor (illustrated in Figure 11.1b). The Förster distance (R_0) for a particular pair of fluorophores depends on their spectral overlap, and this is described by the following equation:

$$R_0 = 0.211 \, [(\kappa^2)(n^{-4}) \, (QY_D) \, (J_\lambda)]^{1/6} \tag{11.2}$$

where
  n is the refractive index (1.4 in aqueous media)
  κ^2 is the orientation factor (0.667 for random orientations)
  J_λ is the spectral overlap integral (Lakowicz 2006)

Thus, the R_0 for any pair of fluorophores used for FRET experiments will depend on the quantum yield of the donor (QY_D) and the J_λ, their spectral overlap integral. For FRET imaging, it is desirable to select a donor fluorophore with a high quantum yield that shares a strong spectral overlap with the acceptor described in more detail in Section 11.4. When these basic requirements for energy transfer are met, the quantification of FRET by microscopy can provide measurements of the spatial relationship between the fluorophores labeling protein inside living cells.

11.3 HOW DID THE FLUORESCENT PROTEINS REVOLUTIONIZE FRET IMAGING?

With the cloning of the *Aequorea* GFP by Douglas Prasher (Prasher et al. 1992), and the demonstration that it still glowed when produced in other organisms (Chalfie et al. 1994), Roger Tsien immediately recognized the astonishing potential of GFP—if only the obvious flaws of the native jellyfish protein could be overcome. From the outset the goal was to develop genetically encoded probes as noninvasive reporters of biological events, and Roger Tsien was particularly interested in developing a genetically encoded sensor for cyclic AMP (cAMP)-mediated activities. In an earlier collaboration with Susan Tayor, Tsien had developed a FRET-based sensor of cAMP activation that exploited the dissociation of the catalytic and regulatory subunits of its cellular binding protein, the cAMP-dependent protein kinase A (PKA). In a technically challenging series of experiments, Tsien covalently labeled the purified subunits of PKA with fluorescein and rhodamine dyes, such that the subunits would still assemble into the holoenzyme, allowing FRET to occur between the fluorescent probes. The labeled subunits were injected into smooth muscle and fibroblast cell lines, and ratio imaging was used to detect the loss of the FRET signal when cAMP bound to PKA, causing its subunits to dissociate and drift apart (Adams et al. 1991).

The technical challenges and limitations of the covalent labeling and cellular injection approach led Roger Tsien to look for genetically encoded fluorescent probes—and the cloning of *Aequorea* GFP was just what was needed. However, to make a useful FRET probe based on GFP required the improvement of its spectral characteristics, as well as the generation of new colors (Zhang 2009; Tsien 2010). Thus, the Tsien laboratory embarked on a series of mutagenesis studies to determine whether different amino acid substitutions might be used to "fine-tune" the spectral characteristics of *Aequorea* GFP. As we will see, this approach has yielded many variations of the *Aequorea* protein with fluorescence emission ranging from the blue to the yellow regions of the visible spectrum (Tsien 1998; Shaner et al. 2005).

11.4 IMPROVEMENT OF *AEQUOREA* GFP

The GFP from the jellyfish *Aequorea* evolved as a FRET acceptor, absorbing the excited-state energy from the chemiluminescent blue light-emitting protein aequorin and emitting longer wavelength green light (Shimomura et al. 1962; Morise et al. 1974). The chromophore responsible for the green light emission is encoded by the primary amino acid sequence of GFP and forms spontaneously without the requirement for cofactors (other than molecular oxygen) or external enzyme components, through a self-catalyzed protein-folding mechanism (for more on this subject, see the earlier chapters). The chromophore forms through the cyclization of the adjacent Ser^{65}–Tyr^{66}–Gly^{67} residues (the number denotes the position of the *Aequorea* GFP sequence; see Cody et al. 1993). The crystal structure for GFP was solved in 1996 (Ormö et al. 1996; Yang et al. 1996), revealing its iconic interwoven eleven-stranded "β-barrel" structure. As the protein folds, the tripeptide chromophore is positioned at the core of the β-barrel, driving the cyclization and dehydration reactions necessary to form the mature chromophore (see Chapter 1). The remarkable cylindrical geometry of GFP is conserved in all the FPs yet discovered and appears ideally suited for providing a protected environment for the chromophore.

The *Aequorea* GFP has a complex absorption spectrum, with maximal excitation occurring at 397 nm and a minor secondary excitation peak at 476 nm (Figure 11.2a). Protonation of the chromophore Tyr[66] residue is responsible for the major absorption peak at 397 nm, while a charged intermediate state in some of the chromophores is responsible for the secondary absorption peak at 476 nm (Chattoraj et al. 1996). Unfortunately, the complex absorption spectrum requiring near-ultraviolet (UV) excitation combined with the low quantum yield of *Aequorea* GFP limited its usefulness for cellular imaging applications. Early mutagenesis studies seeking to change the spectral characteristics of *Aequorea* GFP focused on substitutions of the chromophore residues. These studies demonstrated that mutations that change the first amino acid in the chromophore, Ser[65], yielded proteins with a single-peak excitation ranging from 471 to 489 nm (Heim et al. 1994; Cubitt et al. 1995). A particularly useful variant resulted from changing the Ser[65] to threonine (S65T), which stabilized the hydrogen-bonding network in the chromophore, producing a permanently ionized form of the chromophore absorbing at 489 nm (Brejc et al. 1997; Cubitt et al. 1999; see Figure 11.2a). Not only did GFP[S65T] overcome the problems associated with near-UV excitation, but it also improved the maturation efficiency of the protein and yielded much brighter fluorescence (Heim et al. 1995).

Additional modifications were introduced to improve maturation of the protein in mammalian cells at physiological temperatures. This included the incorporation of optimized codon usage to improve translation efficiency and the substitution of the phenylalanine position 64 with leucine (F64L), yielding what are commonly called the enhanced (E) FPs (Yang et al. 1996; Zolotukhin et al. 1996). Taken together, these mutations have dramatically improved the fluorescence signal obtained from GFP-fusion proteins produced in cells, allowing the detection of fewer than 10,000 GFP molecules (Patterson et al. 1997). More recently, enhanced green fluorescent protein (EGFP) was subjected to directed evolution approaches to select for a brighter variant that might prove a better FRET donor for the orange and red FPs. The resulting FP, called Clover, differs from EGFP at 10 residues, and is currently the brightest of the *Aequorea*-based FPs, although this was achieved at the expense of photostability (Table 11.1; Lam et al. 2012).

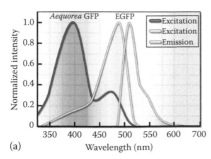
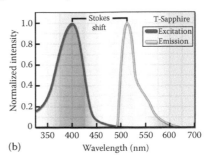

Figure 11.2 (a) The improved spectral characteristics of EGFP result from the S65T substitution. (b) The spectral characteristics of the long Stokes shift FP T-Sapphire. The spectral data are adapted from the Fluorescent Proteins Data Excel file available at http://home.earthlink.net/~pubspectra (see McNamara et al., 2006 for details).

Table 11.1 Properties of selected FPs and their use as FRET probes

FLUORESCENT PROTEIN	Ex (nm)	Em (nm)	EC ($\times 10^{-3}$) $M^{-1}cm^{-1}$	QY	RELATIVE BRIGHTNESS (% OF EGFP)[a]	USE AS FRET PROBE	REFERENCE
Aequorea-based FPs							
EBFP2	383	448	32.0	0.56	53	Donor to GFP/YFPs	Ai et al. (2007)
mCerulean3	433	475	30.0	0.80	71	Donor to YFPs	Markwardt et al. (2011)
mTurquoise	435	477	30.0	0.84	73	Donor to YFPs	Goedhart et al. (2010)
mTurquoise2	434	475	30.0	0.93	82	Donor to YFPs	Goedhart et al. (2012)
EGFP	488	507	56.0	0.60	100	Donor to OFP, RFPs	Ai et al. (2007)
Clover	505	515	111.0	0.76	247	Donor to OFP, RFPs	Lam et al. (2012)
mNeonGreen	506	517	115.0	0.8	271	Acceptor for CFPs, donor to RFPs	Shaner et al. (2013)
mVenus	515	528	92.2	0.57	156	Acceptor for CFPs, donor to RFPs	Nagai et al. (2002)
mCitrine	516	529	77.0	0.76	174	Acceptor for CFP	Griesbeck et al. (2001)
T-Sapphire	399	511	44.0	0.60	79	Long Stokes shift, donor to OFP	Zapata-Hommer and Griesbeck (2003)
mAmetrine	406	526	45.0	0.58	78	Long Stokes shift, donor to OFP	Ai et al. (2008)

Name	Ex	Em	EC	QY	Relative brightness[a]		Reference
REACh	515	528	92.2	0.04	1	Strong absorber, weak emitter, acceptor for FLIM studies	Ganesan et al. (2006) and Murakoshi et al. (2008)
Amber	None	None	0	0	0	Nonabsorbing, nonfluorescent, control for probe environment	Koushik et al. (2006)
Coral FPs							
Midoriishi Cyan	472	495	27.3	0.90	73	Donor to mKO	Karasawa et al. (2004)
mTFP1	462	492	64.0	0.85	162	Donor to YFP, OFP	Day et al. (2008)
Kusabira Orange2	551	565	63.8	0.62	118	Acceptor for CFP	Karasawa et al. (2004)
dTomato	554	581	69.0	0.69	142	Acceptor for mTFP1, YFP Acceptor for mAmetrine	Sun et al. (2009) and Ai et al. (2008)
mCherry	587	610	72.0	0.22	47	Acceptor for GFP	Yasuda et al. (2006)
TagRFP-T	555	584	81.0	0.41	99	Acceptor for GFP	Shcherbo et al. (2009)
mRuby2	559	600	113.0	0.38	126	Acceptor for GFP, Clover	Lam et al. (2012)

Note: The peak excitation (Ex) and emission (Em) wavelengths, molar extinction coefficient (EC), quantum yield (QY), and relative brightness are listed.
[a] Relative brightness is intrinsic brightness (EC × QY)/EGFP.

Importantly, the high intrinsic brightness and spectral overlap with red FPs make Clover a useful donor fluorophore in FRET-based studies (discussed subsequently).

11.4.1 LONG STOKES SHIFT VARIANTS OF *AEQUOREA* GFP

The mutagenesis of *Aequorea* GFP also yielded variants that had just a single absorption peak near 400 nm, producing FPs with an exceptionally large Stokes shift, which refers to the separation between the peak wavelength for excitation and the peak wavelength for emission (see Figure 11.2b). A large Stokes shift can be a useful trait for probes used in FRET studies, since it overcomes some of the problems associated with spectral bleed-through (SBT) signals (described in Section 11.8). The first long Stokes shift FP resulted from the substitution of the isoleucine for threonine at position 203 (T203I; Tsien 1998). The T203 residue of GFP is in one of the β-barrel strands and is positioned close to the chromophore, where it influences the local environment. The T203I substitution produced a GFP with only the single-peak absorption at 399 nm and this variant is called Sapphire (see Figure 11.2). A newer version, called T-Sapphire, was developed for more efficient maturation inside living cells (Zapata-Hommer and Griesbeck 2003). A similar long Stokes shift variant of *Aequorea* GFP, called Ametrine (Ai et al. 2008), was also generated and has been used in novel FRET probes that are described later (see Section 11.7.3).

11.5 NEW COLOR VARIANTS BASED ON *AEQUOREA* GFP

Aside from improving the spectral characteristics of GFP, substitutions around the chromophore also produced the color variants that Roger Tsien needed for his FRET-based sensors. One of the earliest color variants of the *Aequorea* GFP is a blue FP (BFP) that results from substitution of tyrosine[66] with histidine (Heim et al. 1994; Cubitt et al. 1995). In its original form, BFP has a low quantum yield and is very susceptible to photobleaching. Despite its obvious limitations, BFP allowed, for the first time, the opportunity to genetically label proteins with fluorescent probes that could be used to detect FRET. Early demonstrations of this capability used fusion proteins consisting of BFP directly coupled to GFP[S65T] through a protease sensitive linker (Heim and Tsien 1996; Mitra et al. 1996). Illumination of cells expressing the fusion protein with wavelengths to excite BFP resulted in an increased signal in the GFP (acceptor) channel. To demonstrate that this signal resulted from FRET, the linker was cleaved by a protease to separate the FPs. This resulted in the loss of most of the signal in the acceptor channel, clearly demonstrating FRET between the BFP and GFP probes.

In collaboration with Tullio Pozzan, Roger Tsien and Susan Taylor developed a genetically encoded FRET sensor for cAMP by labeling the catalytic and regulatory subunits of PKA with EGFP and EBFP, respectively (Zaccolo et al. 2000). Kinetic changes in the FRET signal were detected from cells expressing the fusion proteins following treatments to increase intracellular cAMP. Many other FRET studies have used BFP in combination with different GFP derivatives to detect cellular events such as transcription factor dimerization (Day 1998; Periasamy and Day 1999), calcium fluctuations (Miyawaki et al. 1997; Romoser et al. 1997), and apoptosis (Xu et al. 1998). Recently, several groups used mutagenesis strategies to develop new BFP variants with much higher quantum yields and improved photostabilities (Mena et al. 2006; Ai et al. 2007; Kremers et al. 2007). EBFP2, generated by Ai et al. (2007), is the brightest and most stable of the BFPs currently available and has been shown to be an excellent donor for FRET studies (see Table 11.1; Day and Davidson 2009).

Another early mutant variant of the *Aequorea* GFP resulted in an FP that emits in the cyan region of the spectrum. The substitution of Tyr[66] with tryptophan (T66W) produced an FP with a broad, bimodal absorption spectrum with peaks at 433 and 445 nm and an equally broad emission profile with maxima at 475 and 503 nm (Heim et al. 1994; Cubitt et al. 1995). The enhanced variant, ECFP, retained the complex excitation spectrum, which indicates the presence of more than one excited-state species (Tramier et al. 2002). This attribute has limited the utility of ECFP as a probe for fluorescence lifetime imaging microscopy (FLIM), an important method for measuring FRET (discussed in Section 11.13.2).

Efforts to address the complex excited-state characteristics of ECFP yielded an improved variant called Cerulean (Rizzo et al. 2004), which resulted from targeted substitutions on the solvent-exposed surface of ECFP (Tyr[145] and His[148]). Cerulean is about 1.5-fold brighter than ECFP, but when expressed in living cells, the fluorescence decay kinetics still indicate the presence of more than one excited-state species (Millington et al. 2007; Goedhart et al. 2010). Additionally, Cerulean has relatively poor photostability (Shaner et al. 2005) and shows some reversible photoswitching behavior (Shaner et al. 2008). Because of these shortcomings, several different groups used directed evolution approaches to improve Cerulean. Mutagenesis targeting the residues that influence the planarity of the chromophore yielded a new cyan FP called Cerulean3 (Markwardt et al. 2011). A similar strategy used by a different group resulted in another improved variant of Cerulean called Turquoise (Goedhart et al. 2010). Both of these cyan FPs are brighter than Cerulean, more photostable, and, importantly, have a single fluorescence lifetime (see Tables 11.1 and 11.2). More recently, a fluorescence lifetime screening approach was used to select for still brighter variants of Turquoise. This approach yielded Turquoise2, the brightest variant of Cerulean yet. Turquoise2 has the T65S and I146F substitutions that allow for tighter packing of the chromophore, resulting in a quantum yield of 0.93 (see Table 11.1; Goedhart et al. 2012). These newer cyan FP variants, with their blue-green emission,

Table 11.2 FD FLIM analysis of the variants of Cerulean

FLUORESCENT PROTEIN[a]	TWO-COMPONENT LIFETIME (FRACTION)	Tau(f)[b] (±SD)	χ^{2c} (±SD)	R_0 VENUS[d] (Å)
Cerulean	2.2 ns (0.66) 4.6 ns (0.34)	3.0 ± 0.05	1.1 ± 0.4	50
Cerulean3	3.9 ns (0.99)	3.9 ± 0.06	5.7 ± 2.5	54
Turquoise	3.9 ns (0.99)	3.9 ± 0.06	3.6 ± 2.0	55
Turquoise2	4.1 ns (0.99)	4.1 ± 0.04	3.2 ± 1.6	56

[a] Expressed in cells at 37°, n = 10 or more.
[b] Tau(f) is the average lifetime.
[c] Chi-square for the fit of measurements at 12 frequencies from 10 to 120 MHz.
[d] Förster distance determined by Equation 11.2.

Applications

share a significant spectral overlap with the longest wavelength mutant variants of *Aequorea* GFP, and have proven useful for FRET microscopy.

The longest wavelength emission variants of *Aequorea* GFP also resulted from substitutions at the T203 residue in the β-barrel, the residue that is also responsible for the long Stokes shift variant, Sapphire, described earlier (described in Section 11.4.1). The T203Y substitution, coupled with the S65G change in the chromophore, generated a bright yellow-green FP (YFP) that is optimally excited at 514 nm and has a peak emission at 527 nm (Ormö et al. 1996; Wachter et al. 1998). The EYFP variant, however, is sensitive to both pH and halides, limiting its usefulness for certain types of studies in living cells. Efforts to improve enhanced yellow FP (EYFP) led to the discovery that substitution of the glutamine at position 69 for methionine (Q69M) that increases the stability of the protein while simultaneously reducing its chloride sensitivity (Griesbeck et al. 2001). This variant, called Citrine is more photostable than many previous yellow FPs. The mutagenesis of EYFP by Nagai and colleagues (Nagai et al. 2002) also demonstrated that substitution of the phenylalanine at position 46 with leucine (F46L) dramatically improved the maturation efficiency and reduced the halide sensitivity of YFP to yield a derivative that they named Venus. Because of their high intrinsic brightness and single exponential decay kinetics, Turquoise or Cerulean3, used in combination with either the Venus or Citrine YFP, remain among the best FRET pairings currently available (discussed in Section 11.13.2). After years of mutagenesis studies to select spectral variants of *Aequorea* GFP, the YFPs remain the most red shifted of the *Aequorea*-color variants (Shaner et al. 2007). The emission spectra for the key color variants derived from *Aequorea* GFP are shown in Figure 11.3.

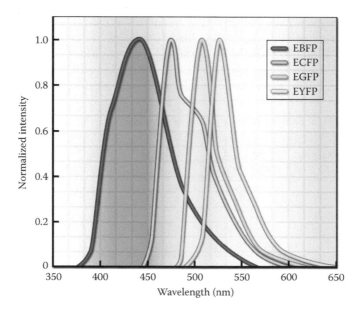

Figure 11.3 The emission spectra for the primary *Aequorea* FP color variants.

11.6 MONOMERIC VARIANTS OF THE *AEQUOREA*-BASED FPs

Nearly all the FPs discovered to date show a strong tendency to self-associate into dimers, tetramers, or oligomers. The comparison of the sequence of many different GFP-like proteins from marine organisms demonstrated that hydrophobic residues, typically at the carboxyl-terminus of the protein, are involved in the formation of interaction interfaces (Verkhusha and Lukyanov 2004). A tendency to self-associate, although weak, was also observed for the *Aequorea* FPs, and a dimer interface was identified in the crystal structure of GFP (Yang et al. 1996). Critically, the replacement of the hydrophobic residues in the dimer interface of the *Aequorea* FPs with positively charged residues eliminates the dimer formation without changing the spectral characteristics (Zacharias et al. 2002; Zhang et al. 2002). The most effective mutation to disrupt the dimer interface in the *Aequorea* proteins is the A206K substitution, where the hydrophilic lysine residue replaces the nonpolar amino acid alanine. The addition of the "monomerizing" A206K mutation to any of the *Aequorea* FPs used in FRET-based studies is highly recommended, since it overcomes possible artifacts resulting from the interactions between the probes themselves.

11.7 CORAL FPs FROM OTHER MARINE ORGANISMS AND THEIR USE IN FRET IMAGING

The observations that many marine organisms produce proteins with fluorescent characteristics led to the mining of these creatures for novel FPs (Matz et al. 1999, 2002; Labas et al. 2002; Karasawa et al. 2004; Shagin et al. 2004; Shcherbo et al. 2007; see Chapter 4). The cloning and engineering of many different GFP-like proteins from these marine organisms dramatically expanded the fluorescence protein palette, covering the spectral range into the deep red and providing important alternatives to the *Aequorea* FPs (reviewed in Day and Davidson 2009). Many of the new FPs share significant spectral overlap, providing the opportunity to develop FRET probes with emission in the orange and red spectral range. However, it is important to emphasize here that the choice of the best FP pairs for FRET-based studies is often not obvious from their spectral and photophysical characteristics alone (Piston et al. 2007) and is best determined by direct comparisons (discussed in Section 11.13.2).

11.7.1 CYAN AND GREEN FPs FROM CORALS

A cyan FP isolated from an *Acropora* stony coral species (Karasawa et al. 2004) is called Midori-ishi Cyan (abbreviated MiCy). Unlike the *Aequorea* ECFP variant, which has a tryptophan residue in the second position of the chromophore (T66W), the MiCy possesses a tyrosine residue in that position, which is typical of the GFPs. This attribute shifts the absorption and emission spectra toward the green, and MiCy FP is the most green shifted ($\lambda_{ex} = 472$, $\lambda_{em} = 495$ nm) of the cyan spectral class. The intrinsic brightness of MiCy is similar to Cerulean (Table 11.1), but the protein shows significant pH sensitivity, which limits its use in some cellular compartments. Another drawback of the MiCy FP is that it forms a homodimeric complex, similar to the GFP-like protein from the bioluminescent sea pansy, *Renilla reniformis* (Ward and Cormier 1979).

Applications

Despite this, MiCy was proven to be useful as the donor in a novel FRET combination with the monomeric Kusabira Orange (KO) FP (see Section 11.7.3). Moreover, MiCy has a single exponential decay with an average lifetime of 3.4 ns, making it a useful probe for FRET measurements by FLIM (discussed in the succeeding text).

Another useful cyan-colored FP was isolated from the coral *Clavularia* and engineered by directed evolution to generate a monomeric (m) teal FP, called mTFP1, with remarkable brightness (Ai et al. 2006). Similar to MiCy, mTFP1 also has a tyrosine residue at the central chromophore position, shifting both the excitation and emission spectra (λ_{ex} = 462, λ_{em} = 495 nm) to the more green wavelengths when compared to CFP (Table 11.1 and Figure 11.4). The mTFP1 protein has a high intrinsic brightness, similar to the brightest of the *Aequorea* FPs, and displays a relatively narrow emission spectrum that strongly overlaps the excitation spectrum of the yellow and orange FPs. The mTFP1 is an excellent donor fluorophore for FRET studies using the Venus FP, and its single fluorescence lifetime (2.8 ns) makes it a useful probe for FRET measurements by FLIM (described later in the chapter; see Day et al. 2008; Padilla-Parra et al. 2009).

The continued directed evolution of mTFP1 led to the selection of a novel green color variant. From the crystal structure of mTFP1, it was known that the histidine residue at position 163 made contact with the chromophore, and this residue was targeted for mutagenesis (Ai et al. 2006). The mTFP1 H163M variant was identified that further shifted its emission spectrum into the green. This protein was then subjected to random mutagenesis, and the result was a protein called mWasabi, with a peak emission of 509 nm (Ai et al. 2008). This *Clavularia*-based protein is among the brightest and most photostable of the GFPs and has the advantage of a relatively narrow excitation and emission spectra (Figure 11.4). Since mTFP1 has proven useful for FRET-based imaging, it seems likely that mWasabi will also be useful for this approach.

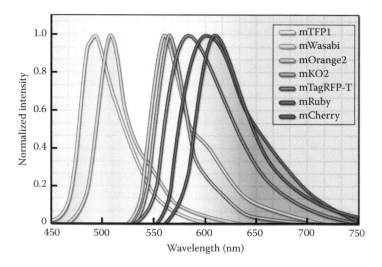

Figure 11.4 The emission spectra for selected FPs isolated from corals.

11.7.2 NOVEL YELLOW FP FROM FISH

The analysis of GFP-like proteins from a cephalochordate, the Amphioxus fish *Branchiostoma floridae*, has yielded a potentially very useful FP (Baumann et al. 2008; Shaner et al. 2013). The fluorescence imaging of the adult fish demonstrated discrete areas of green and red fluorescence surrounding the mouthparts. Genomic sequencing of *B. floridae* indicated the presence of at least 12 GFP-like proteins, which may have arisen by duplication during evolution (Baumann et al. 2008). A yellow fluorescent protein was cloned from *Branchiostoma lanceolatum* (LanYFP, Allele Biotechnology, San Diego, CA), and found to have an unusually high quantum yield and extinction coefficient, making it an attractive candidate for further development. The Shaner laboratory used mutagenesis and directed evolution to select a very bright monomeric form of LanYFP. The result is a novel FP, called mNeonGreen, which has 21 substitutions compared to the parent LanYFP. The mNeonGreen has an exceptionally high extinction coefficient (Table 11.1), and is among the brightest of the monomeric FPs yet described (Shaner et al. 2013). Its spectral characteristics are similar to mVenus, and this, coupled with its very high intrinsic brightness, suggests that mNeonGreen will be a good FRET acceptor for cyan fluorescent proteins (see Section 11.13).

11.7.3 ORANGE AND RED FPs FROM CORALS

There are several important reasons why it was critical to develop FPs that emit in the orange and red spectral region. First, living cells and tissues are more tolerant of illumination with longer wavelength light, which allows imaging for extended time periods. Second, the autofluorescence background from cells and tissues is significantly reduced when illuminated with the longer wavelengths, allowing the probes to be detected deeper inside tissues and organisms (Shcherbo et al. 2009; see Chapter 5). Finally, the addition of orange and red FPs (RFPs) to the available blue to yellow FPs provides tools for multicolor imaging and potentially allows the development of new FRET biosensors with spectral profiles in the longer wavelength regions.

The first widely available coral FP was isolated from the mushroom anemone *Discosoma striata* (Matz et al. 1999). The DsRed FP has a peak absorbance at 558 nm and a maximum emission at 583 nm, providing the first RFP. In its original form, DsRed is not well suited for live-cell imaging applications because it matures slowly into an obligate tetramer, generating a green intermediate as it develops (Baird et al. 2000). These problems were overcome using both random and directed mutagenesis strategies, leading to the development of the first monomeric RFP, mRFP1 (Bevis and Glick 2002; Campbell et al. 2002). Unfortunately, mRFP1 had several shortcomings, including poor quantum yield and photostability, and a large fraction of the protein that never fully developed to the fluorescent state (Hillesheim et al. 2006). Further directed evolution of mRFP1 yielded a variety of FPs with interesting characteristics, including the mCherry FP—a rapid maturing and bright mRFP (Table 11.1; Shaner et al. 2004, 2005; Wang et al. 2004). Because EGFP and mCherry share strong spectral overlap, they have been used for FRET-based imaging studies (Peter et al. 2005; Tramier et al. 2006; Yasuda et al. 2006). As will be discussed in the succeeding text, this pairing has advantages for live-cell imaging, especially when FLIM is used.

The directed evolution of mRFP1 also yielded several different orange FPs (Shaner et al. 2004, 2005). Because of their significant spectral overlap with both the commonly used cyan and green FPs, as well as the RFPs, the orange FPs provide potential alternative fluorophores for FRET studies. For example, the tandem dimer (d) Tomato protein is among the brightest of the current FPs and has proved useful as a FRET acceptor. The major limitation to the dTomato FP for live-cell imaging studies is its size—twice that of the monomeric FPs. However, dTomato has been shown to function well as a fusion partner for many cellular proteins (Shaner et al. 2007; Day and Davidson 2009) and was used in a transgenic mouse model, indicating it is nontoxic and well behaved *in vivo* (Luche et al. 2007). Recently, we used dTomato as an acceptor fluorophore for FRET studies with both the mTFP1 and Venus FPs (Sun et al. 2010).

The dTomato FP was also paired with a long Stokes shift GFP Ametrine (see Section 11.4.1; Ai et al. 2008). Ametrine was identified in a screen for violet-excitable *Aequorea* GFP variant, and directed evolution was used to select a bright yellow fluorescing protein that retained the violet excitation (λ_{ex} = 406 nm). To develop a FRET-based biosensor probe to measure caspase-3 activities, Ametrine was paired with dTomato. Importantly, because of its violet excitation and long Stokes shift, it was possible to also include a second FRET biosensor in the same cellular assay that used the combination of mTFP1 and Citrine FP probes (Sections 11.7.1 and 11.5, respectively). The Ametrine–dTomato and mTFP1–Citrine caspase-3 biosensor probes were each targeted to different subcellular compartments, and ratio imaging was used to monitor activities in the different, nonoverlapping cellular regions. However, since the emission profile of Ametrine and Citrine are very similar, it was not possible to visualize the two probes simultaneously (Ai et al. 2008). With the development of new spectral variants of the FPs, it may be possible to design FRET probes for the simultaneous detection of two different cellular activities.

A monomeric orange FP was also isolated in the directed evolution screens of mRFP1 (Shaner et al. 2004). The original mOrange1 protein had relatively poor photostability, limiting its usefulness for quantitative measurements. Recently, mOrange1 was subjected to further directed evolution, selecting for increased photostability, leading to an improved variant, mOrange2 (Shaner et al. 2008). However, the mOrange proteins exhibit a strong tendency for photoconversion—that is, they change their color upon strong or prolonged illumination, and this greatly limits their utility for FRET-based measurements. For example, using typical laser-scanning illumination at 488 nm, Dave Piston's group (Kremers et al. 2009) showed that these FPs were efficiently photoconverted from orange- to far-red (λ_{em} = 640 nm)-emitting proteins. While this behavior makes mOrange2 an attractive optical highlighter (see Chapter 6), it limits its use for other applications. So while mOrange2 was used in FRET-based assays (Goedhart et al. 2007; Ouyang et al. 2010), the photoconversion to red-emitting fluorophores prevents quantitative measurements, especially when used as a donor fluorophore for red acceptors, such as mCherry.

Another orange FP isolated from the mushroom coral *Fungia concinna* and called KO has spectral characteristics that are very similar to the mOrange proteins derived from mRFP1 (see Figure 11.4). The KO FP was engineered to a bright, photostable monomeric protein, called mKO (Karasawa et al. 2004; see Table 11.1). A fast-folding version containing eight additional mutations, named mKO2, was developed that has improved characteristics for live-cell imaging (Sakaue-Sawano et al. 2008). Under laser excitation, the mKO FP is much less susceptible to photoconversion than the mOrange proteins (Kremers et al. 2009). As was mentioned earlier (Section 11.7.1), mKO was developed as a FRET acceptor for MiCy (Karasawa et al. 2004). More recently, mKO was used in FRET imaging studies with donor probes other than MiCy (Goedhart et al. 2007; Sun et al. 2009).

Several bright RFPs have also been developed that should allow the development of FRET probes with long-wavelength spectral characteristics. For instance, the TagRFP was engineered from a protein isolated from the sea anemone *Entacmaea quadricolor* (Merzlyak et al. 2007) and is among the brightest of the monomeric RFPs currently available. The TagRFP is optimally excited at 555 nm, with an emission peak at 584 nm (Figure 11.4), and it is well tolerated as a fusion partner for many different proteins expressed in a variety of mammalian cell systems (Day and Davidson 2009). The TagRFP has been successfully used as an acceptor for FRET studies (Shcherbo et al. 2009). The original TagRFP, however, had relatively poor photostability compared to other FPs in this spectral class. Directed evolution was used to select more photostable variants of TagRFP, and a single mutation (S158T) was identified that increases the photostability almost tenfold (Shaner et al. 2008). The resulting FP, named TagRFP-T, has spectral properties similar to the parent and is among the most photostable of the FPs yet discovered (see Table 11.1). Another potentially useful RFP, also isolated from *E. quadricolor*, was engineered to a bright, monomeric RFP named mRuby (see Figure 11.4). The mRuby FP contains 29 mutations relative to the parent, and the spectral characteristics (λ_{ex} = 558, λ_{em} = 605 nm) are similar to mCherry (Kredel et al. 2009). The mRuby FP is one of the brightest monomeric RFPs yet developed (see Table 11.1) and has proven to be an effective fusion partner for many different cellular proteins (Day and Davidson 2009). More recently, mRuby was subject to further directed evolution using a selection method based on FRET measurements from tandem fusions with Clover (see Section 11.4). The resulting variant, called mRuby2, differed from mRuby by four substitutions, and had improved brightness and photostability, while retaining a similar emission profile (Lam et al. 2012). The Clover-mRuby2 pair share excellent spectral overlap and the combination was shown to function well in FRET-based biosensor probes (Lam et al. 2012).

11.8 USING THE FPs FOR FRET MEASUREMENTS

When energy is transferred from donor fluorophore to an acceptor fluorophore, the fluorescence emission from the donor is quenched, and there is increased (sensitized) emission from the acceptor that can be detected in the FRET channel (illustrated in Figure 11.5). Recall that a key requirement for FRET is that the donor and acceptor must share a significant spectral overlap (see Figure 11.1). An undesired consequence of the strong spectral overlap is significant background fluorescence (the SBT signal) that contaminates the FRET signal. The SBT results from the direct excitation of the acceptor by the donor excitation wavelengths (arrow, Figure 11.5), and the donor emission signal that bleeds into the FRET detection channel (hatching, Figure 11.5). Therefore, the accurate measurement of FRET signals requires correction methods that define and remove these different SBT components from the sensitized acceptor or FRET signal (Periasamy et al. 2005). Since the SBT signals contaminate the FRET (acceptor) channel, and not the donor channel (see Figure 11.5), an alternative method is to detect energy transfer by measuring the quenched state of the donor fluorophore.

Applications

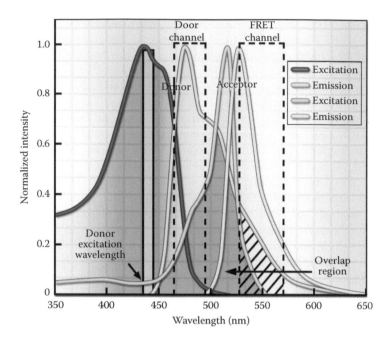

Figure 11.5 Spectral cross talk between donor and acceptor FRET probes limits the detection of sensitized acceptor emission. The excitation and emission spectra for the CFP (donor) and YFP (acceptor) FRET pair is shown. The dashed boxes indicate typical donor and FRET detection channels. The arrow indicates the direct acceptor excitation at the donor excitation wavelength, and the hatching shows donor SBT into the FRET channel.

The methods of acceptor photobleaching FRET (pbFRET) and FLIM both measure the quenched donor signal. Since the measurements are made in the donor channel, they are not affected by SBT signals and can be the most accurate methods to quantify FRET (Periasamy and Clegg, 2010). Below, the advantages and limitations of the SBT correction method (Section 11.11), as well as the methods of pbFRET (Section 11.12) and FLIM (Section 11.13), are discussed.

11.9 IMPORTANCE OF STANDARDS FOR FRET MEASUREMENTS

The wide availability of FPs that emit across the visible spectrum and that have a suitable spectral overlap for FRET-based microscopic measurements has led to broad application of this approach. However, a consequence of this newfound popularity has been the "degradation in the validity of the interpretations" of these experiments (Vogel et al. 2006). As with any highly technical approach, the interpretation of the experimental results can be problematic, and it is often difficult to directly compare the accuracies of different methods. The Vogel laboratory addressed this issue by introducing a set of genetically encoded fusion proteins that can be used to evaluate systems for making FRET measurements (Thaler et al. 2005; Koushik et al. 2006).

They created a series of genetic constructs that encode mCerulean (the donor) directly coupled to mVenus (the acceptor) through protein linkers of different lengths to serve as "FRET standards" (these are now available through Addgene, Cambridge, MA). For example, they generated a genetic construct that encoded mCerulean separated from mVenus by a short five-amino-acid (5 aa) linker (Cerulean–5aa–Venus), producing a fusion protein with consistently high (~40%) FRET efficiency. In addition, they generated a genetic construct encoding a much longer linker from the tumor necrosis factor receptor–associated factor (TRAF) domain separating the mCerulean and mVenus proteins (Cerulean–TRAF–Venus). This produced a fusion protein with low FRET efficiency (~6%). They also developed an important control for FLIM measurements. The fluorescence lifetime of a probe is sensitive to the local probe environment (described in Section 11.13). Therefore, it was important to develop probes that closely mimic the environmental effects of the nearby acceptor protein on the donor fluorophore. Here, a mutant variant of mVenus that neither absorbs nor emits fluorescence was generated, called Amber, which contains the substitution of the chromophore tyrosine with cysteine (Table 11.1). This produces a protein that folds correctly, but cannot function as a FRET acceptor (Koushik et al. 2006). The presence of the folded β-barrel in the context of fusion proteins containing the donor fluorophore replicates the molecular environment of the donor, without quenching the fluorescence.

The Vogel group expressed FRET standard fusion proteins in living cells and then used microscopic techniques to measure the efficiency of energy transfer (Thaler et al. 2005; Koushik et al. 2006). For each of the fusion proteins tested, there was consensus in the results obtained by the different FRET methods, demonstrating that these genetic constructs could serve as "FRET standards." What is more important is that other laboratories can use these same genetic constructs to verify and evaluate FRET measurements obtained in their own experimental systems. The genetically encoded FRET standard proteins are also useful tools for optimizing cell culture conditions for FRET measurements and for evaluating optical systems used for the detection of FRET (discussed in Section 11.13.2). The low FRET efficiency standards are especially useful for assessing the background noise in the system. The linked probes described here also serve as a starting point for the design of biosensor probes. The biosensor proteins use a bioactive linker peptide to separate the donor and acceptor fluorophores and use FRET to report changes in the conformation of the linker resulting from its modification or the binding of a substrate (DiPilato and Zhang 2010).

11.10 FRET-BASED BIOSENSOR PROTEINS

The biosensor proteins use a bioactive linker peptide to separate the donor and acceptor fluorophores, and use FRET to report changes in the conformation of the linker resulting from its modification, or the binding of a substrate (DiPalato and Zhang 2010; Hum et al. 2012). Similar to the FRET standard approach described earlier (Section 11.9), the biosensor proteins incorporate a linker between the donor and acceptor FPs. In this case, however, instead of simply serving as a spacer, the linker protein encodes a detector sequence for bioactivity. For example, the binding of a ligand, or the modification of the protein linker, leads to a change in the conformation of the linker. This changes the spacing between the donor and acceptor FPs, which can be detected as changes in the FRET signal. Since the biosensor probes consist of the donor and acceptor fluorophores directly coupled to one another, ratio imaging of the donor and acceptor signal can be

Applications

used to quantify the FRET signal. The ratio imaging of the linked probes automatically corrects for the SBT background, greatly simplifying FRET measurements.

The early biosensor probes incorporated calcium-sensitive linkers to monitor intracellular calcium fluctuations (Miyawaki et al. 1997; Romoser et al. 1997). These probes have been continually modified to incorporate optimized cyan and yellow FPs, increasing their sensitivity and dynamic range (Tian et al. 2009). Recently, linkers that incorporate a variety of different kinase-sensitive domains that undergo conformational reorganization upon phosphorylation have been developed. The dynamic changes in the FRET signal allow the direct visualization of localized kinase activity inside living cells (Zhang et al. 2002; Miyawaki 2005; Zhang and Allen 2007; Seong et al. 2009; Hum et al. 2012). To demonstrate that the probe responses are specific, it is critical to also make measurements with probes that incorporate mutations in the sensor region—probes with similar structure that do not respond to the activity that is being measured. A general concern regarding the biosensor probe approach is the potential for proteolytic cleavage of the linker over the time course of the experiment, resulting in separation of the FPs, leading to the loss of FRET signals. Another concern is that the transfection approaches for introducing FP-biosensor probes can yield very high levels of the fusion proteins in the target cells, especially when strong promoters are used. The high-level expression of these probes can lead to sequestration of the endogenous effector molecules and result in aberrant signals in inappropriate subcellular domains (Haugh 2012). In addition, although the FRET measurements from biosensor probes, when collected and quantified properly, are remarkably robust, there can be substantial heterogeneity in the measurements. The data must be collected from multiple cells and statistically analyzed to prevent the user from reaching erroneous conclusions from a nonrepresentative measurement. New genetically encoded biosensor probes continue to be developed, combining the features of subcellular targeting, signal pathway specificity, and detection sensitivity that allow the real-time monitoring of cellular events inside living cells (Zhang and Allen 2007; Seong et al. 2009; DiPilato and Zhang 2010; Miyawaki 2011; Zhou et al. 2012). The efforts to incorporate the new FPs are yielding new biosensor probes that can be used in different spectral windows, such as the pairing of Ametrine and dTomato (Ai et al. 2008; described in Section 11.7.3). These FRET-based biosensors have tremendous potential for the development of large-scale screening approaches for the discovery of novel pharmaceuticals and the development of therapeutic strategies (You et al. 2006).

11.11 FRET ANALYSIS BY SPECTRAL BLEED-THROUGH CORRECTION ALGORITHMS

In contrast to the linked FRET-biosensor probes, intermolecular FRET experiments are designed to detect the association of independently produced proteins, each labeled with either the donor or acceptor FPs. In this case, the ratio of donor to acceptor is not fixed and will be highly variable between individual cells within a transfected population. Since the donor–acceptor ratio varies from cell to cell, the SBT background signal will also be different for each transfected cell. As a result, the intensity-based measurements of FRET are highly dependent upon the donor to the acceptor ratio and will work best over a limited range of ratios (Berney and Danuser 2003). Therefore, the quantitative measurement of FRET using the sensitized acceptor emission signal requires accurate methods to identify and remove the background SBT signals. Many computer algorithms have been developed for this purpose, yielding corrected FRET measurements that can be quantified (Berney and Danuser 2003; Periasamy et al. 2005).

The common approach for filter-based imaging is to acquire images of control cells that produce either the donor- or the acceptor-labeled proteins alone, using precisely the same imaging conditions that will be used to detect FRET. The reference images are acquired of the donor using the donor excitation with donor emission (the donor channel), and the donor excitation with acceptor emission (the FRET channel). Similarly, reference images are acquired of the acceptor using the acceptor excitation with acceptor emission (the acceptor channel), and the donor excitation with acceptor emission (the FRET channel). Together, the reference images are used to define the SBT components that arise from the direct excitation of the acceptor (acceptor SBT), as well as the donor emission bleed-through into the FRET channel (donor SBT, see Figure 11.5). A computer algorithm uses pixel intensity matrices to partition the intensity values for the SBT signals, and these same pixel ranges are identified in the FRET images acquired from the experimental cells that coexpress the donor- and acceptor-labeled proteins. The computer algorithm then removes the SBT contributions by pixel-to-pixel subtraction, yielding the corrected FRET signal (Elangovan et al. 2003; Chen et al. 2007).

The strength of this approach is that it is relatively simple to implement, and most current imaging system software packages include an algorithm that will determine FRET efficiency based on SBT correction. However, the computer algorithm–based correction is sensitive to the collection conditions used on a particular instrument, so the reference cell measurements must be made for every experiment. Further, cell movement, focal plane drift, and a lack of precisely registered images used for FRET determinations are potential sources of artifacts that will appear as regions of either very high or negative FRET in energy transfer images and must be critically evaluated.

An alternative SBT correction method is applied when spectral imaging approaches are used. Spectral imaging systems typically use acousto-optic tunable filters (AOTF) to obtain images over a series of discrete wavelength bands, generating what are called lambda stacks. The spectral signatures for different fluorophores or background signals are obtained from these lambda stacks, and the method of linear unmixing is then used to separate the contributions of the individual signals in each pixel of an acquired image (Dickinson et al. 2003). The spectral imaging and linear unmixing method is among the most accurate methods for removing the contribution of the donor SBT from the FRET signal (Chen et al. 2007). However, since the acceptor SBT and the FRET signals have identical spectra, a computer algorithm still must be used to determine the acceptor SBT contribution and remove that component from the spectral FRET data—analogous to methods for SBT correction in filter-based FRET microscopy systems described earlier. For example, a method called ps-FRET (Chen et al. 2007) uses the acceptor spectral data obtained from reference cells expressing the acceptor alone to make corrections for acceptor SBT.

The spectral imaging method is illustrated here using measurements from cells that express the FRET standard proteins consisting of mTFP1 (described in Section 11.7.1) directly coupled to mVenus through linker peptides. Fusion proteins with relatively high FRET (mTFP–5aa–Venus) and low FRET efficiency (mTFP–TRAF–Venus) were produced in living cells, and the emission spectra for the different fusion proteins were obtained using donor excitation with the 458 nm laser line. The representative spectra obtained from cells expressing each of the fusion proteins are shown in Figure 11.6. The results graphically illustrate the relative contributions of sensitized acceptor emission and SBT signals to the emission spectra for the different fusion proteins. The emission spectrum from the low-FRET-efficiency probe (mTFP–TRAF–Venus) shows the donor emission (peak emission at 490 nm), and predominately acceptor SBT in the 530 nm emission range that results from the direct excitation of Venus by the 458 nm laser line (Figure 11.6c). In striking contrast, the emission spectrum for the mTFP–5aa–Venus (Figure 11.6a) clearly shows the strong contribution of the sensitized Venus emission in the 530 nm emission range that is the result of energy transfer. Importantly, we can verify that the acceptor emission signals detected here resulted from FRET by using the pbFRET method, described in the following.

11.12 VERIFYING FRET MEASUREMENTS USING ACCEPTOR PHOTOBLEACHING

The direct transfer of donor excited-state energy to the acceptor quenches the donor emission. Therefore, if the acceptor fluorophore is destroyed, the quenching pathway due to energy transfer is eliminated, and the donor signal will increase. The technique of pbFRET exploits this phenomenon and detects the dequenching of the donor signal when the acceptor fluorophore is destroyed by photobleaching (Bastiaens and Jovin 1996; Bastiaens et al. 1996). Since the same cell is used to determine the quenched and unquenched donor, the method can be very accurate and will reveal regional differences in the FRET signal within individual cells. Importantly, this approach provides a method to verify FRET measurements made using other techniques.

The photobleaching approach requires the selective bleaching of the acceptor, because any bleaching of the donor fluorophore will result in an underestimation of the dequenching. Further, the acceptor must be bleached nearly to completion, since any remaining acceptor will still be available for FRET, again resulting in an underestimation of the donor dequenching. So it is important to select an acceptor FP that is susceptible to photobleaching. Many newer FPs, such as TagRFP and mRuby2, have been engineered for high photostability and may not be amenable to the acceptor photobleaching approach (see Table 11.1). In contrast, the Venus FP is sensitive to photobleaching, making it useful for FRET measurements by acceptor photobleaching (Nagai et al. 2002; Shaner et al. 2005). The major limitation of this approach, however, is that it is an end-point assay and cannot be repeated on the same cells. Furthermore, the application of acceptor photobleaching to living cells requires that protein complexes be relatively stable over the bleach period.

Here, pbFRET is used to verify the ps-FRET measurements discussed earlier. After the spectra for the mTFP–5aa–Venus and mTFP–TRAF–Venus fusion proteins were obtained, the linked Venus fluorophores were selectively photobleached using the 514 nm laser line. The spectra were then reacquired using the 458 nm laser line to excite the mTFP1 donor under identical conditions to the first measurements. Following the photobleaching of the Venus fluorophores in the mTFP1–5aa–Venus fusion protein,

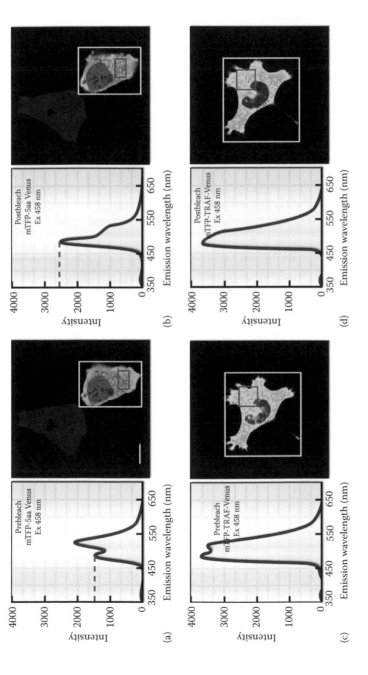

Figure 11.6 The measurement of the changes in the donor signals from either (a, b) mTFP–5aa–Venus or (c, d) mTFP–TRAF–Venus fusion proteins after acceptor photobleaching. Spectral measurements were acquired from cells expressing the indicated fusion proteins (a, c); the calibration bar indicates 10 μm. The linked Venus fluorophore was then photobleached by more than 70% using the 514 nm laser line. The spectral measurements were then reacquired under identical conditions to the first, and changes in the donor signal were measured. The dashed line in (b) indicates the change in the donor signal for the mTFP–5aa–Venus fusion protein after acceptor photobleaching. In contrast, there was little change in the donor signal for mTFP–TRAF–Venus (d) following acceptor photobleaching. (From Day, R. et al., *J. Biomed. Opt.*, 13, 031203, 2008.)

there was a large increase in the donor signal (peak emission 490 nm), reflecting the dequenching of the mTFP (compare Figure 11.6a and b). The comparison of the pre- and post-acceptor bleach spectra for mTFP–5aa–Venus for several different cells revealed a mean increase in the donor signal of about 40%. In contrast, pbFRET for the mTFP–TRAF–Venus fusion protein (Figure 11.6c and d) yielded an average change in the donor signal of about 5% (Day et al. 2008).

11.13 FLUORESCENCE LIFETIME MEASUREMENTS

The fluorescence lifetime is the average time a fluorophore spends in the excited state before returning to the ground state, typically with the emission of a photon. The fluorescence lifetime is an intrinsic property of the fluorophore, and it carries information about events in the local microenvironment of the probe that can affect the photophysical processes. Most probes used in biological studies have lifetimes ranging from one to about ten ns. When a population of fluorophores is excited, there is an exponential decay in fluorescence as the probes relax back to the ground state. The average time the population spends in the excited state is known as the fluorescence lifetime, τ, and it is defined by the following equation:

$$\tau = \frac{1}{\kappa_r + \kappa_{nr}} \tag{11.3}$$

where
    κ_r is the radiative rate constant
    κ_{nr} is the rate constant for all nonradiative processes (Lakowicz 2006)

The nonradiative rate constant, κ_{nr}, is sensitive to environmental factors that affect the excited state. Since energy transfer is a quenching process that depopulates the excited state of the donor fluorophore, the donor fluorescence lifetime is shortened by FRET. There is a direct and inverse relationship between the fluorescent lifetime of the donor fluorophore and the FRET efficiency. When the donor lifetimes are determined in the absence (τ_D) and in the presence of the acceptor (τ_{DA}), the ratio of the lifetimes provides a straightforward method to determine E_{FRET} by the following equation:

$$E_{FRET} = 1 - \frac{\tau_{DA}}{\tau_D} \tag{11.4}$$

11.13.1 FLIM METHODS AND DATA ANALYSIS

The FLIM methods can map the spatial distribution of probe lifetimes inside living cells and can accurately measure the shorter donor lifetimes that result from FRET (Bastiaens and Squire 1999; Peter et al. 2005; Yasuda 2006; reviewed in Periasamy and Clegg 2010). FLIM is particularly useful for biological applications, since measurements made in the time domain (TD) are independent of variations in the probe concentration and are not affected by excitation intensity, factors that can limit steady-state intensity-based measurements. The FRET–FLIM approach uses optical filtering to isolate the donor fluorescence emission signal and then measure the donor fluorescence lifetime. The FLIM techniques are broadly subdivided into the TD and the frequency domain (FD) methods. The physics that underlies TD and FD FLIM methods is identical, but they differ in the way the measurements are analyzed (Clegg 2010). Briefly, the TD method uses a pulsed-light source synchronized to high-speed detectors to measure the fluorescence decay profile at different intervals after each excitation pulse. The fluorescence lifetime of the fluorophore is estimated by analyzing the recorded decay profile. The FD method uses a modulated light source to excite a fluorophore and then measures the modulation and phase of the emission signals. The fundamental modulation frequency is chosen depending on the lifetime of the fluorophore and is usually between 10 and 140 MHz for nanosecond decays. The emission signal is analyzed for changes in phase and amplitude relative to the excitation source to extract the fluorescence lifetime of the fluorophore.

The FD FLIM measurements are commonly analyzed using the phasor or polar plot. This method was originally developed as a way to analyze transient responses to repetitive perturbations, and can be applied to any system with frequency characteristics. The phasor plot is a global representation of the relative modulation fraction and the phase delay of the emission signal from every pixel in an image, allowing the direct determination of the fluorescence lifetime of a fluorophore (Jameson et al. 1984; Redford and Clegg 2005; Hinde et al. 2012). Importantly, the phasor plot does not require a fitting model to determine fluorescence lifetime distributions, but rather expresses the overall decay in each pixel in terms of the polar coordinates on a universal semicircle, with shorter lifetimes to the right and longer lifetimes to the left (see Figure 11.7). Here, phasor analysis for FD FRET measurements is illustrated using cells that expressed either mTFP1-5aa-Amber (unquenched donor) or mTFP1–5aa–Venus (donor quenched by FRET) FRET standard fusion proteins.

Images (256 × 256 pixels) of cells expressing each different protein (right panels, Figure 11.7a) were acquired, and the FD method was used to determine the distribution of donor lifetimes inside each cell. The distribution of lifetimes at every pixel in the images of the two different cells is displayed on a common phasor plot, allowing direct comparison of the donor lifetimes in each cell (Figure 11.7a). The lifetime distribution for the cell expressing the mTFP1-5aa-Amber (unquenched donor, Section 11.9) protein falls directly on the semicircle, indicating it fits well to a single component lifetime. The average lifetime for the unquenched mTFP1 is 2.9 ns. In contrast, the average donor lifetime for the mTFP1–5aa–Venus protein was shortened to about 1.89 ns, which is consistent with quenching due to FRET (Figure 11.7a). To demonstrate that the quenched signal results from FRET, the cell in Figure 11.7a was next subjected to selective acceptor photobleaching of the Venus fluorophore, and the fluorescence lifetime of the donor (mTFP1) was then reacquired (Figure 11.7b). Following the photobleaching of the Venus fluorophores, the intensity of the donor signal was increased, reflecting the donor dequenching (Figure 11.7b, compare right panel scale bars). The donor dequenching is also reflected in the shift of the donor lifetime from 1.9 to 2.9 ns, which is the lifetime of the unquenched mTFP1. Together, these results illustrate the strength of the FRET–FLIM approach for quantitative measurements of protein interactions inside living cells and the value of the acceptor photobleaching FRET approach for verifying those measurements.

Applications

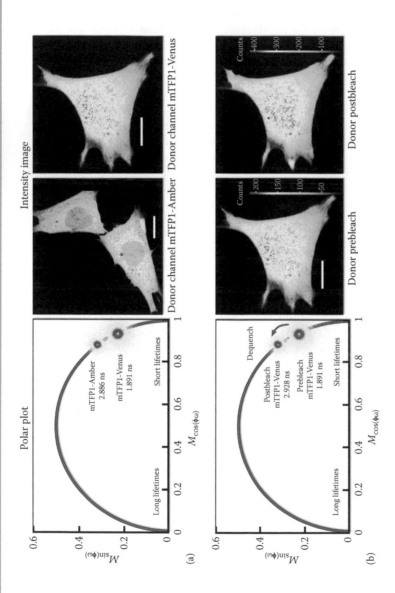

Figure 11.7 The polar plot analysis of the donor (mTFP1) lifetime for cells expressing mTFP1-5aa-Amber (unquenched donor) or mTFP1–5aa–Venus (quenched donor; see text for details). The intensity image for each cell is shown in the right panels; the calibration bar indicates 10 μm. (a) The lifetime distribution for all pixels in each image is displayed on the polar plot, showing the unquenched lifetime of mTFP1-5aa-Amber, and the quenched lifetime when mTFP1 is linked to the acceptor, Venus. (b) The same cell expressing mTFP1–5aa–Venus shown in (a) was then subjected to selective acceptor photobleaching. The bleaching of Venus resulted in the dequenching of mTFP1 and the shift in the lifetime to that of the unquenched donor. The prebleach and postbleach intensity images for the cell are shown in the right panels and were acquired under identical conditions. The average counts per pixel are shown on the scale bar and the size calibration bar indicates 10 μm.

Fluorescence lifetime provides detailed information about any local environmental event that influences the excited state. Therefore, a potential limitation to the approach is that environmental factors, such as changing pH or collisional quenching, will also shorten the measured fluorescence lifetime. Thus, care must be taken in interpreting FRET–FLIM data from living cells. It is critical to identify the sources of noise in FRET–FLIM measurements to determine the reliability of the data analysis. In this regard, the FRET standard proteins are valuable tools, since they should report the same range of FRET signals each time they are used, and will effectively reveal problems in the imaging system. Another limitation to FRET–FLIM is that the acquisition of the data is typically slow. For example, acquiring sufficient photon counts to assign lifetimes using the FD method described above required about 45 seconds, which limit its application for monitoring very dynamic events. There are FLIM methods based on wide-field imaging using a gated image intensifier camera, or spinning disk confocal systems that have the ability to acquire the data in seconds (Periasamy and Clegg 2010; Buranachai et al. 2008). As the technology continues to evolve, it is expected that the time required to obtain FLIM measurements will decrease significantly, making this technique even more useful for measurements of dynamic changes in biosensor activities in single living cells. What is more, since lifetime measurements are made only in the donor channel, this approach has great potential for the simultaneous monitoring of multiple intracellular signaling events through the combination of two different FRET-based biosensor proteins (Carlson and Campbell 2009).

11.13.2 FLUOROPHORE PAIRS FOR FRET–FLIM

The donor fluorophore used for FRET–FLIM should ideally have only one lifetime component to allow the unambiguous assignment of the quenched or unquenched donor lifetime populations. In this regard, the complex decay kinetics of ECFP and Cerulean can be problematic for lifetime analysis (Traimer et al. 2006; Yasuda et al. 2006; Millington et al. 2007; Goedhart et al. 2010). For example, here, FD FLIM measurements of Cerulean were analyzed for multiple lifetime components, revealing two distinct lifetime fractions, with an average lifetime (tau(f)) of 3 ns (Table 11.2). In contrast, the same analysis returned only a single lifetime for Cerulean3, Turquoise, and Turquoise2, so the measured lifetime and average lifetime (tau(f)) are the same (Table 11.2). The measurements also reveal that each of the new variants have longer lifetimes compared to Cerulean, which is consistent with their improved quantum yields (Lakowicz 2006). Additionally, the newer Cerulean3 and the Turquoise variants have markedly improved photostability, as well as reduced photoswitching behavior. Taken together, these results support the choice of the newer Cerulean-based FPs as preferred donors for FRET-based imaging for several reasons. First, the donor quantum yield determines the R_0 for the FRET pair (see Equation 11.2), and each of these variants has a higher quantum yield than original Cerulean, particularly Turquoise2 with a quantum yield of 0.93 (Table 11.1). Second, the single exponential decay (Tables 11.2 and 11.3) allows the unambiguous assignment of quenched and unquenched fractions in FRET–FLIM studies. Third, most imaging systems will already be configured for the use of cyan and yellow FPs.

An underappreciated characteristic of the genetically encoded FRET standards is the ease of substituting sequences encoding other FPs, enabling the direct comparison of different donor and acceptor fluorophore pairs. Here, FD FLIM measurements of a

Table 11.3 FRET standards comparing the different variants of Cerulean as energy transfer donors for Venus

FRET STANDARD[a]	TWO-COMPONENT LIFETIME (FRACTION)	Tau(f)[b] (±SD)	Tau(α)[c] (±SD)	E_{FRET}[d] (%±SD)
Cerulean-5aa-Amber	2.2 ns (0.66) 4.6 ns (0.34)	3.0 ± 0.05	NA	0
Cerulean-5aa-Venus	3.0 ns (0.55) 1.2 ns (0.45)	NA	1.8 ± 0.07	41.1 ± 2.4
Cerulean3-5aa-Amber	3.9 ns (0.99)	3.9 ± 0.06	NA	0
Cerulean3-5aa-Venus	3.3 ns (0.72) 1.2 ns (0.28)	NA	2.2 ± 0.09	41.7 ± 2.4
Turquoise-5aa-Amber	3.8 ns (0.99)	3.8 ± 0.04	NA	0
Turquoise-5aa-Venus	3.3 ns (0.69) 1.3 ns (0.31)	NA	2.3 ± 0.07	40.8 ± 1.8
Turquoise2-5aa-Amber	4.1 ns (0.99)	4.1 ± 0.04	NA	0
Turquoise2-5aa-Venus	3.3 ns (0.77) 1.1 ns (0.23)	NA	2.3 ± 0.05	44.6 ± 1.2

[a] Expressed in cells at 37°.
[b] Tau(f) is the average lifetime.
[c] Tau(α) is the amplitude-weighted lifetime.
[d] Determined by $E_{FRET} = (1 - DA_{tau(\alpha)}/D_{tau(f)})$.

series of different FRET standards are used to evaluate the newer variants of Cerulean as donor fluorophores for Venus (Table 11.3). The average unquenched donor lifetime (tau(f)) was determined for each of the donor fluorophores linked to Amber, the mutant variant of Venus that folds but does not absorb or emit fluorescence (Section 11.9). Then the amplitude-weighted lifetimes (tau(α)) of the quenched donors were determined for each of the FRET standards containing the Venus acceptor (Lakowicz 2006). Equation 11.4 is used to determine the FRET efficiency (E_{FRET}) for each of the standard fusion proteins expressed in living cells. The results demonstrate similar E_{FRET} for Cerulean, Cerulean3, and Turquoise (Table 11.3). However, the single lifetime for Cerulean3 and Turquoise variants makes them the preferred donors for FLIM studies. Further, as was expected based on the higher quantum yield of Turquoise2, the use of this FP resulted in an improvement in the E_{FRET} compared to Cerulean3 and Turquoise (Table 11.3).

This same approach is then used to characterize two of the newer yellow FPs, Clover (Section 11.4) or NeonGreen (Section 11.7.2), as acceptors for Turquoise. Based on their high extinction coefficients and strong spectral overlap with the cyan FPs, they could represent an improvement over Venus as an acceptor fluorophore. Here, the FRET standards used have a longer 10aa linker sequence (-SGLRSPPVAT-) developed in the Davidson laboratory. Predictably, the longer linker between Turquoise and Venus reduced the E_{FRET} (compare Tables 11.3 and 11.4). The substitution of NeonGreen for Venus resulted in a marked increase in the E_{FRET}. Interestingly, despite the similarity of Clover and NeonGreen in their photophysical and spectral characteristics, the substitution of Clover for Venus resulted in a reduction in the E_{FRET} (Table 11.4).

Table 11.4 FRET standards comparing Venus, NeonGreen, and Clover as energy transfer acceptors for Turquoise

FRET STANDARD[a]	TWO-COMPONENT LIFETIME (FRACTION)	Tau(α)[b] (±SD)	E$_{FRET}$[c] (%±SD)
Turquoise-10aa-Venus	3.3 ns (0.73) 1.6 ns (0.27)	2.3 ± 0.07	36.5 ± 2.4
Turquoise-10aa-NeonGreen	3.1 ns (0.74) 1.1 ns (0.26)	2.1 ± 0.03	44.9 ± 2.0
Turquoise-10aa-Clover	3.8 ns (0.58) 2.0 ns (0.42)	2.8 ± 0.1	27.4 ± 2.6

[a] Expressed in cells at 37°.
[b] Tau(α) is the amplitude-weighted lifetime.
[c] Determined by E$_{FRET}$ = (1 – DA$_{tau(α)}$/ D$_{tau(f)}$); see Table 11.3 for D$_{tau(f)}$.

This could indicate that Clover is not the best choice of a fluorophore as an acceptor for the cyan FPs. This also underscores the importance of direct comparisons using the FRET standard approach to quantify the performance of different FPs in FRET-based measurements.

For live-cell imaging, there are important advantages for using probes that are excited at longer wavelengths. The longer wavelength spectral windows decrease phototoxicity in the living specimens and also reduce autofluorescence background from the sample. We would expect the new orange and RFPs (Table 11.1) to have advantages for FRET studies but published studies using these probes for intensity-based FRET measurements are limited. The reason for this is likely that sensitized acceptor emission measurements favor fluorophores with a high quantum yield, and most of the RFPs have relatively low intrinsic brightness (Table 11.1). In contrast, the acceptor quantum yield is irrelevant if FLIM is used to detect the lifetime of the donor. Recently, Clover was proven to be a useful donor fluorophore for the improved red FP Ruby2 (Section 11.7.3). Compared to the cyan and yellow FPs, the combination of Clover and Ruby2 has an increased Förster distance and also an improved dynamic range (the difference between the minimum and maximum signal) when used in FRET biosensor probes monitored by ratiometric imaging (Lam et al. 2012).

Using fluorophores with even more spectral overlap will also increase the distance over which FRET can be detected (Patterson et al. 2000). For example, EGFP shares substantial spectral overlap with Venus (Table 11.1), which would make this an effective FRET pair. However, this exceptionally strong spectral overlap causes a profound increase in the background SBT signals, including the back bleed-through of the Venus signal into the donor (EGFP) channel.

This would appear to eliminate the GFP/YFP pair for FRET–FLIM measurements, but a strategy was developed that exploits the strong spectral overlap of this pair, while overcoming the problem of acceptor SBT. Novel YFPs were developed that have a high absorbance coefficient but have extremely low quantum yield. This class of chromophore, called resonance-energy-accepting chromoproteins (REACh) (see Table 11.1), permits the optimal use of GFP as a donor for FRET–FLIM (Ganesan et al. 2006; Murakoshi et al. 2008). Their very low quantum yield overcomes the potential problem of acceptor back-bleed-through emission into the donor channel. This allows the use of filters with a wider donor spectral window to collect optimally the donor signal. The measurement

Applications

of a double-exponential fluorescence lifetime decay curve for EGFP in the presence of the dark chromoproteins will now accurately reflect the populations of free donor and donor quenched by the REACh probe (Ganesan et al. 2006). What is more, the absence of fluorescence from REACh probes means that the spectral window normally occupied by the acceptor is now available for the detection of another probe. This opens the possibility of correlating the protein–protein interactions detected by FRET with the behavior of another labeled protein expressed inside the same living cells, the cellular biochemical network (Ganesan et al. 2006; Murakoshi et al. 2008).

11.13.3 GENERAL LIMITATIONS OF FRET-BASED IMAGING WITH THE FLUORESCENT PROTEINS

It is important to recognize that any amount of exogenous protein that is produced in a cell is, by definition, overexpressed relative to its endogenous counterpart. The transfection approach can yield very high levels of the fusion proteins in the target cells, especially when strong promoters are used. This can result in improper protein distribution and protein dysfunction that could lead to erroneous interpretations of protein activities. It is critical to verify that the fusion proteins retain the functions of the endogenous protein and that they have the expected subcellular localization. However, even with careful assessment of FP-fusion protein function and biochemical demonstrations of protein interactions, false-negative results are common in FRET imaging studies, because it can be difficult to achieve the required spatial relationships for FRET. Furthermore, positive FRET results from single cells are, by themselves, not sufficient to characterize the associations between proteins in living cells. Although the FRET measurements, when collected and quantified properly, are remarkably robust, there is still heterogeneity in the measurements. There can be substantial cell-to-cell heterogeneity for some types of interactions, and it is possible that only a subpopulation of cells responds to a particular stimulus. Therefore, data must be collected and statistically analyzed from multiple cells to prevent the user from reaching false conclusions based on a nonrepresentative measurement.

11.14 CONCLUSION AND PERSPECTIVES

Over the past few years, we have witnessed a remarkable expansion in the FP palette, which now spans the entire visible spectrum (reviewed in Day and Davidson 2009). Most FPs in common use today have been extensively modified through mutagenesis to optimize their characteristics for expression and imaging in living cells. Furthermore, some of these FPs have been developed specifically as probes for FRET imaging, allowing ångstrom-scale measurements of the spatial relationship between the fluorophores labeling proteins inside living cells. However, it must be emphasized that the best FP pairs for FRET-based studies is often not obvious from their spectral and photophysical characteristics alone, and there are limitations to many of the available FPs for FRET-based imaging studies.

It is also important to consider that live-cell imaging is a trade-off between acquiring adequate signal from the expressed FPs, while limiting cell damage that might be caused by the illumination of the fluorophores. Living systems are more tolerant of longer wavelength illumination, and the newer generations of bright, photostable yellow to red FPs offer alternatives to the blue and cyan variants that require near-UV excitation.

Many of the newer generation FPs are very bright and can be detected with minimal exposure of the living cells to the excitation illumination. Furthermore, if probes in the blue spectrum are necessary, long-wavelength excitation of these probes can be achieved with two-photon microscopy, offering a less damaging alternative to near-UV excitation (Drobishev et al. 2011). The many new FP probes are allowing noninvasive imaging techniques to complement and extend the results that are obtained by the biochemical analysis of the endogenous cellular proteins. Importantly, the measurements obtained from proteins labeled with the FPs in the natural environment inside living cells provide the most physiologically relevant information about protein behavior currently available.

REFERENCES

Adams, S. R., A. T. Harootunian, Y. J. Buechler, S. S. Taylor, and R. Y. Tsien. 1991. Fluorescence ratio imaging of cyclic AMP in single cells. *Nature* 349:694–697.

Ai, H. W., J. N. Henderson, S. J. Remington, and R. E. Campbell. 2006. Directed evolution of a monomeric, bright and photostable version of *Clavularia* cyan fluorescent protein: Structural characterization and applications in fluorescence imaging. *Biochem J* 400:531–540.

Ai, H. W., N. C. Shaner, Z. Cheng, R. Y. Tsien, and R. E. Campbell. 2007. Exploration of new chromophore structures leads to the identification of improved blue fluorescent proteins. *Biochemistry* 46:5904–5910.

Ai, H. W., S. G. Olenych, P. Wong, M. W. Davidson, and R. E. Campbell. 2008. Hue-shifted monomeric variants of *Clavularia* cyan fluorescent protein: Identification of the molecular determinants of color and applications in fluorescence imaging. *BMC Biol* 6:13.

Baird, G. S., D. A. Zacharias, and R. Y. Tsien. 2000. Biochemistry, mutagenesis, and oligomerization of DsRed, a red fluorescent protein from coral. *Proc Natl Acad Sci USA* 97:11984–11989.

Bastiaens, P. I. and A. Squire. 1999. Fluorescence lifetime imaging microscopy: Spatial resolution of biochemical processes in the cell. *Trends Cell Biol* 9:48–52.

Bastiaens, P. I., I. V. Majoul, P. J. Verveer, H. D. Soling, and T. M. Jovin. 1996. Imaging the intracellular trafficking and state of the AB5 quaternary structure of cholera toxin. *EMBO J* 15:4246–4253.

Bastiaens, P. I. and T. M. Jovin. 1996. Microspectroscopic imaging tracks the intracellular processing of a signal transduction protein: Fluorescent-labeled protein kinase C beta I. *Proc Natl Acad Sci USA* 93:8407–8412.

Baumann, D., M. Cook, L. Ma, A. Mushegian, E. Sanders, J. Schwartz, and C. R. Yu. 2008. A family of GFP-like proteins with different spectral properties in lancelet *Branchiostoma floridae*. *Biol Direct* 3:28.

Berney, C. and G. Danuser. 2003. FRET or no FRET: A quantitative comparison. *Biophys J* 84:3992–4010.

Bevis, B. J. and B. S. Glick. 2002. Rapidly maturing variants of the *Discosoma* red fluorescent protein (DsRed). *Nat Biotechnol* 20:83–87.

Brejc, K., T. K. Sixma, P. A. Kitts, S. R. Kain, R. Y. Tsien, M. Ormo, and S. J. Remington. 1997. Structural basis for dual excitation and photoisomerization of the *Aequorea victoria* green fluorescent protein. *Proc Natl Acad Sci USA* 94:2306–2311.

Buranachai, C., D. Kamiyama, A. Chiba, B. D. Williams, and R. M. Clegg. 2008. Rapid frequency-domain FLIM spinning disk confocal microscope: Lifetime resolution, image improvement and wavelet analysis. *J Fluoresc* 18:929–942.

Campbell, R. E., O. Tour, A. E. Palmer, P. A. Steinbach, G. S. Baird, D. A. Zacharias, and R. Y. Tsien. 2002. A monomeric red fluorescent protein. *Proc Natl Acad Sci USA* 99:7877–7882.

Carlson, H. J. and R. E. Campbell. 2009. Genetically encoded FRET-based biosensors for multiparameter fluorescence imaging. *Curr Opin Chem Biol* 20:19–27.

Chalfie, M., Y. Tu, G. Euskirchen, W. W. Ward, and D. C. Prasher. 1994. Green fluorescent protein as a marker for gene expression. *Science* 263:802–805.

Chattoraj, M., B. A. King, G. U. Bublitz, and S. G. Boxer. 1996. Ultra-fast excited state dynamics in green fluorescent protein: Multiple states and proton transfer. *Proc Natl Acad Sci USA* 93:8362–8367.

Chen, Y., J. P. Mauldin, R. N. Day, and A. Periasamy. 2007. Characterization of spectral FRET imaging microscopy for monitoring nuclear protein interactions. *J Microsc* 228:139–152.

Clegg, R. M. 2010. Fluorescence lifetime-resolved imaging what, why, how—A prologue. In *FLIM Microscopy in Biology and Medicine*, A. Periasamy and R. M. Clegg, eds., CRC Press, Boca Raton, FL, pp. 3–34.

Cody, C. W., D. C. Prasher, W. M. Westler, F. G. Prendergast, and W. W. Ward. 1993. Chemical structure of the hexapeptide chromophore of the *Aequorea* green-fluorescent protein. *Biochemistry* 32:1212–1218.

Cubitt, A. B., L. A. Woollenweber, and R. Heim. 1999. Understanding structure-function relationships in the *Aequorea victoria* green fluorescent protein. *Methods Cell Biol* 58:19–30.

Day, R. N. 1998. Visualization of Pit-1 transcription factor interactions in the living cell nucleus by fluorescence resonance energy transfer microscopy. *Mol Endocrinol* 12:1410–1419.

Day, R. N., C. F. Booker, and A. Periasamy. 2008. Characterization of an improved donor fluorescent protein for Forster resonance energy transfer microscopy. *J Biomed Opt* 13:031203.

Day, R. N. and M. W. Davidson. 2009. The fluorescent protein palette: Tools for cellular imaging. *Chem Soc Rev* 38:2887–2921.

Dickinson, M. E., E. Simbuerger, B. Zimmermann, C. W. Waters, and S. E. Fraser. 2003. Multiphoton excitation spectra in biological samples. *J Biomed Opt* 8:329–338.

DiPilato, L. M. and J. Zhang. 2010. Fluorescent protein-based biosensors: Resolving spatiotemporal dynamics of signaling. *Curr Opin Chem Biol* 14:37–42.

Drobizhev, M., N. S. Makarov, S. E. Tillo, T. E. Hughes, and A. Rebane. 2011. Two-photon absorption properties of fluorescent proteins. *Nat Methods* 8:393–399.

Elangovan, M., H. Wallrabe, Y. Chen, R. N. Day, M. Barroso, and A. Periasamy. 2003. Characterization of one- and two-photon excitation fluorescence resonance energy transfer microscopy. *Methods* 29:58–73.

Förster, T. 1965. Delocalized excitation and excitation transfer. In *Modern Quantum Chemistry*, Vol. 3, pp. 93–137. O. Sinanoglu, ed., Academic Press Inc., New York.

Ganesan, S., S. M. Ameer-Beg, T. T. Ng, B. Vojnovic, and F. S. Wouters. 2006. A dark yellow fluorescent protein (YFP)-based resonance energy-accepting chromoprotein (REACh) for Forster resonance energy transfer with GFP. *Proc Natl Acad Sci USA* 103:4089–4094.

Giepmans, B. N., S. R. Adams, M. H. Ellisman, and R. Y. Tsien. 2006. The fluorescent toolbox for assessing protein location and function. *Science* 312:217–224.

Goedhart, J., D. von Stetten, M. Noirclerc-Savoye, M. Lelimousin, L. Joosen, M. A. Hink, L. van Weeren, T. W. Gadella, Jr., and A. Royant. 2012. Structure-guided evolution of cyan fluorescent proteins towards a quantum yield of 93%. *Nat Commun* 3:751–759.

Goedhart, J., J. E. Vermeer, M. J. Adjobo-Hermans, L. van Weeren, and T. W. Gadella, Jr. 2007. Sensitive detection of p65 homodimers using red-shifted and fluorescent protein-based FRET couples. *PLoS One* 2:e1011.

Goedhart, J., L. van Weeren, M. A. Hink, N. O. Vischer, K. Jalink, and T. W. Gadella, Jr. 2010. Bright cyan fluorescent protein variants identified by fluorescence lifetime screening. *Nat Methods* 7:137–139.

Griesbeck, O., G. S. Baird, R. E. Campbell, D. A. Zacharias, and R. Y. Tsien. 2001. Reducing the environmental sensitivity of yellow fluorescent protein. Mechanism and applications. *J Biol Chem* 276:29188–29194.

Gryczynski, Z., Gryczynski, I., and Lakowicz, J. R. 2005. Basics of fluorescence and FRET. In *Molecular Imaging: FRET Microscopy and Spectroscopy*, Vol. 2, pp. 21–56. A. Periasamy and R. N. Day, eds., Oxford University Press, New York.

Applications

Haugh, J. M. 2012. Live-cell fluorescence microscopy with molecular biosensors: What are we really measuring? *Biophys J* 102:2003–2011.

Heim, R., A. B. Cubitt, and R. Y. Tsien. 1995. Improved green fluorescence. *Nature* 373:663–664.

Heim, R., D. C. Prasher, and R. Y. Tsien. 1994. Wavelength mutations and posttranslational autoxidation of green fluorescent protein. *Proc Natl Acad Sci USA* 91:12501–12504.

Heim, R. and R. Y. Tsien. 1996. Engineering green fluorescent protein for improved brightness, longer wavelengths and fluorescence resonance energy transfer. *Curr Biol* 6:178–182.

Hillesheim, L. N., Y. Chen, and J. D. Muller. 2006. Dual-color photon counting histogram analysis of mRFP1 and EGFP in living cells. *Biophys J* 91:4273–4284.

Hinde, E., M. A. Digman, C. Welch, K. M. Hahn, and E. Gratton. 2012. Biosensor Förster resonance energy transfer detection by the phasor approach to fluorescence lifetime imaging microscopy. *Microsc Res Tech* 75:271–281.

Hum, J. M., A. P. Siegel, F. M. Pavalko, and R. N. Day. 2012. Monitoring biosensor activity in living cells with fluorescence lifetime imaging microscopy. *Int J Mol Sci* 13:14385–14400.

Inoué, S. 2006. Foundations of confocal scanned imaging in light microscopy. In *Handbook of Biological Confocal Microscopy*, Vol. 1, pp. 1–17. J. B. Pawley, ed., Springer, New York.

Inouye, S. and F. I. Tsuji. 1994. *Aequorea* green fluorescent protein. Expression of the gene and fluorescence characteristics of the recombinant protein. *FEBS Lett* 341:277–280.

Jameson, D. M., E. Gratton, and R. D. Hall. 1984. The measurement and analysis of heterogeneous emissions by multifrequency phase and modulation fluorometry. *Appl Spectrosc Rev* 20:55–106.

Karasawa, S., T. Araki, T. Nagai, H. Mizuno, and A. Miyawaki. 2004. Cyan-emitting and orange-emitting fluorescent proteins as a donor/acceptor pair for fluorescence resonance energy transfer. *Biochem J* 381:307–312.

Koushik, S. V., H. Chen, C. Thaler, H. L. Puhl, 3rd, and S. S. Vogel. 2006. Cerulean, Venus, and VenusY67C FRET reference standards. *Biophys J* 91:L99–L101.

Kredel, S., F. Oswald, K. Nienhaus, K. Deuschle, C. Rocker, M. Wolff, R. Hcilkcr, G. U. Nienhaus, and J. Wiedenmann. 2009. mRuby, a bright monomeric red fluorescent protein for labeling of subcellular structures. *PLoS One* 4:e4391.

Kremers, G. J., J. Goedhart, D. J. van den Heuvel, H. C. Gerritsen, and T. W. Gadella, Jr. 2007. Improved green and blue fluorescent proteins for expression in bacteria and mammalian cells. *Biochemistry* 46:3775–3783.

Kremers, G. J., K. L. Hazelwood, C. S. Murphy, M. W. Davidson, and D. W. Piston. 2009. Photoconversion in orange and red fluorescent proteins. *Nat Methods* 6:355–358.

Labas, Y. A., N. G. Gurskaya, Y. G. Yanushevich, A. F. Fradkov, K. A. Lukyanov, S. A. Lukyanov, and M. V. Matz. 2002. Diversity and evolution of the green fluorescent protein family. *Proc Natl Acad Sci USA* 99:4256–4261.

Lam, A. J., F. St-Pierre, Y. Gong, J. D. Marshall, P. J. Cranfill, M. A. Baird, M. R. McKeown, J. Wiedenmann, M. W. Davidson, M. J. Schnitzer, R. Y. Tsien, and M. Z. Lin. 2012. Improving FRET dynamic range with bright green and red fluorescent proteins. *Nat Methods* 9:1005–1012.

Lakowicz, J. R. 2006. *Principles of Fluorescence Spectroscopy*. Springer, New York.

Lippincott-Schwartz, J., E. Snapp, and A. Kenworthy. 2001. Studying protein dynamics in living cells. *Nat Rev Mol Cell Biol* 2:444–456.

Luche, H., O. Weber, T. Nageswara Rao, C. Blum, and H. J. Fehling. 2007. Faithful activation of an extra-bright red fluorescent protein in "knock-in" Cre-reporter mice ideally suited for lineage tracing studies. *Eur J Immunol* 37:43–53.

Markwardt, M. L., G. J. Kremers, C. A. Kraft, K. Ray, P. J. Cranfill, K. A. Wilson, R. N. Day, R. M. Wachter, M. W. Davidson, and M. A. Rizzo. 2011. An improved cerulean fluorescent protein with enhanced brightness and reduced reversible photoswitching. *PLoS One* 6:e17896.

Matz, M. V., A. F. Fradkov, Y. A. Labas, A. P. Savitsky, A. G. Zaraisky, M. L. Markelov, and S. A. Lukyanov. 1999. Fluorescent proteins from nonbioluminescent Anthozoa species. *Nat Biotechnol* 17:969–973.

Matz, M. V., K. A. Lukyanov, and S. A. Lukyanov. 2002. Family of the green fluorescent protein: Journey to the end of the rainbow. *Bioessays* 24:953–959.

McNamara, G., A. Gupta, J. Reynaert, T. D. Coates, and C. Boswell. 2006. Spectral imaging microscopy web sites and data. *Cytometry A* 69:863–871.

Mena, M. A., T. P. Treynor, S. L. Mayo, and P. S. Daugherty. 2006. Blue fluorescent proteins with enhanced brightness and photostability from a structurally targeted library. *Nat Biotechnol* 24:1569–1571.

Merzlyak, E. M., J. Goedhart, D. Shcherbo, M. E. Bulina, A. S. Shcheglov, A. F. Fradkov, A. Gaintzeva, K. A. Lukyanov, S. Lukyanov, T. W. Gadella, and D. M. Chudakov. 2007. Bright monomeric red fluorescent protein with an extended fluorescence lifetime. *Nat Methods* 4:555–557.

Millington, M., G. J. Grindlay, K. Altenbach, R. K. Neely, W. Kolch, M. Bencina, N. D. Read, A. C. Jones, D. T. Dryden, and S. W. Magennis. 2007. High-precision FLIM-FRET in fixed and living cells reveals heterogeneity in a simple CFP-YFP fusion protein. *Biophys Chem* 127:155–164.

Mitra, R. D., C. M. Silva, and D. C. Youvan. 1996. Fluorescence resonance energy transfer between blue-emitting and red-shifted excitation derivatives of the green fluorescent protein. *Gene* 173:13–17.

Miyawaki, A. 2011. Development of probes for cellular functions using fluorescent proteins and fluorescence resonance energy transfer. *Ann Rev Biochem* 80:357–373.

Miyawaki, A., J. Llopis, R. Heim, J. M. McCaffery, J. A. Adams, M. Ikura, and R. Y. Tsien. 1997. Fluorescent indicators for Ca2+ based on green fluorescent proteins and calmodulin. *Nature* 388:882–887.

Morise, H., O. Shimomura, F. H. Johnson, and J. Winant. 1974. Intermolecular energy transfer in the bioluminescent system of *Aequorea*. *Biochemistry* 13:2656–2662.

Murakoshi, H., S. J. Lee, and R. Yasuda. 2008. Highly sensitive and quantitative FRET-FLIM imaging in single dendritic spines using improved non-radiative YFP. *Brain Cell Biol* 36:31–42.

Nagai, T., K. Ibata, E. S. Park, M. Kubota, K. Mikoshiba, and A. Miyawaki. 2002. A variant of yellow fluorescent protein with fast and efficient maturation for cell-biological applications. *Nat Biotechnol* 20:87–90.

Ormö, M., A. B. Cubitt, K. Kallio, L. A. Gross, R. Y. Tsien, and S. J. Remington. 1996. Crystal structure of the *Aequorea victoria* green fluorescent protein. *Science* 273:1392–1395.

Ouyang, M., H. Huang, N. C. Shaner, A. G. Remacle, S. A. Shiryaev, A. Y. Strongin, R. Y. Tsien, and Y. Wang. 2010. Simultaneous visualization of protumorigenic Src and MT1-MMP activities with fluorescence resonance energy transfer. *Cancer Res* 70:2204–2212.

Padilla-Parra, S., N. Auduge, H. Lalucque, J. C. Mevel, M. Coppey-Moisan, and M. Tramier. 2009. Quantitative comparison of different fluorescent protein couples for fast FRET-FLIM acquisition. *Biophys J* 97:2368–2376.

Patterson, G. H., D. W. Piston, and B. G. Barisas. 2000. Forster distances between green fluorescent protein pairs. *Anal Biochem* 284:438–440.

Periasamy, A. and R. M. Clegg. 2010. *FLIM Microscopy in Biology and Medicine*. Taylor & Francis, Boca Raton, FL.

Periasamy, A. and R. N. Day. 1999. Visualizing protein interactions in living cells using digitized GFP imaging and FRET microscopy. *Methods Cell Biol* 58:293–314.

Periasamy, A., R. N. Day, and American Physiological Society (1887-). 2005. *Molecular Imaging: FRET Microscopy and Spectroscopy*. Oxford University Press: Published for the American Physiological Society, Oxford; New York.

Peter, M., S. M. Ameer-Beg, M. K. Hughes, M. D. Keppler, S. Prag, M. Marsh, B. Vojnovic, and T. Ng. 2005. Multiphoton-FLIM quantification of the EGFP-mRFP1 FRET pair for localization of membrane receptor-kinase interactions. *Biophys J* 88:1224–1237.

Piston, D. W. and G. J. Kremers. 2007. Fluorescent protein FRET: The good, the bad and the ugly. *Trends Biochem Sci* 32:407–414.

Prasher, D. C., V. K. Eckenrode, W. W. Ward, F. G. Prendergast, and M. J. Cormier. 1992. Primary structure of the *Aequorea victoria* green-fluorescent protein. *Gene* 111:229–233.

Redford, G. I. and R. M. Clegg. 2005. Polar plot representation for frequency-domain analysis of fluorescence lifetimes. *J Fluoresc* 15:805–815.

Rizzo, M. A., G. H. Springer, B. Granada, and D. W. Piston. 2004. An improved cyan fluorescent protein variant useful for FRET. *Nat Biotechnol* 22:445–449.

Romoser, V. A., P. M. Hinkle, and A. Persechini. 1997. Detection in living cells of Ca2+-dependent changes in the fluorescence emission of an indicator composed of two green fluorescent protein variants linked by a calmodulin-binding sequence. A new class of fluorescent indicators. *J Biol Chem* 272:13270–13274.

Sakaue-Sawano, A., H. Kurokawa, T. Morimura, A. Hanyu, H. Hama, H. Osawa, S. Kashiwagi, K. Fukami, T. Miyata, H. Miyoshi, T. Imamura, M. Ogawa, H. Masai, and A. Miyawaki. 2008. Visualizing spatiotemporal dynamics of multicellular cell-cycle progression. *Cell* 132:487–498.

Seong, J., S. Lu, M. Ouyang, H. Huang, J. Zhang, M. C. Frame, and Y. Wang. 2009. Visualization of Src activity at different compartments of the plasma membrane by FRET imaging. *Chem Biol* 16:48–57.

Shagin, D. A., E. V. Barsova, Y. G. Yanushevich, A. F. Fradkov, K. A. Lukyanov, Y. A. Labas, T. N. Semenova, J. A. Ugalde, A. Meyers, J. M. Nunez, E. A. Widder, S. A. Lukyanov, and M. V. Matz. 2004. GFP-like proteins as ubiquitous metazoan superfamily: Evolution of functional features and structural complexity. *Mol Biol Evol* 21:841–850.

Shaner, N. C., R. E. Campbell, P. A. Steinbach, B. N. Giepmans, A. E. Palmer, and R. Y. Tsien. 2004. Improved monomeric red, orange and yellow fluorescent proteins derived from *Discosoma* sp. red fluorescent protein. *Nat Biotechnol* 22:1567–1572.

Shaner, N. C., G. G. Lambert, A. Chammas, Y. Ni, P. J. Cranfill, M. A. Baird, B. R. Sell, R. N. Day, M. W. Davidson, and J. Wang. 2013. A bright monomeric green fluorescent protein derived from *Branchiostoma lanceolatum*. *Nat Methods* 10:407–409.

Shaner, N. C., G. H. Patterson, and M. W. Davidson. 2007. Advances in fluorescent protein technology. *J Cell Sci* 120:4247–4260.

Shaner, N. C., M. Z. Lin, M. R. McKeown, P. A. Steinbach, K. L. Hazelwood, M. W. Davidson, and R. Y. Tsien. 2008. Improving the photostability of bright monomeric orange and red fluorescent proteins. *Nat Methods* 5:545–551.

Shaner, N. C., P. A. Steinbach, and R. Y. Tsien. 2005. A guide to choosing fluorescent proteins. *Nat Methods* 2:905–909.

Shcherbo, D., E. A. Souslova, J. Goedhart, T. V. Chepurnykh, A. Gaintzeva, I. I. Shemiakina, T. W. Gadella, S. Lukyanov, and D. M. Chudakov. 2009. Practical and reliable FRET/FLIM pair of fluorescent proteins. *BMC Biotechnol* 9:24.

Shcherbo, D., E. M. Merzlyak, T. V. Chepurnykh, A. F. Fradkov, G. V. Ermakova, E. A. Solovieva, K. A. Lukyanov, E. A. Bogdanova, A. G. Zaraisky, S. Lukyanov, and D. M. Chudakov. 2007. Bright far-red fluorescent protein for whole-body imaging. *Nat Methods* 4:741–746.

Shimomura, O., F. H. Johnson, and Y. Saiga. 1962. Extraction, purification and properties of aequorin, a bioluminescent protein from the luminous hydromedusan, *Aequorea*. *J Cell Comp Physiol* 59:223–239.

Stryer, L. 1978. Fluorescence energy transfer as a spectroscopic ruler. *Annu Rev Biochem* 47:819–846.

Sun, Y., C. F. Booker, S. Kumari, R. N. Day, M. Davidson, and A. Periasamy. 2009. Characterization of an orange acceptor fluorescent protein for sensitized spectral fluorescence resonance energy transfer microscopy using a white-light laser. *J Biomed Opt* 14:054009.

Sun, Y., H. Wallrabe, C. F. Booker, R. N. Day, and A. Periasamy. 2010. Three-color spectral FRET microscopy localizes three interacting proteins in living cells. *Biophys J* 99:1274–1283.

Thaler, C., S. V. Koushik, P. S. Blank, and S. S. Vogel. 2005. Quantitative multiphoton spectral imaging and its use for measuring resonance energy transfer. *Biophys J* 89:2736–2749.

Tian, L., S. A. Hires, T. Mao, D. Huber, M. E. Chiappe, S. H. Chalasani, L. Petreanu, J. Akerboom, S. A. McKinney, E. R. Schreiter, C. I. Bargmann, V. Jayaraman, K. Svoboda, and L. L. Looger. 2009. Imaging neural activity in worms, flies and mice with improved GCaMP calcium indicators. *Nat Methods* 6:875–881.

Tramier, M., I. Gautier, T. Piolot, S. Ravalet, K. Kemnitz, J. Coppey, C. Durieux, V. Mignotte, and M. Coppey-Moisan. 2002. Picosecond-hetero-FRET microscopy to probe protein-protein interactions in live cells. *Biophys J* 83:3570–3577.

Tramier, M., M. Zahid, J. C. Mevel, M. J. Masse, and M. Coppey-Moisan. 2006. Sensitivity of CFP/YFP and GFP/mCherry pairs to donor photobleaching on FRET determination by fluorescence lifetime imaging microscopy in living cells. *Microsc Res Tech* 69:933–939.

Tsien, R. Y. 1998. The green fluorescent protein. *Annu Rev Biochem* 67:509–544.

Tsien, R. Y. 2010. The 2009 Lindau Nobel Laureate Meeting: Roger Y. Tsien, Chemistry 2008. *J Vis Exp* e1575.

Verkhusha, V. V. and K. A. Lukyanov. 2004. The molecular properties and applications of *Anthozoa* fluorescent proteins and chromoproteins. *Nat Biotechnol* 22:289–296.

Vogel, S. S., C. Thaler, and S. V. Koushik. 2006. Fanciful FRET. *Sci STKE* 2006:re2.

Vogel, S. S., T. A. Nguyen, B. W. van der Meer, and P. S. Blank. 2012. The impact of heterogeneity and dark acceptor states on FRET: Implications for using fluorescent protein donors and acceptors. *PLoS One* 7:e49593.

Wachter, R. M., M. A. Elsliger, K. Kallio, G. T. Hanson, and S. J. Remington. 1998. Structural basis of spectral shifts in the yellow-emission variants of green fluorescent protein. *Structure* 6:1267–1277.

Wang, L., W. C. Jackson, P. A. Steinbach, and R. Y. Tsien. 2004. Evolution of new nonantibody proteins via iterative somatic hypermutation. *Proc Natl Acad Sci USA* 101:16745–16749.

Ward, W. W. and M. J. Cormier. 1979. An energy transfer protein in coelenterate bioluminescence. Characterization of the Renilla green-fluorescent protein. *J Biol Chem* 254:781–788.

Xu, X., A. L. Gerard, B. C. Huang, D. C. Anderson, D. G. Payan, and Y. Luo. 1998. Detection of programmed cell death using fluorescence energy transfer. *Nucleic Acids Res* 26:2034–2035.

Yang, F., L. G. Moss, and G. N. Phillips, Jr. 1996. The molecular structure of green fluorescent protein. *Nat Biotechnol* 14:1246–1251.

Yasuda, R. 2006. Imaging spatiotemporal dynamics of neuronal signaling using fluorescence resonance energy transfer and fluorescence lifetime imaging microscopy. *Curr Opin Neurobiol* 16:551–561.

Yasuda, R., C. D. Harvey, H. Zhong, A. Sobczyk, L. van Aelst, and K. Svoboda. 2006. Supersensitive Ras activation in dendrites and spines revealed by two-photon fluorescence lifetime imaging. *Nat Neurosci* 9:283–291.

You, X., A. W. Nguyen, A. Jabaiah, M. A. Sheff, K. S. Thorn, and P. S. Daugherty. 2006. Intracellular protein interaction mapping with FRET hybrids. *Proc Natl Acad Sci USA* 103:18458–18463.

Zaccolo, M., F. De Giorgi, C. Y. Cho, L. Feng, T. Knapp, P. A. Negulescu, S. S. Taylor, R. Y. Tsien, and T. Pozzan. 2000. A genetically encoded, fluorescent indicator for cyclic AMP in living cells. *Nat Cell Biol* 2:25–29.

Zacharias, D. A., J. D. Violin, A. C. Newton, and R. Y. Tsien. 2002. Partitioning of lipid-modified monomeric GFPs into membrane microdomains of live cells. *Science* 296:913–916.

Zapata-Hommer, O. and O. Griesbeck. 2003. Efficiently folding and circularly permuted variants of the Sapphire mutant of GFP. *BMC Biotechnol* 3:5.

Zhang, J. 2009. The colorful journey of green fluorescent protein. *ACS Chem Biol* 4:85–88.

Zhang, J. and M. D. Allen. 2007. FRET-based biosensors for protein kinases: Illuminating the kinome. *Mol Biosyst* 3:759–765.

Zhang, J., R. E. Campbell, A. Y. Ting, and R. Y. Tsien. 2002. Creating new fluorescent probes for cell biology. *Nat Rev Mol Cell Biol* 3:906–918.

Zhou, X., K. J. Herbst-Robinson, and J. Zhang. 2012. Visualizing dynamic activities of signaling enzymes using genetically encodable FRET-based biosensors from designs to applications. *Methods Enzymol* 504:317–340.

Zolotukhin, S., M. Potter, W. W. Hauswirth, J. Guy, and N. Muzyczka. 1996. A "humanized" green fluorescent protein cDNA adapted for high-level expression in mammalian cells. *J Virol* 70:4646–4654.

Superresolution techniques using fluorescent protein technology

John R. Allen and *Michael W. Davidson*
Florida State University

Stephen T. Ross
Nikon Instruments, Inc.

Contents

12.1 INTRODUCTION

For decades, fluorescence microscopy has proved a powerful tool for visualizing subcellular structures. Due to its minimally invasive nature, fluorescence microscopy can be used to visualize the temporal dynamics of cells, tissues, and often entire organisms *in vivo*. However, as with all forms of optical microscopy, resolution is ultimately limited by the diffraction of light waves occurring between the sample and the detector. This limit, popularly known as the diffraction barrier, was first characterized by Ernst Abbe and Lord Rayleigh in the nineteenth century and had continued to define a maximum resolution of roughly 200 and 500 nm in the lateral and axial dimensions, respectively. However, there has been a recent surge in methods designed to circumvent the diffraction limit, some theoretically capable of diffraction-unlimited resolution. These powerful new techniques, some approaching single-molecule resolution, are collectively referred to as "superresolution" techniques. We will briefly survey some of the most promising superresolution microscopies and explore the unique requirements imposed on fluorescent probe selection, with an emphasis on the use of fluorescent proteins (FPs). But first, it is of the utmost importance to understand the diffraction barrier and other factors that traditionally limit resolution.

When using a microscope to image a point source of light, the result is a large diffracted focal spot known as an Airy disk, after the British astronomer George Airy. The 3D diffraction pattern analog of the Airy disk is the point-spread function (PSF), which (as serially viewed through a microscope objective) appears as an ellipsoid. An Airy disk appears as a single bright spot surrounded by concentric rings of decreasing intensity, and its size governs the minimum distance at which two separate point sources of light can be discerned. This is represented by the classic Abbe equations for lateral and axial resolutions:

$$\text{Resolution}_{x,y} = \frac{\lambda}{2[\eta \sin(\alpha)]}$$

$$\text{Resolution}_z = \frac{2\lambda}{[\eta \sin(\alpha)]^2}$$

where

λ is the wavelength of excitation light

η is the refractive index of the imaging medium

the combination of $\eta \cdot \sin(\alpha)$ represents the numerical aperture (NA) of the objective lens

Other popular resolution metrics include the Rayleigh and Sparrow criterions. Each metric, in essence, defines an arbitrary minimum distance required between two point source emitters such that they can be considered optically resolved from each other. In an ideal configuration, using a high-NA objective and high-frequency ultraviolet (UV)-violet excitation light, the greatest theoretically obtainable resolution is about 150 nm laterally and 400 nm axially. However, in reality, these values are closer to about 200 and 500 nm, respectively, rendering structures separated by distances less than these values unresolved. Thus, while traditional fluorescence microscopy techniques (e.g., wide-field, spinning disk, laser scanning, total internal reflection fluorescence [TIRF], and multiphoton) are invaluable tools for visualizing larger cellular constructs such as mitochondria, cytoskeletal components, and the endoplasmic reticulum, they ultimately fall short when attempting to characterize activities involving smaller cellular components (e.g., ribosomes, synaptic vesicles, and single molecules) where fine structures and close interactions are obfuscated by the overlapping Airy disks created by simultaneous excitation of each point source. This limitation has led to the development of a myriad of new techniques and approaches with the ultimate goal of circumventing the resolution limit imposed by the diffraction of light.

12.2 SUPERRESOLUTION MICROSCOPIES

The first true optical superresolution technique was actually a near-field approach developed in 1986 known as near-field scanning optical microscopy (NSOM; Betzig et al. 1986). NSOM works by placing a small probe equipped with a detector a distance shorter than a single excitation wavelength from the sample and serially scanning it to create a superresolution image. By imaging this close to the sample, NSOM can detect normally inaccessible evanescent fields (which decay exponentially with increasing distance from their origin). However, this technique is very difficult to use on biologically relevant samples with variable topography and is largely confined to surface studies. Starting in the early 1990s, new far-field techniques began to emerge capable of circumventing the diffraction barrier of light microscopy. The first, standing wave fluorescence microscopy (SWFM; Bailey et al. 1993), uses the interference of a pair of counterpropagating laser beams to create standing nodes and antinodes of illumination along the optical (z) axis. These nodes, approximately 50 nm in width, can be used to selectively excite fluorescence from very thin optical sections. However, proper utilization of this technique requires extremely thin samples that do not occupy multiple illumination planes. Not long after the introduction of SWFM, axial resolution saw an approximate sevenfold increase with the introduction of 4Pi (Hell and Stelzer 1992, Hell and Nagorni 1998, Gugel et al. 2004) and I^5M (Gustafsson et al. 1995, 1999). Both techniques utilize objectives opposed on either side of the specimen (sandwiched between a pair of coverslips) to manipulate the shape of the PSF. The most sophisticated iterations of these related techniques use both interference of the excitation illumination to create near-isotropic illumination PSFs and constructive interference of emission light at the camera image plane to sharpen axial resolution.

The first true leap forward in lateral resolution came with the introduction of stimulated emission depletion (STED) fluorescence microscopy in 1995 by Stefan Hell and colleagues (Hell and Wichmann 1994, Westphal and Hell 2005). STED is one of many closely related techniques based on the concept of reversibly saturable optical fluorescence transitions (RESOLFT) proposed by Hell (2005) but the most prominent, being fully realized as a viable, widespread, and commercially available superresolution imaging methodology. RESOLFT techniques rely upon fluorophores that can be stably switched between a bright fluorescent "on" state and a dark "off" state or, in theory, any two states A and B that can be reversibly switched between. An example of such a transition is *cis–trans* isomerization of the chromophore in a photoswitchable FP (PS-FP), allowing it to reversibly switch between an "on" and "off" state (Figure 12.2c). In the case of STED, the bright state is the excited singlet state S_1 of the fluorophore, and the dark state is the electronic ground state S_0; stimulated emission of a molecule in the S_1 state prevents normal fluorescent relaxation to the ground state, as illustrated by a simplified Jablonski energy diagram (Figure 12.2a). Another example would be the transition between the excited singlet state and the metastable dark triplet states used in ground state depletion (GSD; Hell and Kroug 1995, Bretschneider et al. 2007) and GSD individual molecule return (GSDIM; Folling et al. 2008) microscopy. The resolving power of various superresolution techniques, such as STED, are compared against those of more traditional optical microscopies, such as wide-field and confocal, in Figure 12.1.

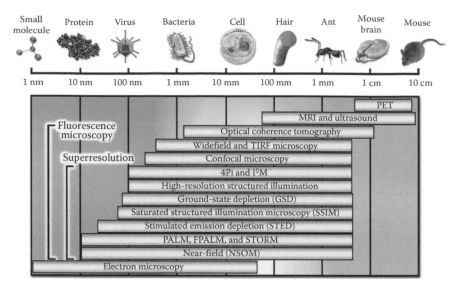

Figure 12.1 Comparison of different biological imaging techniques, showing spatial resolution ranges and corresponding biological features of similar size. The bracketed superresolution-type techniques allow for the observation of features smaller than approximately 200 nm in the lateral directions. (Reprinted with permission from Allen, J.R., Ross, S.T., and Davidson, M.W., Single-molecule localization microscopy for superresolution, *J. Opt.*, 15, 2013, 094001. IOP Publishing Ltd.)

12.2.1 STIMULATED EMISSION DEPLETION MICROSCOPY

When a molecule photoswitches to an on or off state, the probability that it will remain in that state decreases exponentially with increasing excitation intensity. The saturation intensity (I_{sat}) is thus defined as the light intensity at which switching will occur in 50% of molecules. The hypothetical STED beam profiles corresponding to varying intensities exceeding I_{sat} are illustrated by Figure 12.2b. STED utilizes a wide, donut-shaped, mode-locked depletion laser, termed a "STED beam," with an intensity greatly exceeding the saturation intensity of the fluorophore, resulting in stimulated emission of irradiated molecules. The donut shape of the STED beam is illustrated by a wireframe rendering in Figure 12.2d. Stimulated emission is a process by which a molecule in an excited state may drop to a lower energy level through interaction with an electromagnetic wave (photon in STED) of a certain frequency. This results in the creation of a photon in the same phase and direction of travel as the incident photon, thus preventing normal fluorescent emission. Each point of the viewfield is irradiated with a very short pulse from an excitation laser, followed immediately by another short pulse (~10–300 ps each) from the STED beam to stimulate emission prior to fluorescent emission. However, the STED beam has a narrow, central zero node where stimulated emission occurs in a much smaller proportion of molecules, allowing the excitation laser to induce fluorescence from molecules excited in this area. This technique effectively reduces the PSF of the excitation beam to a subdiffraction-limited area, where size is directly determined by the intensity of the STED beam, as illustrated by Figure 12.3b. Figure 12.3a shows a simplified STED light path, specifically the donut-shaped STED beam encompassing the excitation beam. By raster-scanning each position in the image

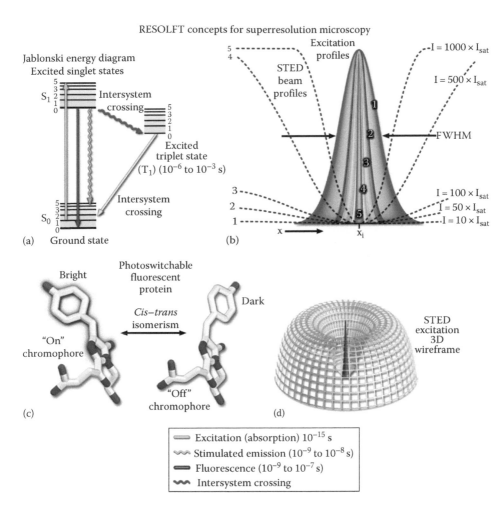

RESOLFT concepts for superresolution microscopy

Figure 12.2 Physical concepts governing probe saturation and optical resolution for STED- and RESOLFT-type optical microscopies. (a) Simplified Jablonski energy diagram illustrating energized states and timescales associated with excitation, fluorescence, and stimulated emission. Following excitation to the S_1 singlet state, stimulated emission is achieved by illumination with certain frequencies of light and results in two photons of the same energy, phase, and direction as the incident photon triggering stimulated emission. (b) STED beam and excitation profiles with varying degrees of STED beam intensity (I_{sat} = the fluorophore-specific saturation intensity). Note that with stronger illumination, STED beam profiles are steeper and thus result in narrower excitation profiles. The lowest illumination intensity ($I = 10 \times I_{sat}$) corresponds with excitation profile #1, while the highest intensity ($I = 1000 \times I_{sat}$) corresponds with excitation profile #5. (c) Cis–trans isomerization of the chromophore is the mechanism employed by many reversibly PS-FPs to cycle between fluorescent and nonfluorescent (dark) states. This is important because it allows one to turn the probe "off" in a manner that requires much lower laser power than required for stimulated emission of a conventional fluorophore, making them popular probes for RESOLFT-type alternatives to traditional STED microscopy. (d) Wireframe illustration of the STED beam and resulting excitation profiles. Note the donut-shaped intensity distribution of the STED beam, featuring a zero node at the origin to allow for excitation.

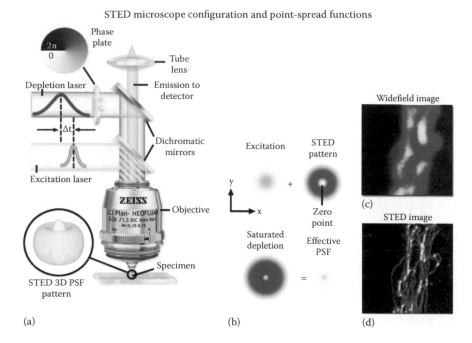

STED microscope configuration and point-spread functions

Figure 12.3 Simplified diagram of a STED microscope configuration, illustrating the improved excitation PSF size and corresponding gain in image resolution. (a) Light path and relative phase of the depletion and excitation lasers. A helical phase plate is used to shape the STED beam into its characteristic donut shape, after which an excitation laser of normal Gaussian-type distribution is introduced into the light path, forming the necessary STED 3D PSF pattern at the specimen plane. (Reproduced from Klar, T.A., Jakobs, S., Dyba, M., Egner, A., and Hell, S.W., Fluorescence microscopy with diffraction resolution barrier broken by stimulated emission, *Proc. Natl. Acad. Sci. USA*, 97, 8206–8210, Copyright 2000 National Academy of Sciences, U.S.A. With permission.) (b) STED microscope PSFs, illustrating the excitation PSF, the STED beam PSF, an overlay of the two, and finally the effective excitation PSF due to saturated depletion effects. (Reprinted by permission from Macmillan Publishers Ltd., *Nature*, Willig, K.I., Rizzoli, S.O., Westphal, V. et al., STED microscopy reveals that synaptotagmin remains clustered after synaptic vesicle exocytosis, *Nature*, 440, 935–939, Copyright 2006.) (c) High-resolution wide-field image of a bundle of microtubules. (d) Superresolution STED image of the same bundle of microtubules from (c), clearly resolving individual tubules that were unresolved in the wide-field image.

field using this configuration, a superresolution image can be acquired. An excellent review of STED-type microscopy and work that has been done in the field is provided by Tobias Müller and colleagues (Mueller et al. 2012).

Though it is theoretically possible to achieve diffraction-unlimited resolution using this technique, it is not realized for various reasons. Firstly, the investigator must balance the need for increased resolution with the need to prevent significant photobleaching and/or phototoxic effects. Also, the zero node of the STED beam is not perfectly rectangular, causing some amount of stimulated emission at the central zero node and reducing the emission intensity to about 25% of that obtained with a standard confocal microscope, which already suffers from relatively low signal-to-noise levels. The intensities required to create an excitation PSF with subnanometer dimensions would destroy or quickly photobleach most fluorophores. Still, STED is routinely used to acquire images with 20–50 nm lateral and sub-100 nm axial resolution using a variety of probes (Klar et al. 2000, Schmidt et al. 2009), with sub-10 nm resolution having been demonstrated (Rittweger et al. 2009). A wide assortment of probes have been tested and characterized for STED microscopy and are not subject to the same constraints as those used for single-molecule imaging, as described later. This is primarily due to the fact that stimulated emission can be achieved with probes that do not necessarily exhibit predictable on–off-switching behavior in the absence of a depletion laser. 3D STED has been demonstrated using a 4Pi-style opposed objective configuration; the technique yields a nearly isotropic (spherical) PSF and, for this reason, has been termed iso-STED (Dyba and Hell 2002). Iso-STED has yielded sub-40 nm isotropic resolution (Schmidt et al. 2009).

12.2.2 SIM AND SSIM MICROSCOPIES

STED and its variations rely upon sharpening the excitation PSF in order to obtain subdiffraction resolution. Structured illumination microscopy (SIM) (Heintzmann and Cremer 1999, Frohn et al. 2000, Gustafsson 2000) is a superresolution technique that relies upon patterned modulation of the fluorescence signal rather than reducing the dimensions of a standard PSF. The classic resolution limit of light microscopy discussed earlier specifies not only a minimum distance that objects can be resolved due to diffraction but similarly a corresponding maximum observable spatial frequency (k_0) given by the following equation:

$$k_0 = \frac{2NA}{\lambda_{em}}$$

where
    NA is the numerical aperture
    λ_{em} is the observed emission wavelength

Information with a spatial frequency greater of magnitude greater than k_0 cannot be directly observed in a standard fluorescence microscope. However, it is possible to artificially relocate high-frequency spatial information into observable frequency space. Given an illumination frequency represented by k_1 and sample frequency by k, moiré fringes will occur at the difference and sum frequencies $k \pm k_1$. Modulated frequencies are observable given that their magnitude is less than or equal to that of k_0. Thus, frequency values greater than k_0 can artificially be brought into observable frequency space and subsequently replaced to their proper location with postacquisition computational analysis. If we were to visualize specimen structure in reciprocal space (Fourier space), it would appear as a circle with low-frequency information residing closer to the origin and higher-frequency information toward the periphery, with a maximum radius equal to k_0 (approximating the diffraction limit). However, by using structured illumination of frequency k, the origin is effectively moved to k and all information within a radius k_0 becomes observable. By performing several iterations with different k, all information with frequency equal to or less than $2k_0$ becomes observable; this process is explored in Figure 12.4. Postacquisition mathematical processing can then be used to restore information about sample features to its true place in reciprocal space. This process effectively doubles the NA of the objective lens, making resolutions down to ~100 nm possible in the lateral dimensions.

Applications

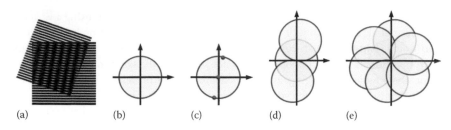

Figure 12.4 Illustration of resolution enhancement via SIM. (a) Moiré fringes (visible as near-vertical black bars) resulting from overlay of two fine patterns in a multiplicative fashion. Moire effects allow for the observation of frequencies resulting from the mixing of high-frequency spatial information with a known pattern of illumination. High-frequency spatial information can then be identified and assigned to its proper place in frequency space during postacquisition computational analysis. (b) Representations of 2D observable region of frequency space and observable frequencies are located inside of the blue circle, whose radius is equal to the magnitude of k_0 (the maximum observable spatial frequency). (c) The OTF support can be expanded using structured illumination, where additional harmonics are introduced (represented by the red dots) that are each independently convolved with the sample frequencies. The OTF support expansion can be maximized by using an illumination spatial frequency of magnitude k_0. (d) The new OTF support realized using a single rotation of the grid pattern. (e) A circular OTF support is circumscribed using three or more rotations of the grid pattern, allowing for a doubling of conventional diffraction-limited resolution. (Gustafsson, M.G.L.: Surpassing the lateral resolution limit by a factor of two using structured illumination microscopy. *J. Microsc.* 2000. 198. 82–87. Copyright Wiley-VCH Verlag GmbH & Co. KGaA. Reprinted with permission.)

SIM uses spatially structured and linearly polarized laser light in combination with a wide-field fluorescence microscope. Laser light is diffracted into several orders, of which only the 0th and 1st are unblocked. When a sample is illuminated by sinusoidally structured laser light (generated by interference of the 1st diffraction orders), moiré fringes will occur, representing spatial information that has changed position in reciprocal space. By capturing a sequence of images using different grid orientations and phases, spatial information from an area twice the size of the normal observable region of reciprocal space becomes available. Commercial instruments are generally capable of three or five grid rotations (of 120° or 72°, respectively), allowing one to optimize an imaging experiment for either imaging speed or resolution. For each rotation of the grid pattern, most instruments will capture images at three different phases. Acquisition of SIM images is rapid, especially in comparison to other superresolution techniques such as STED and single-molecule localization methods (to be discussed). Acquisition times of about 600 ms per SIM frame are possible with commercially available SIM instruments, making it capable of live-cell imaging of less dynamic cellular structures (e.g., Golgi apparatus, mitochondria, and actin). It is also possible to utilize interference of the zeroth diffraction order to produce a 3D interference pattern, where intensity varies sinusoidally axially as well as laterally (Gustafsson et al. 2008). This variation yields a twofold increase in axial resolution (to approximately 300 nm) as compared to resolutions obtained with standard techniques such as confocal microscopy and can be used to capture high-resolution z-series. Similar to iso-STED, the two-objective SIM analog I^5S uses a second opposing objective in a 4Pi-type geometry to further increase axial resolution, down to approximately 100 nm (Shao et al. 2008). A great review of structured illumination techniques is provided by Matthias Langhorst and colleagues (Langhorst et al. 2009).

Saturated structured illumination microscopy (SSIM) (Heintzmann et al. 2002, Gustafsson 2005) is an SIM derivative that relies upon the introduction of additional high-frequency harmonics, resulting in even further resolution gains. As we have explored, in SIM, high-frequency spatial characteristics are combined with structured illumination to create observable lower-frequency moiré patterns. The resolution enhancement of SIM is ultimately limited by the spatial frequency of the illumination pattern, which itself is limited by the diffraction of light. However, if the fluorescent probe exhibits a nonlinear response to excitation illumination, higher-frequency harmonics are introduced into the illumination pattern, making resolutions past those obtainable using SIM alone possible. Photoswitchable synthetic dyes and certain FPs with optical highlighting capabilities exhibit nonlinear responses to excitation illumination. The simplest approach to obtaining a nonlinear response is to use excitation intensities in excess of the saturation intensity; the result is a nonlinear response to increasing illumination intensities. This is because once an area is saturated in the fluorescent state, increased excitation intensities do not result in proportional increases in fluorescent emission. This technique has been shown to have a resolving power of about 50 nm in the lateral directions. SSIM was first applied toward the imaging of biologically relevant structures in 2012; the FP Dronpa was used to image purified microtubules, as well as cellular actin (Figure 12.5) and nuclear pore complexes (Rego et al. 2012).

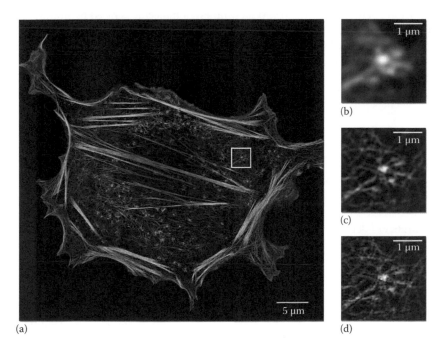

Figure 12.5 Resolution enhancement through use of SSIM. (a) SSIM image of a cell with actin labeled by the FP Dronpa. (b) Zoomed-in wide-field image of the region of interest defined in (a). (c) SIM image of the same area from (b). (d) SSIM image of same area from (b). (Reprinted with permission from Rego, E.H., Shao, L., Macklin, J.J. et al., Nonlinear structured-illumination microscopy with a photoswitchable protein reveals cellular structures at 50-nm resolution, *Proc. Natl. Acad. Sci. USA*, 109, E135–E143, Copyright 2012, National Academy of Sciences, U.S.A.)

Applications

12.2.3 SINGLE-MOLECULE LOCALIZATION MICROSCOPIES

The superresolution techniques described to this point fundamentally rely upon spatially patterned illumination. SIM acquisition times are relatively quick, requiring only enough time to acquire a small set of approximately 9–15 images using different illumination patterns, but is fundamentally limited to providing a twofold increase in resolution. STED is theoretically capable of single nanometer resolution but requires long acquisition times and high laser intensities and is difficult to implement with multicolor imaging. The family of single-molecule superresolution techniques, including stochastic optical reconstruction microscopy (STORM) (Rust et al. 2006), photoactivated localization microscopy (PALM) (Betzig et al. 2006), fluorescence PALM (FPALM) (Hess et al. 2006), and their related variants, rely upon the sequential imaging of sparse subsets of the entire fluorophore ensemble in an imaging field. There are several methods for accomplishing this, but the common goal of each technique is to temporally separate which molecules exist in the bright state so that individual (nonoverlapping) PSFs can be sequentially identified and fitted to a subdiffraction-limited area. The single-molecule localization procedure is briefly explored in Figure 12.6. These techniques have their origins in older single-molecule tracking experiments, where fluorophores were so diffuse that researchers did not have to compensate for overlapping signal. For the purpose of simplicity, such pointillistic techniques may be collectively referred to as single-molecule localization microscopy (SMLM). At its simplest, a molecule is fitted by finding the centroid position of the detected PSF over several imaging frames and subsequently determining the uncertainty based upon the position data (the localization precision). A simplified metric for characterizing the uncertainty of a single localization is given by the following equation:

$$\sigma = \frac{s}{\sqrt{N}}$$

where
  σ is the uncertainty
  s is the standard deviation of a 2D Gaussian function approximating the PSF of the emitter
  N is the number of photons detected

The number of detected photons can be thought of as the number of measurements of the fluorophore position. In theory, an emission of 10,000 photons will yield a localization precision of approximately 1–2 nm. This equation is only an approximation and does not take into account variables such as photon distribution, pixel size, background emission, or detector noise. It is of the utmost importance that individual PSFs are separated by a distance greater than that defined by the Abbe resolution limit to allow for independent identification and localization of each fluorescent signal. Data are generally acquired over several hundreds to several tens of thousands of frames and are used to create a composite reconstruction image.

STORM, in its original form, relies upon the use of specialized pairs of fluorescent dyes that are conjugated to the same molecule (e.g., Cy3 and Cy5 conjugated to the same antibody). These specialized dye pairs, termed "molecular switches," consist of an "activator" and "reporter" fluorophore; excitation of the activator facilitates increased recovery of the reporter from a dark state. Detection involves localizing individual

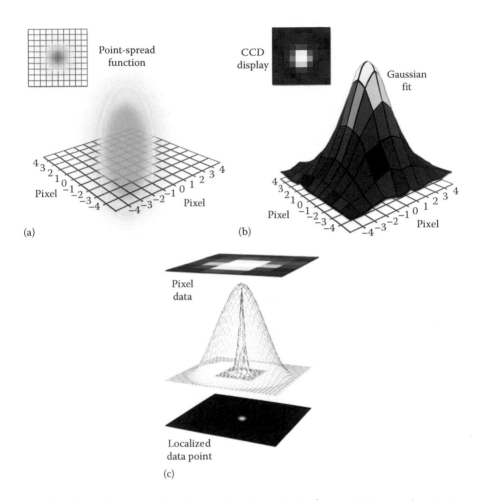

Figure 12.6 Generalized procedure for localizing the emission from single (nonoverlapping) molecules to a subdiffraction-limited area. (a) Idealized PSF showing the distribution of signal from a single emitter as collected through a single microscope objective. (b) Real charge-coupled device (CCD) image of an isolated emitter and corresponding height map of signal intensity. Note that pixel size is smaller than the optical resolution of the microscope. (c) A 2D Gaussian fit is applied to the pixel data, allowing the identification of the centroid position of emitter and the ultimate dimensions of the data point determined by the position uncertainty of the centroid. (Reprinted with permission from Allen, J.R., Ross, S.T., and Davidson, M.W., Single molecule localization microscopy for superresolution, *J. Opt.*, 15, 2013. IOP Publishing Ltd.)

reporter dye molecules: an initial bleaching step is used to drive the vast majority of reporters into a dark state, followed by activation of sparse subsets of reporters to the fluorescent state by the use of weaker laser light corresponding with the absorption spectrum of the activator (e.g., green light for the Cy3–Cy5 switch) and imaged using a higher-power reporter laser corresponding to the absorption spectrum of the reporter. The reporter laser also accomplishes the task of switching the reporter back to a dark state. Low activation powers ensure that only a fraction of molecules are activated, which is necessary for identifying individual PSFs when imaging. In this configuration, multiple switches tagged to different structures but with the same reporter fluorophore

Superresolution techniques using fluorescent protein technology

can be imaged. For example, if imaging mitochondria tagged with Cy2–Cy5 and tubulin tagged with Cy3–Cy5, one would use weak blue activation, image the activated Cy5 tagged to the mitochondria, use green activation light, and then image the Cy5-tagged tubulin. This process can be repeated several thousand times to create a reconstruction image; the acquisition procedure is illustrated by Figure 12.7. The primary advantage of this method is that one can use a high-performance dye such as Cy5 to tag multiple structures in the same specimen; synthetics outside of the far-red emission spectrum that perform at the same level have yet to be identified. Consequently, this method often results in high amounts of cross talk, complicating its use for quantitative measurements such as colocalization, even after the application of cross-talk correction operations. The use of an activator only ensures that the majority of the imaged reporters are tagged to the structure being imaged; nonspecific and spontaneous activation remains an issue.

Another technique, alternatively termed GSDIM or direct STORM (*d*STORM) (Heilemann et al. 2008), relies upon the use of reporter fluorophores not paired with an activator. In this setup, the reporter laser accomplishes all three tasks necessary

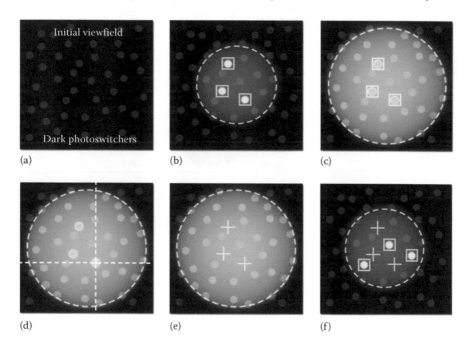

Figure 12.7 Generalized procedure for detecting single molecules in an SMLM-type setup. (a) Begin with an initial viewfield where all molecules are or have been driven into a dark state. (b) Photoactivate a small nonoverlapping population of emitters to the fluorescent state through the use of an activation beam. (c) Image molecules activated in (b) using a readout beam to excite fluorescence. (d) Localize individual molecules excited in (c) to a subdiffraction-limited area. (e) Continue imaging until all activated molecules have either been driven back into a dark state or have been permanently photobleached. (f) Activate a new subset of molecules to be imaged, continuously repeating this process until a satisfactory number of molecules are detected or the entire population is depleted. (Reprinted from *Methods Cell Biol.*, 89, Gould, T.J. and Hess, S.T., Nanoscale biological fluorescence imaging: Breaking the diffraction barrier, 329–358, Copyright 2008, with permission from Elsevier.)

for STORM imaging: activation to the fluorescent state, stimulation of fluorescence, and deactivation to the dark state. Spectrally distinct reporter dyes are generally imaged sequentially using this method, resulting in much lower levels of cross talk but introducing the need to account for possible chromatic aberration and requiring the use of less-than-optimal reporter fluorophores. There are no currently any known reporters outside of the far-red spectrum that can match the high photon outputs and number of switching cycles afforded by carbocyanine probes such as Alexa Fluor 647 or Cy5, ultimately resulting in a decrease in resolution for structures tagged with alternative probes. Probes for single-molecule superresolution imaging will be more thoroughly explored later. It should be noted that when using synthetic dyes in either a STORM or dSTORM configuration, specialized buffer conditions are generally necessary. A "STORM buffer" has several components that help ensure favorable photoswitching kinetics. An oxygen-scavenging system, such as an aliphatic thiol, helps some probes enter a dark state. Many fluorophores that had previously thought to be photostable under high intensity illumination have been shown to enter dark states in low-oxygen conditions.

PALM and FPALM are the original SMLMs designed for use in conjunction with optical highlighting FPs. To avoid confusion, it should be noted that each SMLM is essentially the same, with the greatest difference between most techniques being the probe of choice. SMLMs are often used in conjunction with a TIRF microscope (Axelrod 1981). Though, not strictly required, epifluorescence is used in a PALM with independently running acquisition (PALMIRA) setup (Egner et al. 2007), TIRF allows for the exclusive visualization of evanescent waves originating less than ~200 nm from the coverglass–medium interface, helping to significantly reduce background noise. This type of setup is great for imaging cellular features occurring in close proximity to the coverglass, such as plasma membrane-bound proteins and focal adhesion complexes. The use of a TIRF microscope in PALM is especially helpful due to the higher levels of background and weaker overall signal associated with FP expression as compared to synthetic dye staining. It should be noted that one of the greatest drawbacks to all single-molecule superresolution techniques is the requirement of long acquisition times; many thousands of frames of data are required in order to create a meaningful reconstruction. A high-quality STORM reconstruction often requires 30 min or longer of continuous imaging. PALM imaging of FPs can take even longer, with some early works describing imaging times in excess of 12 h (Betzig et al. 2006). The vast number of required imaging frames may also present data storage challenges; single datasets are often several gigabytes in size.

A 3D SMLM imaging has been demonstrated using a wide variety of techniques. It should be noted that 3D SMLM does not impose any extraneous requirements on probe selection. However, commonsense measures, such as choosing far-red or near-infrared probes for imaging deep in tissues, continue to be applicable. Some of the most popular and accessible methods for 3D SMLM involve coding z-information into the shape of the PSF; this has been demonstrated using two different methods. The first is to introduce astigmatism into the optical path using a cylindrical lens by Huang et al. (2008) (Figure 12.8). Depending on the axial position of a single emission event, the PSF will appear elliptical, stretched either in the x- or y-direction (Figure 12.8a). With proper calibration, this ellipticity can be used to determine axial position with an accuracy of about 40–50 nm over an approximate 800 nm range. The other method

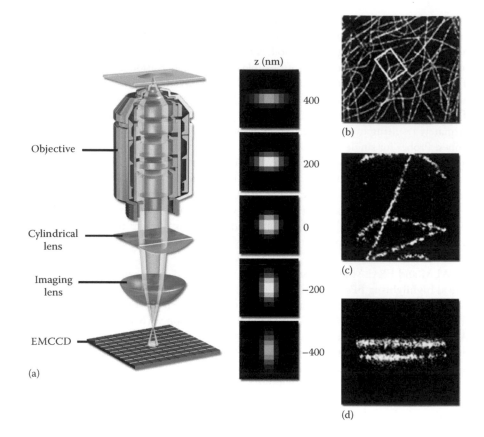

Figure 12.8 Overview of 3D SMLM imaging using astigmatism. (a) Optical path of excitation and emission light, emission light passes through a cylindrical lens introducing astigmatism, prior to passing through the imaging lens and signal being recorded on the electron multiplying charge-coupled device (EMCCD) detector. The appearance of the PSF of single emitters is shown; features at the focal point (z = 0 nm) appear circular, as one would expect without astigmatism. Features further away from the focal point of the objective (z = 0–400 nm) appear stretched in the x-direction, while those closer to the objective (z = −400–0 nm) appear stretched in the y-direction instead. (b) Normal 2D x–y view of microtubules pseudocolored to show differences in z-position. (c) Zoomed in view of the region of interest marked in (b). (d) x–z view of microtubules from (c), clearly showing separation of individual tubules along the z axis. (From Huang, B., Wang, W., Bates, M., and Zhuang, X., Three-dimensional super-resolution imaging by stochastic optical reconstruction microscopy, *Science*, 319, 810–813, 2008. Reprinted with permission of AAAS.)

relies upon using a spatial light modulator to reshape the PSF into a double-helix form (Pavani et al. 2009). In two dimensions, the double-helix PSF appears as a pair of circles that rotate about a central point with varying z-position. SMLMs have been introduced that, similar to iso-STED and I⁵S, utilize opposed objectives in a 4Pi type of geometry to sharpen the PSF and gain 10–20 nm resolution in both the lateral and axial directions (Shtengel et al. 2009, Xu et al. 2012). An excellent review of SMLM and superresolution microscopy in general is provided by Derek Toomre and Jeorg Bewersdorf (Toomre and Bewersdorf 2010).

Applications

12.3 FLUORESCENT PROBES FOR SUPERRESOLUTION

There are several desirable probe qualities for superresolution imaging; by and large, they are identical to those for conventional fluorescence imaging. In general, probes should have high brightness (product of extinction coefficient and quantum yield), photostability, and contrast levels, and the active and inactive species should have well-separated spectral profiles. Included here is a brief survey of the types of probes available for superresolution imaging, as well as their respective benefits and shortcomings. Specifically, the use of synthetic dyes, quantum dots, and FPs will be explored. The relative size of these types of probes, as well as immunoglobulin G, which is commonly used to target probes to specific structures in immunofluorescent experiments, are compared in Figure 12.9. The primary focus will be on the application of FP probes toward SMLM-type imaging; due to the more stringent and unique standards, probes must meet for this type of imaging. STED and SIM have requirements similar to those of more traditional fluorescence microscopies, with photostability arguably being the most important attribute. With STED, a probe must be able to go

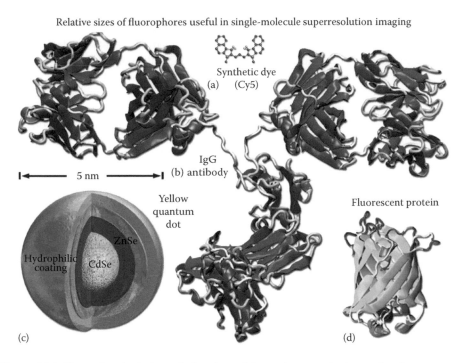

Relative sizes of fluorophores useful in single-molecule superresolution imaging

Figure 12.9 Illustration depicting relative sizes of fluorophores and other molecules commonly used for immunofluorescence of cellular details, roughly to scale. (a) Depiction of a small-molecule organic synthetic dye; in this instance, the structure belongs to the carbocyanine dye Cy5. (b) Ribbon structure of an immunoglobulin G (IgG) antibody, which is commonly used for both direct and indirect immunofluorescent staining. (c) Generalized structure of a quantum dot, illustrating the central cadmium selenide core and the zinc selenide and hydrophilic coating. (d) Ribbon structure of an FP.

Applications

undergo continuous stimulated emission so that it can be imaged without first being photobleached and similarly must have low rates of alternative transitions to molecular states that could potentially compete with stimulated emission.

With SIM, the reconstruction algorithms perform optimally when signal levels are relatively constant between imaging frames; a large decrease in signal between frames (e.g., due to photobleaching) can seriously decrease the realized resolution and, in extreme cases, cause the algorithms to fail altogether. SSIM is more complicated; the primary concern is to find probes that exhibit nonlinear responses to excitation intensity. In many ways, this technique is still in its infancy; the only probe that it has been demonstrated with in a biologically relevant sample is the FP Dronpa (Rego et al. 2012). Dronpa is excited and subsequently turned off by blue (488 nm) light and activated with UV–violet (405 nm) light (Habuchi et al. 2005). For SSIM, the off state is saturated using 488 nm illumination and, in combination with sinusoidal intensity variations, results in clearly defined illumination minima where Dronpa molecules are not driven to a dark state. It is also possible to use saturation of the on state to achieve the necessary nonlinear response. In theory, it is possible to use a wide variety of optical highlighter FPs in this manner; it simply has not been demonstrated to this point.

As stated, SMLMs have significantly different probe requirements as compared to other superresolution methods. Firstly, probes should spend very little time in the fluorescent state (low duty cycle) in order to facilitate both spatial and temporal separations of fluorophore emission. Next, probes should emit a large number of photons with each switching cycle; remember that localization precision scales with the inverse square of the number of detected photons. Finally, the investigator should be aware of the approximate number of switching cycles a probe will experience prior to photobleaching. If one is trying to quantify the amount of a given molecule, then a single switching cycle is ideal. Similarly, a single fluorophore should be associated with each molecule of interest. However, if quantification is not necessarily the goal, one will obtain a more detailed reconstruction if the probe can undergo a large number of switching cycles (localization measurements).

12.3.1 SYNTHETIC DYES

Synthetic fluorophores are very small molecules (about 1–2 nm in width) with high quantum yields (often in excess of 0.90), extinction coefficients (often in excess of 100,000 M^{-1} cm^{-1}), and excellent photostability. There is a large and continually developing variety of synthetic dyes with emission spectra spanning from the blue to near-infrared region. There are many commercially available high-performance synthetic dyes to choose from; popular choices include the Alexa Fluors, cyanines (Cy dyes), ATTO dyes, photochromic rhodamine, and caged derivatives of traditionally popular dyes such as fluorescein and rhodamine. Due to these ideal properties, synthetic dyes have proven to be the probe of choice for STED microscopy, with over 30 different spectrally varying dyes having been demonstrated as suitable for STED imaging.

Many synthetic dyes have very desirable properties for SMLM, including high photon emission counts and almost infinitely high contrast ratios between the dark and fluorescent states. The major downside of synthetics for SMLM is that the majority require some type of independent targeting mechanism, with perhaps the most widespread method being indirect antibody labeling. Indirect immunofluorescence protocols are tedious, must be carefully optimized, and result in a separation of several nanometers of the fluorophore from the target. Using a typical indirect antibody staining protocol, the dye is separated by about 10–20 nm from the actual structure of interest. This value exceeds the approximate 10 nm resolution afforded by high-performance synthetics such as Alexa Fluor 647 or Cy5 (Figure 12.9a). Performing SMLM on living cells using synthetic dyes is especially difficult. Though some dyes will localize to specific subcellular structures (e.g., the MitoTracker series, Hoechst dyes), most do not produce very high levels of background staining or do not possess the necessary on–off switching behavior required for STORM imaging. Furthermore, due to its relatively large size, a dye-labeled antibody cannot easily permeate a cell membrane. Cells must be fixed and somehow permeabilized (generally through the use of a precipitant fixative or a detergent solution) to allow antibodies free access to their intended targets. It should be noted that it is possible to label antigens present on the surface of living cells. This presents its own challenges though, for example, many antibodies have low affinity for their target antigen in its native, nonfixed conformation, and factors such as endocytosis of applied antibodies may cause erroneous localization artifacts. A popular alternative is the use of self-labeling proteins such as SNAP, CLIP, and Halo tags. A SNAP-tag, for example, is expressed in a manner similar to an FP; it is encoded to be expressed fused to the protein of interest. The mature SNAP protein, in this case, a derivative of human O^6-alkylguanine-DNA alkyltransferase (hAGT), is fused to the protein of interest and covalently binds to the fluorophore (conjugated to a guanine leaving group via a benzyl linker) (Keppler et al. 2004). This can be done in both live and fixed cells; however, biological membranes may be impermeable to certain fluorophores. For a comprehensive evaluation of more than twenty different synthetic fluorophores for SMLM, refer to the work of Graham Dempsey and colleagues (Dempsey et al. 2011).

12.3.2 QUANTUM DOTS

Quantum dots (Alivisatos 1996, Chan and Nie 1998) are inorganic semiconductor nanocrystals formed by a cadmium selenide (CdSe) core enveloped by a zinc sulfide shell (Figure 12.9c) and displaying remarkably discrete and symmetrical emission spectra of approximately 30 nm full width at half maximum (FWHM). Excitation spectra are unusually broad and similar between quantum dots of different emission spectra, generally exhibiting an excitation maximum in the UV–violet spectrum and a large shoulder stretching into the blue-green region. Interestingly, the emission spectrum of a quantum dot is directly related to the size of its CdSe core, with larger dots having more red-shifted emission spectra and thus making it possible to synthesize quantum dots with tailor-made emission profiles. When labeling cells, the addition of a passivation layer and hydrophilic coating is required, especially if live-cell labeling is the goal. Additionally, quantum dots enjoy advantages for superresolution imaging similar to those of synthetic fluorophores, including high brightness, contrast, and almost unlimited fatigue resistance.

However, labeling with quantum dots can be difficult; like synthetic dyes, they must be tagged to the structure of interest using some type of independent targeting system, making live-cell imaging currently unfeasible. Though it is possible to use a hybrid system, such as a targeting peptide bound to the quantum dot to bind the structure of interest, a delivery system (e.g., microinjection) is still necessary given the large size of even a single dot. Furthermore, quantum dots have been shown to be toxic in living cells, potentially leading to localization artifacts and/or changes in normal cell function. Not all quantum dots exhibit the necessary switching kinetics required for single-molecule superresolution, and those that do often have very short off-state lifetimes, making imaging densely labeled structures very difficult. However, resolutions of about 12 nm using Qdot 705 (Invitrogen) applied to microtubules with an indirect antibody staining protocol were demonstrated in a dSTORM-type setup (Hoyer et al. 2011). This method relies on the blueing property displayed by some quantum dots where, upon illumination with high-intensity lasers, their emission spectra are slowly and stochastically blue-shifted. This allows one to image sparse subsets of the total population and, with time, over 90% of the entire ensemble.

12.3.3 FLUORESCENT PROTEINS

Genetically encoded FPs have emerged as invaluable tools for imaging subcellular structures with SMLM. For the remainder of this discussion, we will focus on the application and development of FPs for SMLM. As discussed, SIM, STED, and traditional fluorescence microscopies have the same general requirements of FP probes, including high photostability, high brightness, and good maturation and folding characteristics. Specifically, high-performance FPs such as Citrine have been shown to work well for STED imaging (Hein et al. 2008). Other FPs used for STED imaging will be discussed. However, probes for SMLM have an extremely unique set of standards that must be met, making the identification of well-performing varieties more cumbersome. There are several advantages afforded by using FPs for SMLM: they are relatively small, only about 4 nm in length and 2 nm in width and depth (Figure 12.9d). Though synthetic dyes are smaller, as previously discussed, they are generally conjugated to a primary or secondary antibody, which can span about 12–15 nm across, exceeding the 10 nm precision offered by some synthetic probes. Even though FPs do not provide the same localization precision as synthetic dyes, it is simpler to locate an FP closer to the structure of interest. One of the greatest advantages afforded by using FPs is the ability to label cellular components *in vitro* with minimal invasiveness, often for long-term live-cell experimentation. Reliance upon antibodies for intracellular labeling additionally requires fixation and permeabilization steps. Properly characterized FP fusion chimeras do not suffer from this limitation, enabling long-term single-molecule superresolution studies of living cells. However, whenever using FPs, the investigator must keep in mind that the native endogenous protein is not being visualized.

Many FPs that exhibit optical highlighting abilities can be rendered virtually useless by other factors. Probes must have high brightness, quick chromophore maturation rates, high contrast between native and photo-changed species, good photostability, and so forth. It is desirable for FPs to be monomeric; other oligomeric states often result in poor localization and the creation of aggregation artifacts and can be problematic in quantitative SMLM studies. Tandem dimers (e.g., tdEos, tdTomato) seem to behave as monomers; however, they are twice the size of their monomeric counterparts and associate more than one chromophore with each protein of interest. As with most biological samples, autofluorescence can be an issue and can be exacerbated through the use of many commonly applied transfection reagents. FPs generally emit less photons per molecule than synthetic dyes. For example, tdEos, a popular probe for PALM imaging, emits approximately 750 photons per molecule per switching cycle (Betzig et al. 2006), as compared to about 6000 photons from Alexa Fluor 647 or Cy5 (Rust et al. 2006). Many optical highlighters emit far less, such as Photoactivatable GFP (PA-GFP) which only emits about 300 photons upon activation. Thus, it is of the utmost importance that the effects of autofluorescence are minimized and the signal-to-noise ratio maximized. Dual-color PALM using FP fusions presents even more problems. Performing two-color PALM is simplest when using probes with well-separated emission spectra; however, FPs generally have very wide fluorescence emission bandwidths (roughly 40–100 nm), especially when compared to synthetic dyes (30–50 nm) or quantum dots (30 nm). Additionally, most optical highlighter FPs are activated using similar wavelengths of light (most often UV–violet). Another potential complication is that red-emitting probes (e.g., Eos fluorescent protein [EosFP], mKikGR, Dendra2) have preactivation fluorescence emission profiles that overlap with the postactivation emission profiles of green-emitting photoactivators or photoconverters (e.g., Dronpa, PA-GFP), making it difficult to visualize structures independently prior to SMLM imaging.

12.4 FLUORESCENT PROTEINS FOR SUPERRESOLUTION

Before discussing individual FPs for use in PALM, it is important that the differences between the three primary groups of optical highlighters are distinguished. Photoactivatable FPs (PA-FPs) irreversibly switch from a dark to a bright fluorescent state upon irradiation with certain wavelengths of light. For example, to highlight a region containing photo-activatible GFP (PA-GFP) (Patterson and Lippincott-Schwartz 2002), it must first be photoactivated by the use of UV–violet light (DAPI-type filter set or 405 nm laser line) before stimulating green emission using blue light (fluorescein isothiocyanate [FITC]-type filter set or 488 nm laser line). The next group is the photoconvertible FPs (PC-FPs), whose members undergo a permanent red shift in their emission spectra upon activation with certain wavelengths of light. A popular member of this group is tandem-dimer Eos (tdEos) (Nienhaus et al. 2006), which is a green-emitting FP in its native state but, upon activation with UV–violet light, irreversibly switches into a red-emitting species. The final group includes the PS-FPs; these FPs are capable of reversible photoswitching from a dark to a fluorescent state upon activation. One example from this class that we have already explored is Dronpa, which is excited to fluoresce and subsequently driven to a dark state using blue light but reactivated to a potentially fluorescent state using UV–violet light.

12.4.1 PHOTOACTIVATABLE FLUORESCENT PROTEINS

We have already discussed the only known green PA-FP: PA-GFP; however, there are several PA-FPs in other spectral classes, especially among the red FPs. Popular examples of red PA-FPs include PAmCherry1 (Subach et al. 2009) and PAmRFP1 (Verkhusha and Sorkin 2005), both derived from DsRed and exhibiting irreversible switching from a dark to a red fluorescent state upon activation with violet light. PAmCherry1 was introduced as one of the three potentially useful photoactivatable mCherry derivatives; it was found to be brighter than PAmCherry3, more photostable than PAmCherry2, and easier to photoactivate than both. Though both PAmCherry1 and PAmRFP1 are monomeric, PAmCherry1 has a faster maturation rate, higher quantum yield, and a higher contrast ratio between the dark and fluorescent states. PA-GFP and PAmCherry1 have been successfully used together in two-color PALM experiments (Subach et al. 2009). Initially both probes exist in a dark state and following photoactivation,

Applications

which is achieved for both probes using UV–violet light; the fluorescence emission profiles are well separated and the probes exhibit minimal nonspecific excitation. At no point does either probe exist in a state where their excitation bandwidths could potentially overlap. Similar to PAmCherry1, PATagRFP (Subach et al. 2010a) is a bright, monomeric dark-red photoactivatable FP that has been used in conjunction with PA-GFP for PALM imaging. Specifically, it was used in two-color single-particle tracking PALM (sptPALM) to show that single-molecule transmembrane proteins colocalize with clathrin light chain-enriched areas on the plasma membrane (Subach et al. 2010a). Recently, photoactivatable mKate (PAmKate) was introduced (Gunewardene et al. 2011), extending the availability of photoactivatable FPs into the far-red region. PAmKate was demonstrated as a suitable SMLM probe in three-color PALM imaging experiments where it was coexpressed with Dendra2 and PAmCherry fusion constructs (Gunewardene et al. 2011).

12.4.2 PHOTOCONVERTIBLE FLUORESCENT PROTEINS

PC-FPs irreversibly convert from one emission bandwidth to another (most commonly green to red) upon activation with a certain color of light (generally UV–violet). The Eos family of FPs performs very well and has been extensively characterized for use in PALM imaging. The EosFP derivatives tdEos, mEos2 (McKinney et al. 2009), and mEos3.1 and mEos3.2 (Zhang et al. 2012) are all well suited to SMLM, with monomeric Eos varieties often preferred because of their superior performance in fusion chimeras. Other green-to-red photoconverters include Kaede (Ando et al. 2002), mClavGR1 and mClavGR2 (Hoi et al. 2010), mMaple (McEvoy et al. 2012), Kikume green–red (KikGR1) (Tsutsui et al. 2005) and its monomeric variant mKikGR1 (Habuchi et al. 2008), and Dendra and Dendra-2 (Gurskaya et al. 2006). Kaede was used in the original PALM paper for SMLM imaging of the lysosomal transmembrane protein CD63 (Betzig et al. 2006). It should be noted that Dendra-2 is unique in that activation can be achieved with blue light, which is less phototoxic than the typical UV–violet wavelengths required for photoconversion; this is especially desirable in live-cell imaging scenarios. There are also photoconverters that switch between wavelengths differing from the green-to-red motif discussed so far. The photo-switchable cyan FP (PS-CFP) and its improved variant PS-CFP2 (Chudakov et al. 2004) convert from a cyan to green species upon activation with violet light. Additionally, PS-CFP2 has a very high contrast ratio (>2000) and photon output exceeding that of green photoswitchers such as Dronpa, making it viable for use in dual-color color PALM experiments with green-to-red photoconverters (Shroff et al. 2007). The FP Phamret (Matsuda et al. 2008) also undergoes conversion from a cyan to a green species upon violet activation but with a significantly different mechanism. Phamret is actually a combination of two different FPs: PA-GFP and an enhanced cyan FP (ECFP) variant. Illuminations of the 458 nm type allow visualization of cyan fluorescence; however, upon violet illumination, the PA-GFP moiety is activated and FRET from the ECFP donor results in green fluorescence upon 458 nm illumination. PSmOrange (Subach et al. 2011) and its recently introduced variant PSmOrange2 (Subach et al. 2012) are capable of photoconverting from an orange to a far-red state upon activation with blue light. This may prove to be of benefit to multicolor PALM experiments since PSmOrange is sufficiently red-shifted to avoid any overlap when used simultaneously with another photoconverter such as PS-CFP2.

12.4.3 PHOTOSWITCHABLE FLUORESCENT PROTEINS

The final class of optical highlighters is the PS-FPs, which are capable of reversible on/off switching when exposed to the proper wavelength of light. One of the first to be used for PALM imaging was Dronpa; Dronpa-tagged α-actinin has been used in two-color PALM to check for colocalization with tdEos-tagged vinculin in focal adhesions (Shroff et al. 2007). This imaging scheme relies upon first exhausting the population of tdEos in the green-fluorescent state by photoconversion into, and the subsequent imaging of, the red state. Though it is possible to successfully adopt this imaging scheme, it is not ideal since there is a high probability of signal from a green–red photoconverter in its native state being observed during imaging of a green probe such as PA-GFP, Dronpa, or mTFP0.7 (another dark-green photoswitcher) (Henderson et al. 2007). Dronpa and many other FPs in this class suffer from low photon outputs and suboptimal switching kinetics. The development of improved Dronpa variants has been a major research focus in recent years. Several Dronpa variants are available that can be used for SMLM, including Dronpa-2 and Dronpa-3 (Ando et al. 2007), bsDronpa and Padron (Andresen et al. 2008), and rsFastLime (Stiel et al. 2007). rsFastLime has proven especially useful in PALMIRA experiments, where spontaneous off–on–off events are recorded asynchronously by a detector running at a very high frame rate, around 500 Hz with rsFastLime (Egner et al. 2007). Additionally two-color PALMIRA-type SMLM has been demonstrated with both rsFastLime and Cy5 together (Bock et al. 2007). The fast off-switching character combined with the high rate of spontaneous conversion into the bright state allows rsFastLime to be, unlike most photoswitchers, activated by, excited to fluoresce, and driven back to a dark state using only a single wavelength of light (blue). Additional violet activation increases the number of molecules switched on, helping expedite imaging if necessary.

Padron is a Dronpa variant that is driven into a dark state with UV illumination. Both Dronpa-2 and Padron have both been demonstrated as viable for dual-color STED imaging (Willig et al. 2011). The reversibly switchable enhanced GFP (rsEGFP) (Grotjohann et al. 2011) and its new variant rsEGFP2 (Grotjohann et al. 2012) were demonstrated with STED-type RESOLFT imaging upon their introduction, though not yet with SMLM. rsEGFP, like Dronpa, is activated with violet light and imaged and driven to a dark state using blue light. However, rsEGFP demonstrates remarkable fatigue resistance, able to complete about 1200 full on–off cycles prior to exhaustion, with on–off cycling requiring only about 20 ms, over 10-fold faster than Dronpa. mGeos-M (Chang et al. 2012) is one of several recently introduced "mGeos" variants derived from the green–red photoconverter mEos2 and has been demonstrated to possess the highest known photon budget of any green photoswitcher (about 400 photons per switching cycle) and an off-switching time 152% longer than Dronpa. Additionally, mGeos-M has been demonstrated as a viable probe for two-color PALM imaging in combination with PAmCherry1 (Chang et al. 2012). Other promising mGeos variants introduced along with mGeos-M include mGeos-X. Perhaps one of the most exciting new dark-green photoswitchers is the Citrine derivative Dreiklang (Brakemann et al. 2011), whose fluorescence excitation profile is completely decoupled from its activation and deactivation spectra. Fluorescence is excited by blue 488 nm-type illumination, activation is achieved using 405 nm-type light, and inactivation by UV 365 nm-typelight. SMLM is performed using continuous 488 nm irradiation and continuously adjusting the intensities of the 405 and 365 nm to achieve proper single-molecule switching densities. Dreiklang has been shown to be capable of approximately 150 switching cycles.

Not all photoswitchers are of the dark-green variety; rsCherry and rsCherryRev are dark-red photoswitchers (Stiel et al. 2008). rsCherry is activated by yellow light and deactivated by blue light, but in rsCherryRev, the effects of the switching lights are reversed. Because of its relatively poor contrast ratio, rsCherry has not been found suitable for single-molecule localization, but rsCherryRev does not suffer from the same limitation and has been successfully used in a live-cell PALMIRA setup to track movements of the endoplasmic reticulum. rsTagRFP (Subach et al. 2010b) is a dark-red photoswitcher that behaves similar to rsCherryRev, being activated by 430–450 nm light and inactivated by 550–570 nm light, but with greater brightness and about sixfold greater photoswitching contrast. Kindling FP (KFP1) (Chudakov et al. 2003) is an especially interesting red photoswitcher; upon illumination with low-intensity green light, it will exhibit red fluorescence, which can be immediately quenched using blue light, allowing for very fine temporal control of fluorescence.

12.4.4 OTHER FLUORESCENT PROTEINS FOR SUPERRESOLUTION

As one can see, there exists a huge variation between optical highlighter FPs, even of the same class. The contrast ratios of representative FPs from among the photoactivators, photoconverters, and photoswitchers are compared in Figure 12.10. However, not all optical highlighter FPs fit neatly within any one of those categories. Perhaps one of the most interesting FPs for PALM imaging is IrisFP (Adam et al. 2008) and its monomeric variant mIrisFP (Fuchs et al. 2010), which do not fit into a single optical highlighter category. IrisFP is a mutant of the original tetrameric EosFP and as such displays irreversible green-to-red photoconversion upon illumination with violet light of moderate intensity. However, in the green state, IrisFP can be photoswitched to and from a dark state with weak violet illumination similar to Dronpa and, once photoconverted to the red state, can still be reversibly photoswitched to and from a dark state. As a result of this unique combination of highlighting properties, mIrisFP has recently been reported as the first fluorophore to be used in a combination pulse-chase and PALM experiment (Fuchs et al. 2010). mIrisFP tagged to α-actinin was imaged using photoswitching of the green species, and then a small population was photoconverted to the red species, at which point the migration of the red-converted population could be tracked by performing PALM of the photoswitching red species. NijiFP (Adam et al. 2011) is also capable of both irreversible green–red photoconversion and reversible dark-green/dark-red photoswitching with violet activation. Derived from Dendra-2, NijiFP (the name is from the Japanese word for "rainbow") is advertised as a more efficiently photoconvertible alternative to mIrisFP and, upon its debut, was demonstrated with PALM imaging of actin in fixed cells.

Though ideal for SMLM imaging, optical highlighter FPs are not required for all forms of superresolution imaging (including SMLM). In fact, the majority of FPs applied toward STED imaging are not optical highlighters, including Citrine (Hein et al. 2008), enhanced yellow fluorescent protein (EYFP) (Naegerl et al. 2008), EGFP (Willig et al. 2006a), TagRFP657 (Morozova et al. 2010), and E2-Crimson (Strack et al. 2009). E2-Crimson, a tetrameric DsRed variant, was the first far-red FP

Fluorescent protein contrast ratios

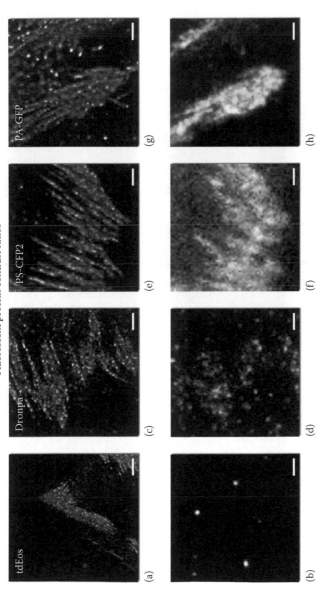

Figure 12.10 Comparison of the contrast ratios of four different optical highlighter FPs fused to the focal adhesion protein paxillin. For each FP, the SMLM reconstruction and the 1000th frame of the wide-field single-molecule data are presented. tdEos SMLM (a) and wide-field (b) images; tdEos clearly has a very high contrast ratio, making it easy to distinguish individual against the background and resulting in a crisp SMLM reconstruction. Dronpa SMLM (c) and wide-field (d) images; showing good contrast, but not to the same degree as tdEos. PS-CFP2 SMLM (e) and wide-field (f) images, demonstrating poorer contrast than the tdEos or Dronpa equivalents, making single-molecule identification more difficult and resulting in a blurrier reconstruction. PA-GFP SMLM (g) and wide-field (h) images; PA-GFP also demonstrates poor contrast compared to high-performance FPs such as tdEos. Scale bar = 2 μm. (From Shroff, H., Galbraith, C.G., Galbraith, J.A. et al., Dual-color superresolution imaging of genetically expressed probes within individual adhesion complexes, *Proc. Natl. Acad. Sci. USA*, 104, 20308–20313. Copyright 2007, National Academy of Sciences, U.S.A. With permission.)

applied toward STED image. However, its tetrameric character is not ideal, with the monomeric mKate derivative TagRFP657 being more recently applied toward far-red STED imaging. SMLM has been demonstrated with conventional FPs in a manner analogous to *d*STORM imaging, having been referred to as stochastic single-molecule superresolution (SSMS) imaging (Shaner et al. 2013). Firstly, molecules are quickly driven into a dark state using high-powered lasers corresponding to the excitation spectrum of the conventional FP, and individual molecules returning to the fluorescent state are subsequently identified while imaging at much lower intensities. This was first

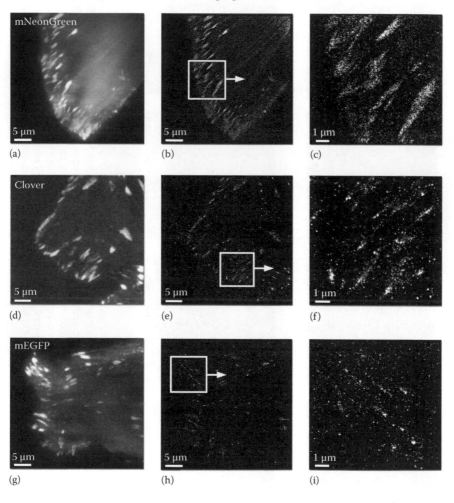

Figure 12.11 Comparison of different conventional FPs fused to the C-terminus of zyxin in live HeLa cells for SSMS-type SMLM imaging. (a), (d), and (g) are the wide-field TIRF images of mNeonGreen, Clover, and mEGFP fused to zyxin, respectively. (b), (e), and (h) are the SMLM reconstructions corresponding to the wide-field images shown in (a), (d), and (g), respectively. (c), (f), and (i) are high-resolution zoomed-in areas from the rectangular insets shown in (b), (e), and (h), respectively. Scale bars = 5 μm in (a), (b), (d), (e), (g), and (h); 1 μm in (c), (f), and (i). (Reprinted by permission from Macmillan Publishers Ltd., *Nat. Methods*, Shaner, N.C., Lambert, G.G., Chammas, A. et al., A bright monomeric green fluorescent protein derived from *Branchiostoma lanceolatum*, 10, 407–409, Copyright 2013.)

demonstrated with EYFP fused to the bacterial actin protein MreB in live *Caulobacter crescentus* cells (Biteen et al. 2008). However, EYFP requires an activation step with UV–violet light in order to recover to the fluorescent state. Several green conventional FPs, including mEGFP, mEmerald, Clover, and mNeonGreen (which performed the best), tagged to several different structures, including β-actin, zyxin, keratin, and myosin IIa, have been used for both live- and fixed-cell SMLM (Figure 12.11) (Shaner et al. 2013).

12.5 CONCLUSIONS

The ability of the superresolution techniques explored to elucidate fine spatial details of subcellular structures will likely lead to a deeper understanding of cellular function and ultrastructure than ever before. With the advancements being made in superresolution technology, it would not be entirely unreasonable to expect single nanometer resolution from commercial instruments within the foreseeable future. Specifically, the use of FP technology in conjunction with these methodologies promises dynamic live-cell superresolution microscopy of cellular processes that, to this point, have been rendered unobservable by the diffraction limit of light microscopy. Thus, proper development and characterization of FPs for superresolution imaging is of the utmost importance.

REFERENCES

Adam, V., Lelimousin, M., Boehme, S. et al. 2008. Structural characterization of IrisFP, an optical highlighter undergoing multiple photo-induced transformations. *Proceedings of the National Academy of Sciences of the United States of America*, 105, 18343–18348.

Adam, V., Moeyaert, B., David, C. C. et al. 2011. Rational design of photoconvertible and biphotochromic fluorescent proteins for advanced microscopy applications. *Chemistry and Biology*, 18, 1241–1251.

Alivisatos, A. P. 1996. Semiconductor clusters, nanocrystals, and quantum dots. *Science*, 271, 933–937.

Allen, J. R., Ross, S. T., and Davidson, M. W. 2013. Single molecule localization microscopy for superresolution. *Journal of Optics*, 15, 094001.

Ando, R., Flors, C., Mizuno, H., Hofkens, J., and Miyawaki, A. 2007. Highlighted generation of fluorescence signals using simultaneous two-color irradiation on Dronpa mutants. *Biophysical Journal*, 92, L97–L99.

Ando, R., Hama, H., Yamamoto-Hino, M., Mizuno, H., and Miyawaki, A. 2002. An optical marker based on the UV-induced green-to-red photoconversion of a fluorescent protein. *Proceedings of the National Academy of Sciences of the United States of America*, 99, 12651–12656.

Andresen, M., Stiel, A. C., Folling, J. et al. 2008. Photoswitchable fluorescent proteins enable monochromatic multilabel imaging and dual color fluorescence nanoscopy. *Nature Biotechnology*, 26, 1035–1040.

Axelrod, D. 1981. Cell-substrate contacts illuminated by total internal-reflection fluorescence. *Journal of Cell Biology*, 89, 141–145.

Bailey, B., Farkas, D. L., Taylor, D. L., and Lanni, F. 1993. Enhancement of axial resolution in fluorescence microscopy by standing-wave excitation. *Nature*, 366, 44–48.

Betzig, E., Lewis, A., Harootunian, A., Isaacson, M., and Kratschmer, E. 1986. Near-field scanning optical microscopy (NSOM)—Development and biophysical applications. *Biophysical Journal*, 49, 269–279.

Betzig, E., Patterson, G. H., Sougrat, R. et al. 2006. Imaging intracellular fluorescent proteins at nanometer resolution. *Science*, 313, 1642–1645.

Applications

Biteen, J. S., Thompson, M. A., Tselentis, N. K., Bowman, G. R., Shapiro, L., and Moerner, W. E. 2008. Super-resolution imaging in live *Caulobacter crescentus* cells using photoswitchable EYFP. *Nature Methods*, 5, 947–949.

Bock, H., Geisler, C., Wurm, C. A. et al. 2007. Two-color far-field fluorescence nanoscopy based on photoswitchable emitters. *Applied Physics B-Lasers and Optics*, 88, 161–165.

Brakemann, T., Stiel, A. C., Weber, G. et al. 2011. A reversibly photoswitchable GFP-like protein with fluorescence excitation decoupled from switching. *Nature Biotechnology*, 29, 942–947.

Bretschneider, S., Eggeling, C., and Hell, S. W. 2007. Breaking the diffraction barrier in fluorescence microscopy by optical shelving. *Physical Review Letters*, 98, 218103.

Chan, W. C. W. and Nie, S. M. 1998. Quantum dot bioconjugates for ultrasensitive nonisotopic detection. *Science*, 281, 2016–2018.

Chang, H., Zhang, M. S., Ji, W. et al. 2012. A unique series of reversibly switchable fluorescent proteins with beneficial properties for various applications. *Proceedings of the National Academy of Sciences of the United States of America*, 109, 4455–4460.

Chudakov, D. M., Belousov, V. V., Zaraisky, A. G. et al. 2003. Kindling fluorescent proteins for precise in vivo photolabeling. *Nature Biotechnology*, 21, 191–194.

Chudakov, D. M., Verkhusha, V. V., Staroverov, D. B., Souslova, E. A., Lukyanov, S., and Lukyanov, K. A. 2004. Photoswitchable cyan fluorescent protein for protein tracking. *Nature Biotechnology*, 22, 1435–1439.

Dempsey, G. T., Vaughan, J. C., Chen, K. H., Bates, M., and Zhuang, X. 2011. Evaluation of fluorophores for optimal performance in localization-based super-resolution imaging. *Nature Methods*, 8, 1027–1036.

Dyba, M. and Hell, S. W. 2002. Focal spots of size lambda/23 open up far-field florescence microscopy at 33 nm axial resolution. *Physical Review Letters*, 88, 163901.

Egner, A., Geisler, C., Von Middendorff, C. et al. 2007. Fluorescence nanoscopy in whole cells by asynchronous localization of photoswitching emitters. *Biophysical Journal*, 93, 3285–3290.

Folling, J., Bossi, M., Bock, H. et al. 2008. Fluorescence nanoscopy by ground-state depletion and single-molecule return. *Nature Methods*, 5, 943–945.

Frohn, J. T., Knapp, H. F., and Stemmer, A. 2000. True optical resolution beyond the Rayleigh limit achieved by standing wave illumination. *Proceedings of the National Academy of Sciences of the United States of America*, 97, 7232–7236.

Fuchs, J., Bohme, S., Oswald, F. et al. 2010. A photoactivatable marker protein for pulse-chase imaging with superresolution. *Nature Methods*, 7, 627–630.

Gould, T. J. and Hess, S. T. 2008. Nanoscale biological fluorescence imaging: Breaking the diffraction barrier. *Methods in Cell Biology*, 89, 329–358.

Grotjohann, T., Testa, I., Leutenegger, M. et al. 2011. Diffraction-unlimited all-optical imaging and writing with a photochromic GFP. *Nature*, 478, 204–208.

Grotjohann, T., Testa, I., Reuss, M. et al. 2012. rsEGFP2 enables fast RESOLFT nanoscopy of living cells. *eLife*, 1, e00248.

Gugel, H., Bewersdorf, J., Jakobs, S., Engelhardt, J., Storz, R., and Hell, S. W. 2004. Cooperative 4Pi excitation and detection yields sevenfold sharper optical sections in live-cell microscopy. *Biophysical Journal*, 87, 4146–4152.

Gunewardene, M. S., Subach, F. V., Gould, T. J. et al. 2011. Superresolution imaging of multiple fluorescent proteins with highly overlapping emission spectra in living cells. *Biophysical Journal*, 101, 1522–1528.

Gurskaya, N. G., Verkhusha, V. V., Shcheglov, A. S. et al. 2006. Engineering of a monomeric green-to-red photoactivatable fluorescent protein induced by blue light. *Nature Biotechnology*, 24, 461–465.

Gustafsson, M. G. L. 2000. Surpassing the lateral resolution limit by a factor of two using structured illumination microscopy. *Journal of Microscopy-Oxford*, 198, 82–87.

Gustafsson, M. G. L. 2005. Nonlinear structured-illumination microscopy: Wide-field fluorescence imaging with theoretically unlimited resolution. *Proceedings of the National Academy of Sciences of the United States of America*, 102, 13081–13086.

Gustafsson, M. G. L., Agard, D. A., and Sedat, J. W. 1995. Sevenfold improvement of axial resolution in 3D widefield microscopy using 2 objective lenses. *3-Dimensional Microscopy—Image Acquisition and Processing II Conference*, February 9–10, 1995, San Jose, CA, pp. 147–156.

Gustafsson, M. G. L., Agard, D. A., and Sedat, J. W. 1999. (IM)-M-5: 3D widefield light microscopy with better than 100 nm axial resolution. *Journal of Microscopy-Oxford*, 195, 10–16.

Gustafsson, M. G. L., Shao, L., Carlton, P. M. et al. 2008. Three-dimensional resolution doubling in wide-field fluorescence microscopy by structured illumination. *Biophysical Journal*, 94, 4957–4970.

Habuchi, S., Ando, R., Dedecker, P. et al. 2005. Reversible single-molecule photoswitching in the GFP-like fluorescent protein Dronpa. *Proceedings of the National Academy of Sciences of the United States of America*, 102, 9511–9516.

Habuchi, S., Tsutsui, H., Kochaniak, A. B., Miyawaki, A., and Van Oijen, A. M. 2008. mKikGR, a monomeric photoswitchable fluorescent protein. *PLoS One*, 3, e3944.

Heilemann, M., Van de Linde, S., Schuttpelz, M. et al. 2008. Subdiffraction-resolution fluorescence imaging with conventional fluorescent probes. *Angewandte Chemie-International Edition*, 47, 6172–6176.

Hein, B., Willig, K. I., and Hell, S. W. 2008. Stimulated emission depletion (STED) nanoscopy of a fluorescent protein-labeled organelle inside a living cell. *Proceedings of the National Academy of Sciences of the United States of America*, 105, 14271–14276.

Heintzmann, R. and Cremer, C. 1999. Laterally modulated excitation microscopy: Improvement of resolution by using a diffraction grating. *Conference on Optical Biopsies and Microscopic Techniques III*, September 9–11, 1998, Stockholm, Sweden, pp. 185–196.

Heintzmann, R., Jovin, T. M., and Cremer, C. 2002. Saturated patterned excitation microscopy—A concept for optical resolution improvement. *Journal of the Optical Society of America A-Optics Image Science and Vision*, 19, 1599–1609.

Hell, S. and Stelzer, E. H. K. 1992. Properties of a 4Pi confocal fluorescence microscope. *Journal of the Optical Society of America A-Optics Image Science and Vision*, 9, 2159–2166.

Hell, S. W. 2005. Fluorescence nanoscopy: Breaking the diffraction barrier by the RESOLFT concept. *NanoBiotechnology*, 1, 296–297.

Hell, S. W. and Kroug, M. 1995. Ground-state-depletion fluorescence microscopy—A concept for breaking the diffraction resolution limit. *Applied Physics B-Lasers and Optics*, 60, 495–497.

Hell, S. W. and Nagorni, M. 1998. 4Pi confocal microscopy with alternate interference. *Optics Letters*, 23, 1567–1569.

Hell, S. W. and Wichmann, J. 1994. Breaking the diffraction resolution limit by stimulated-emission—Stimulated-emission-depletion fluorescence microscopy. *Optics Letters*, 19, 780–782.

Henderson, J. N., Ai, H. W., Campbell, R. E., and Remington, S. J. 2007. Structural basis for reversible photobleaching of a green fluorescent protein homologue. *Proceedings of the National Academy of Sciences of the United States of America*, 104, 6672–6677.

Hess, S. T., Girirajan, T. P. K., and Mason, M. D. 2006. Ultra-high resolution imaging by fluorescence photoactivation localization microscopy. *Biophysical Journal*, 91, 4258–4272.

Hoi, H. F., Shaner, N. C., Davidson, M. W., Cairo, C. W., Wang, J. W., and Campbell, R. E. 2010. A monomeric photoconvertible fluorescent protein for imaging of dynamic protein localization. *Journal of Molecular Biology*, 401, 776–791.

Hoyer, P., Staudt, T., Engelhardt, J., and Hell, S. W. 2011. Quantum dot blueing and blinking enables fluorescence nanoscopy. *Nano Letters*, 11, 245–250.

Huang, B., Wang, W., Bates, M., and Zhuang, X. 2008. Three-dimensional super-resolution imaging by stochastic optical reconstruction microscopy. *Science*, 319, 810–813.

Keppler, A., Kindermann, M., Gendreizig, S., Pick, H., Vogel, H., and Johnsson, K. 2004. Labeling of fusion proteins of O-6-alkylguanine-DNA alkyltransferase with small molecules in vivo and in vitro. *Methods*, 32, 437–444.

Klar, T. A., Jakobs, S., Dyba, M., Egner, A., and Hell, S. W. 2000. Fluorescence microscopy with diffraction resolution barrier broken by stimulated emission. *Proceedings of the National Academy of Sciences of the United States of America*, 97, 8206–8210.

Langhorst, M. F., Schaffer, J., and Goetze, B. 2009. Structure brings clarity: Structured illumination microscopy in cell biology. *Biotechnology Journal*, 4, 858–865.

Matsuda, T., Miyawaki, A., and Nagai, T. 2008. Direct measurement of protein dynamics inside cells using a rationally designed photoconvertible protein. *Nature Methods*, 5, 339–345.

McEvoy, A. L., Hoi, H., Bates, M. et al. 2012. mMaple: A photoconvertible fluorescent protein for use in multiple imaging modalities. *PLoS One*, 7, e51314.

McKinney, S. A., Murphy, C. S., Hazelwood, K. L., Davidson, M. W., and Looger, L. L. 2009. A bright and photostable photoconvertible fluorescent protein. *Nature Methods*, 6, 131–133.

Morozova, K. S., Piatkevich, K. D., Gould, T. J., Zhang, J., Bewersdorf, J., and Verkhusha, V. V. 2010. Far-red fluorescent protein excitable with red lasers for flow cytometry and superresolution STED nanoscopy. *Biophysical Journal*, 99, L13–L15.

Mueller, T., Schumann, C., and Kraegeloh, A. 2012. STED microscopy and its applications: New insights into cellular processes on the nanoscale. *Chemphyschem*, 13, 1986–2000.

Naegerl, U. V., Willig, K. I., Hein, B., Hell, S. W., and Bonhoeffer, T. 2008. Live-cell imaging of dendritic spines by STED microscopy. *Proceedings of the National Academy of Sciences of the United States of America*, 105, 18982–18987.

Nienhaus, G. U., Nienhaus, K., Holzle, A. et al. 2006. Photoconvertible fluorescent protein EosFP: Biophysical properties and cell biology applications. *Photochemistry and Photobiology*, 82, 351–358.

Patterson, G. H. and Lippincott-Schwartz, J. 2002. A photoactivatable GFP for selective photolabeling of proteins and cells. *Science*, 297, 1873–1877.

Pavani, S. R. P., Deluca, J. G., and Piestun, R. 2009. Polarization sensitive, three-dimensional, single-molecule imaging of cells with a double-helix system. *Optics Express*, 17, 19644–19655.

Rego, E. H., Shao, L., Macklin, J. J. et al. 2012. Nonlinear structured-illumination microscopy with a photoswitchable protein reveals cellular structures at 50-nm resolution. *Proceedings of the National Academy of Sciences of the United States of America*, 109, E135–E143.

Rittweger, E., Han, K. Y., Irvine, S. E., Eggeling, C. , and Hell, S. W. 2009. STED microscopy reveals crystal colour centres with nanometric resolution. *Nature Photonics*, 3, 144–147.

Rust, M. J., Bates, M., and Zhuang, X. W. 2006. Sub-diffraction-limit imaging by stochastic optical reconstruction microscopy (STORM). *Nature Methods*, 3, 793–795.

Schmidt, R., Wurm, C. A., Punge, A., Egner, A., Jakobs, S., and Hell, S. W. 2009. Mitochondrial cristae revealed with focused light. *Nano Letters*, 9, 2508–2510.

Shaner, N. C., Lambert, G. G., Chammas, A. et al. 2013. A bright monomeric green fluorescent protein derived from *Branchiostoma lanceolatum*. *Nature Methods*, 10, 407–409.

Shao, L., Isaac, B., Uzawa, S., Agard, D. A., Sedat, J. W., and Gustafsson, M. G. L. 2008. I(5)S: Wide-field light microscopy with 100-nm-scale resolution in three dimensions. *Biophysical Journal*, 94, 4971–4983.

Shroff, H., Galbraith, C. G., Galbraith, J. A. et al. 2007. Dual-color superresolution imaging of genetically expressed probes within individual adhesion complexes. *Proceedings of the National Academy of Sciences of the United States of America*, 104, 20308–20313.

Shtengel, G., Galbraith, J. A., Galbraith, C. G. et al. 2009. Interferometric fluorescent super-resolution microscopy resolves 3D cellular ultrastructure. *Proceedings of the National Academy of Sciences of the United States of America*, 106, 3125–3130.

Stiel, A. C., Andresen, M., Bock, H. et al. 2008. Generation of monomeric reversibly switchable red fluorescent proteins for far-field fluorescence nanoscopy. *Biophysical Journal*, 95, 2989–2997.

Stiel, A. C., Trowitzsch, S., Weber, G. et al. 2007. 1.8 angstrom bright-state structure of the reversibly switchable fluorescent protein Dronpa guides the generation of fast switching variants. *Biochemical Journal*, 402, 35–42.

Strack, R. L., Hein, B., Bhattacharyya, D., Hell, S. W., Keenan, R. J., and Glick, B. S. 2009. A rapidly maturing far-red derivative of DsRed-expres2 for whole-cell labeling. *Biochemistry*, 48, 8279–8281.

Subach, F. V., Patterson, G. H., Manley, S., Gillette, J. M., Lippincott-Schwartz, J., and Verkhusha, V. V. 2009. Photoactivatable mCherry for high-resolution two-color fluorescence microscopy. *Nature Methods*, 6, 153–159.

Subach, F. V., Patterson, G. H., Renz, M., Lippincott-Schwartz, J., and Verkhusha, V. V. 2010a. Bright monomeric photoactivatable red fluorescent protein for two-color super-resolution sptPALM of live cells. *Journal of the American Chemical Society*, 132, 6481–6491.

Subach, F. V., Zhang, L., Gadella, T. W. J., Gurskaya, N. G., Lukyanov, K. A., and Verkhusha, V. V. 2010b. Red fluorescent protein with reversibly photoswitchable absorbance for photochromic FRET. *Chemistry and Biology*, 17, 745–755.

Subach, O. M., Entenberg, D., Condeelis, J. S., and Verkhusha, V. V. 2012. A FRET-facilitated photoswitching using an orange fluorescent protein with the fast photoconversion kinetics. *Journal of the American Chemical Society*, 134, 14789–14799.

Subach, O. M., Patterson, G. H., Ting, L. M., Wang, Y. R., Condeelis, J. S., and Verkhusha, V. V. 2011. A photoswitchable orange-to-far-red fluorescent protein, PSmOrange. *Nature Methods*, 8, 771–777.

Toomre, D. and Bewersdorf, J. 2010. A new wave of cellular imaging. *Annual Review of Cell and Developmental Biology*, 26, 285–314.

Tsutsui, H., Karasawa, S., Shimizu, H., Nukina, N., and Miyawaki, A. 2005. Semi-rational engineering of a coral fluorescent protein into an efficient highlighter. *EMBO Reports*, 6, 233–238.

Verkhusha, V. V. and Sorkin, A. 2005. Conversion of the monomeric red fluorescent protein into a photoactivatable probe. *Chemistry and Biology*, 12, 279–285.

Westphal, V. and Hell, S. W. 2005. Nanoscale resolution in the focal plane of an optical microscope. *Physical Review Letters*, 94, 143903.

Willig, K. I., Kellner, R. R., Medda, R., Hein, B., Jakobs, S., and Hell, S. W. 2006a. Nanoscale resolution in GFP-based microscopy. *Nature Methods*, 3, 721–723.

Willig, K. I., Rizzoli, S. O., Westphal, V. et al. 2006b. STED Microscopy reveals that synaptotagmin remains clustered after synaptic vesicle exocytosis. *Nature*, 440, 935–939.

Willig, K. I., Stiel, A. C., Brakemann, T., Jakobs, S., and Hell, S. W. 2011. Dual-Label STED nanoscopy of living cells using photochromism. *Nano Letters*, 11, 3970–3973.

Xu, K., Babcock, H. P., and Zhuang, X. W. 2012. Dual-objective STORM reveals three-dimensional filament organization in the actin cytoskeleton. *Nature Methods*, 9, 185–188.

Zhang, M., Chang, H., Zhang, Y. et al. 2012. Rational design of true monomeric and bright photoactivatable fluorescent proteins. *Nature Methods*, 9, 727–729.

In vivo imaging revolution made by fluorescent proteins

Robert M. Hoffman
AntiCancer, Inc.
University of California, San Diego

Contents

13.1 DISCOVERY AND PROPERTIES OF GFP

Green fluorescent protein (GFP) was discovered in the bioluminescent jellyfish *Aequorea victoria* by Shimomura (2009). The GFP gene was cloned from *A. victoria* by Doug Prasher, which enabled GFP to become the most powerful tool in cell biology (Prasher et al. 1992). The GFP cDNA encodes a 283-amino-acid polypeptide with a mol. wt. of 27 kd (Prasher et al. 1992, Yang et al. 1996).

The monomeric GFP requires no other *Aequorea* proteins, substrates, or cofactors to fluoresce (Cody et al. 1993). GFP gene gain-of-function mutants have been generated by various techniques (Cormack et al. 1996, Crameri et al. 1996, Delagrave et al. 1995, Heim et al. 1995). For example, the GFPS-65T clone has the serine-65 codon substituted with a threonine codon, which results in a single excitation peak at 490 nm (Heim et al. 1995). Moreover, to develop higher expression in human and other mammalian cells, a humanized hGFP-S65T clone was isolated (Zolotukhin et al. 1996).

In 2008, the Nobel Prize for chemistry was awarded for the discovery and modification of GFP (Tsien 2009). The Nobel announcement cited two uses of GFP (Nobel background), one of which was to track cancer cells *in vivo*, which was pioneered in our laboratory (Chishima et al. 1997a, Hayashi et al. 2007, Hoffman 2005, 2011, Yang et al. 2000a, Yamauchi et al. 2005, 2006).

Fluorescent proteins have very high extinction coefficients ranging from 6,500 up to 95,000 (Morin and Hastings 1971). In addition, they have very high quantum yields ranging from 0.24 up to 0.8 (Prasher et al. 1992). These properties make fluorescent proteins very bright. The large two-photon absorption of GFP is important for deep imaging *in vivo* (Koenig 2000). Another important feature among the family of fluorescent proteins is that there are many colors that can be used simultaneously for multifunctional *in vivo* imaging. These properties make fluorescent proteins optimal for cellular imaging *in vivo* (Yang et al. 2006a, 2007).

Matz et al. (1999) cloned six fluorescent proteins homologous to GFP. The proteins were isolated from the coral *Discosoma*. A red fluorescent protein (RFP) from *Discosoma* has been further developed for expression in mammalian cells and is known as DsRed-2. Monomer variants of the *Discosoma* red protein have been developed with distinguishable colors from yellow-orange to red-orange with fruit names similar to their color, mCherry, mTomato, etc. (Shaner et al. 2004). Fluorescent proteins that can be converted from one color to another or become fluorescent upon light activation have also been developed (Ando et al. 2002, Verkhusha and Lukyanov 2004, Lukyanov 2005).

13.2 NONINVASIVE IMAGING WITH GFP

13.2.1 BRIEF HISTORY OF IMAGING OF SMALL ANIMALS

Methods of external imaging of small animals growing tumors include x-rays, MRI, and ultrasonography. Although these methods are well suited for noninvasive imaging (Tearney et al. 1997), they have limitations in the investigation of internally- growing tumors. In particular, monitoring growth and metastatic dissemination over time by these methods are limited because they either use potentially harmful irradiation or require harsh contrast agents and, therefore, should not be repeated on a frequent, real-time basis. In addition, these methods do not have cellular or subcellular resolution (Yang et al. 2000a). Optical imaging of cancers has been challenging because cancer cells are not clearly distinguished from normal cells. Optical imaging has also been limited by strong nonspecific autofluorescence (Taubes 1997, Yang et al. 2000a).

Previous attempts to mark tumors have had some success. These included labeling with monoclonal antibodies (mAbs) and other high-affinity molecules targeted against tumor-associated markers (Kaushal et al. 2008, McElroy et al. 2008a,b, 2009, Metildi et al. 2012a,b, 2013a,b). However, results are limited by the fact that mAbs are not passed onto the progeny of the marked cancer cells.

Intravital videomicroscopy (IVVM) is another approach to optical imaging of tumor cells. IVVM allows direct observation of cancer cells (Chambers 1995, Naumov et al. 1999). However, IVVM is not well suited for longitudinal studies of tumor growth, progression, and internal metastasis in a live, intact animal (Yang et al. 2000a).

An important advance in optical imaging was making the tumor emit light. One approach inserted the luciferase gene into tumors so that they emit light (Sweeney et al. 1999). However, luciferase enzymes genetically transferred to mammalian cells require the exogenous delivery of their luciferin substrate, which is difficult to control and expensive to use in an intact animal and results in weak signal that does not produce an image and therefore can be detected only by photon counting (Hoffman 2005, Yang et al. 2000a).

A more practical approach to tumor luminance is to make the target tissue selectively fluorescent. One approach is to target tumors with protease-activated, near-infrared (NIR) fluorescent probes (Weissleder et al. 1999). Tumors with appropriate proteases could activate the probes and be imaged externally. However, the system proved to have severe restrictions. The selectivity was limited because most normal tissues have significant protease activity. In fact, the normal activity in liver is so high as to preclude imaging in this most important of metastatic sites. The short lifetime of the fluorescence probes would appear to rule out growth and efficacy studies. The requirement of appropriate, tumor-specific protease activity and effective tumor delivery of the probes also limits this approach (Weissleder et al. 1999, Yang et al. 2000a).

13.2.2 BEGINNING OF NONINVASIVE *IN VIVO* IMAGING WITH GFP

Our laboratory developed an approach to genetically label tumors with GFP whose fluorescence is sufficiently strong to be viewed externally and noninvasively in intact animals (Figure 13.1) (Hoffman 2005, Hoffman and Yang 2006b, Yang et al. 2000a). This was an extension of our previous work, which used stable GFP expression in cancer cells as an extremely effective tumor cell marker in the live animal (Chishima et al. 1997a, Hoffman 2005). The GFP fluorescence illuminated tumor progression and allowed detection of metastases in intact animals. GFP labeling enabled visualization of metastases in soft organs and bone (Chishima et al. 1997a,b,c,d,e, Yang et al. 1998, 1999a,b). A major advantage of GFP-expressing tumor cells is that imaging requires no preparative procedures and, therefore, is uniquely suited for visualizing in the live animal. Using stable, high-GFP-expression tumor cells, external, noninvasive, whole-body, real-time fluorescence optical imaging of internally-growing tumors and metastases was readily carried out.

13.2.3 NONINVASIVE *IN VIVO* IMAGING WITH RFP

RFP possesses enhanced optical qualities that enable it to be used instead of, or as an adjunct to, GFP-based systems. For example, orthotopic implantation of highly red fluorescent human pancreatic tumor fragments onto the pancreas spontaneously

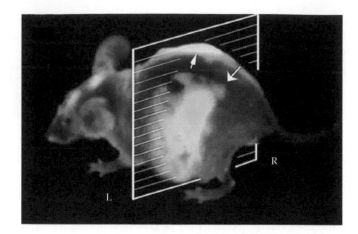

Figure 13.1 Lateral, whole-body image of GFP-expressing metastatic liver lesions. Cancer in the left (thick arrow) and right lobes (fine arrow) of a live nude mouse at day 21 after surgical orthotopic transplantation (SOI). (Reprinted by permission from *Proc. Natl. Acad. Sci. USA*, Yang, M., Baranov, E., Jiang, P., et al., Whole-body optical imaging of green fluorescent protein-expressing tumors and metastases, 97, 1206–1211, Copyright 2000.)

yields extensive, locoregional, primary tumor growth and the development of distant metastases. The primary and metastatic tumors can be visualized, tracked, and imaged in real time, using selective tumor RFP fluorescence (Figure 13.2). Treatment with two well-described therapeutic agents, gemcitabine and CPT-11, was performed to show the ability to compare therapeutic efficacy among drugs using this unique preclinical model (Katz et al. 2003). RFP has been shown to be very suitable for imaging time course growth of human tumors in the brain of the nude mice (Figure 13.3) (Hoffman and Yang 2006b).

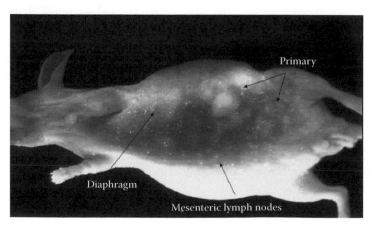

Figure 13.2 External image of a nude mouse on day 17 after SOI of a human pancreatic cancer expressing RFP Extensive locoregional and metastatic growth is visualized by selectively exciting DsRed2-expressed in the tumors. (From Katz, M. et al., *J. Surg. Res.*, 113, 151, 2003.)

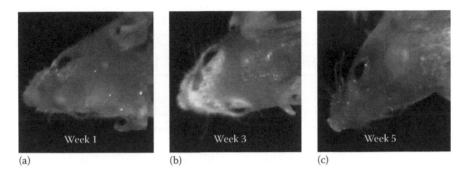

(a)   (b)   (c)

Figure 13.3 Whole-body imaging of a brain tumor. Real-time whole-body imaging of a U87-GFP human glioma growing in the brain of a nude mouse at (a) 1 week, (b) 3 weeks, and (c) 5 weeks after SOI. (From Hoffman, R.M., Yang, M. Whole-body imaging with fluorescent proteins. *Nat. Protoc.*, 1, 1429–1438, 2006. Copyright 2006.)

13.2.4 USING SIMPLE INSTRUMENTS FOR *IN VIVO* IMAGING WITH GFP AND RFP

We have used a blue LED flashlight (LDP LLC, Woodcliff Lake, NJ, United States; www.maxmax.com/OpticalProducts.htm) with an excitation filter (midpoint wavelength peak of 470 nm) and an emission D470/40 filter (Chroma Technology, Brattleboro, VT, United States) for noninvasive, whole-body imaging of mice with GFP- and RFP-expressing tumors growing in or on internal organs (Yang et al. 2005). For example, whole-body imaging with the LED flashlight visualized two different colored tumors, one expressing GFP and the other expressing RFP, implanted in the brain of a nude mouse. GFP and RFP tumors are simultaneously excited with the blue LED flashlight and readily imaged without interfering autofluorescence (Yang et al. 2005).

In another example, a GFP-expressing tumor implanted on the colon was whole-body imaged with the blue LED flashlight. The animal was also opened and imaged in the same way with the blue LED flashlight. There was high accuracy of the whole-body image compared to the image of the opened animal. Whole-body imaging of an RFP-expressing tumor implanted on the liver and a GFP-expressing tumor on the pancreas of the same mouse was performed. As with the RFP and GFP tumors implanted in the brain, the RFP tumor implanted on the liver and GFP tumor implanted on the pancreas were simultaneously excited with the blue LED flashlight. The size of the imaged tumor was comparable for both the whole-body and open images. Even more striking is that the intensity of the whole-body image was 70% of the open image. It has previously been shown that whole-body imaging correlates with actual tumor volume (4). These results lead to the following conclusions: (1) very strong signals emit from GFP- and RFP-expressing tumors inside the animal, (2) the images are readily quantifiable, (3) there is negligible interference from autofluorescence with proper filters, and (4) very simple and low-cost instruments can be used for GFP and RFP whole-body macroimaging (Yang et al. 2005). The data also corrected serious misconceptions in the literature stating "limits" of *in vivo* fluorescent protein imaging (Gross and Piwnica-Worms 2005, Weissleder and Ntziachristos 2003, Kocher and Piwnica-Worms 2013).

13.3 USING TRANSGENIC NUDE MICE EXPRESSING FLUORESCENT PROTEINS FOR COLOR-CODED IMAGING OF TUMOR–HOST INTERACTION

13.3.1 TRANSGENIC GFP NUDE MOUSE

Okabe et al. (1997) produced transgenic mice with GFP under the control of a chicken β-actin promoter and cytomegalovirus enhancer. All of the tissues from these transgenic mice, with the exception of erythrocytes and hair, fluoresce green. We then developed a transgenic GFP nude mouse with ubiquitous GFP expression. The GFP nude mouse was obtained by crossing nontransgenic nude mice with the transgenic C57/B6 mouse in which the β-actin promoter drives GFP expression. In the adult mice, the organs all brightly expressed GFP, including the heart, lungs, spleen, pancreas, esophagus, stomach, and duodenum. RFP-expressing human cancer cell lines, including PC-3-RFP prostate cancer, HCT-116-RFP colon cancer, MDA-MB-435-RFP breast cancer, and HT1080-RFP fibrosarcoma, were transplanted to the transgenic GFP nude mice. Dual-color fluorescence imaging enabled visualization of human tumor–host interaction by whole-body imaging and at the cellular level in fresh and frozen tissues (Yang et al. 2004).

13.3.2 TRANSGENIC RFP NUDE MOUSE

Using the transgenic RFP mouse described by Nagy's group (Vintersten et al. 2004), we developed a transgenic RFP nude mouse that could serve as a host for GFP or GFP–RFP-labeled human cancer cells. A nude mouse expressing RFP is also an appropriate host for transplantation of GFP-expressing stromal cells as well as double-labeled cancer cells expressing GFP in the nucleus and RFP in the cytoplasm (Yamamoto et al. 2004), thereby creating a three-color imaging model of the tumor microenvironment (TME). The RFP nude mouse was obtained by crossing nontransgenic nude mice with the transgenic C57/B6 mouse in which the beta-actin promoter drives RFP (DsRed2) expression in essentially all tissues. In the RFP nude mouse, the organs all brightly expressed RFP, including the heart, lungs, spleen, pancreas, esophagus, stomach, duodenum, and the male and female reproductive systems; brain and spinal cord; and the circulatory system, including the heart and major arteries and veins. The skinned skeleton highly expressed RFP. The bone marrow and spleen cells were also RFP positive. GFP-expressing human cancer cell lines, including HCT-116-GFP colon cancer and MDA-MB-435-GFP breast cancer, were orthotopically transplanted to the transgenic RFP nude mice. These human tumors grew extensively in the transgenic RFP nude mouse. Dual-color fluorescence imaging enabled visualization of human tumor–host interaction (Yang et al. 2009).

13.3.3 TRANSGENIC CFP NUDE MOUSE

Additional colored mice would allow even more processes to be imaged simultaneously. In the cyan fluorescent protein (CFP) mouse developed in Nagy's laboratory (Hadjantonakis et al. 2002), CFP is driven by the β-actin promoter similar to the GFP and RFP transgenic mice. In the CFP mouse, there is differential expression of blue fluorescence among the various organs. However, the pancreas stands out from the rest of the GI tract, displaying the strongest fluorescence of all organs in the mouse. Fluorescence microscopy demonstrated that the CFP fluorescence was expressed in the acinar cells of the pancreas and not the islet cells (Tran Cao et al. 2009).

Applications

XPA-1 human pancreatic cancer cells expressing RFP or GFP in the nucleus and RFP in the cytoplasm were orthotopically implanted in female nude CFP mice. Color-coded fluorescence imaging of these human pancreatic cancer cells implanted into the bright blue fluorescent pancreas of the CFP nude mouse gave novel insight into the interaction of the pancreatic tumor and the normal pancreas, in particular the strong desmoplastic reaction of the tumor (Tran Cao et al. 2009).

13.3.4 RED-SHIFTED FLUORESCENT PROTEINS

RFPs were first described in the late 1990s as mentioned previously. The first such protein was isolated and cloned from the coral *Discosoma* sp. that was obtained from an aquarium shop in Moscow (Matz et al. 1999) and termed DsRed as mentioned previously. After extensive modification by mutagenesis, a bright red protein was eventually isolated, termed DsRed2, with an emission wavelength peak of 588. DsRed2 can be used for whole-body imaging and has been used to noninvasively follow cancer metastasis in real time (Katz et al. 2003) in nude mice as described previously. DsRed2 has also been used to whole-body image tumors growing in transgenic GFP nude mice, allowing for the color coding of cancer and host cells (Hoffman 2008, Yang et al. 2003, 2004, 2006c), as described previously.

In 2004, a report appeared (Shaner et al. 2004) that described a series of red-shifted proteins obtained by mutating DsRed. These proteins, termed mCherry, mRaspberry, mPlum, and mTomato, had emission maxima as long as 649 nm. However, these mutants have low quantum yields, thereby reducing their brightness (Hoffman 2008).

A very bright, red-shifted fluorescent protein was described by Shcherbo et al. (2007). This protein, named Katushka, originated from the sea anemone *Entacmaea quadricolor*. Katushka has an excitation peak at 588 mm and an emission peak at 635 nm, both of which are relatively nonabsorbed by tissues and hemoglobin. Importantly, Katushka has an extinction coefficient of 65,000 M^{-1} cm^{-1}, and a quantum yield of 0.34 which makes Katushka the brightest fluorescent protein with an emission maximum beyond 620 nm making it an ideal candidate for *in vivo* imaging (Shcherbo et al. 2007).

Subsequent mutagenesis experiments generated a monomeric version of Katushka, which is a dimer. The monomer, highly suitable for fusions with other proteins, is termed mKate. Both Katushka and mKate demonstrate high photostability (Shcherbo et al. 2007).

13.4 PERSPECTIVES

13.4.1 METHOD OF CHOICE FOR WHOLE-BODY IMAGING

The features of fluorescent-protein-based imaging, such as a strong and stable signal, enable noninvasive whole-body imaging down to the subcellular level (Yang et al. 2006a, 2007). These properties make fluorescent-protein-based imaging (especially with red-shifted fluorescent proteins) far superior to luciferase-based imaging. Luciferase-based imaging, with its weak signal (Ray et al. 2004), which precludes image acquisition and enables only photon counting with pseudocolor-generated images, has limited applications (Hoffman 2005, Hoffman and Yang 2006b). For example, cellular imaging *in vivo* is not presently possible with luciferase. The dependence on circulating luciferin makes the signal from luciferase imaging unstable (Hoffman 2005, Hoffman and Yang 2006b). The one possible advantage of luciferase-based imaging is that no excitation light is necessary. However, far-red absorbing proteins, such as Katushka, greatly reduce any problems with excitation, even in deep tissues, as shown by Shcherbo et al. (2007). Proteins, such as Katushka,

Applications

as well as photoactivatable- (Lukyanov et al. 2005) and photoconvertible-fluorescent proteins (Ando et al. 2002), provide powerful tools for future whole-body imaging experiments.

Imaging instrumentation, such as those with variable magnification (Yamauchi et al. 2006) or scanning lasers (Yang et al. 2007) and multiphoton microscopy (Condeelis and Segall 2003), makes fluorescent proteins tools of choice for whole-body imaging. Whole-body imaging with fluorescent proteins can now reach the subcellular level using cells labeled in the nucleus with GFP and RFP in the cytoplasm (Yang et al. 2006a, 2007). However, there are misconceptions in the literature suggesting that fluorescent-protein-based imaging is inferior to luciferase (Gross and Piwnica-Worms 2005, Ntziachristos et al. 2005, Weissleder and Ntziachristos 2003, Kocher and Piwnica-Worms 2013). The results described in this chapter should clarify this subject greatly.

Transgenic mice engineered with bright proteins also offer many new possibilities for whole-body imaging. Future human use is also possible, for example, using specifically engineered viruses to label existing tumors *in vivo* for diagnostic and surgical-navigation applications (Kishimoto et al. 2009).

13.5 NONINVASIVE SUBCELLULAR IMAGING WITH GFP AND RFP

We genetically engineered dual-color fluorescent cells with one color in the nucleus and the other in the cytoplasm that enables real-time nuclear–cytoplasmic dynamics to be visualized in living cells *in vivo* as well as *in vitro*. To obtain the dual-color cells, RFP was expressed in the cytoplasm of cancer cells, and GFP linked to histone H2B was expressed in the nucleus. Nuclear GFP expression enabled visualization of nuclear dynamics, whereas simultaneous cytoplasmic RFP expression enabled visualization of nuclear–cytoplasmic ratios as well as simultaneous cell and nuclear shape changes. Thus, total cellular dynamics can be visualized in the living dual-color cells in real time (Figure 13.4) (Yamamoto et al. 2004, Yamauchi et al. 2005, Hoffman and Yang 2006a).

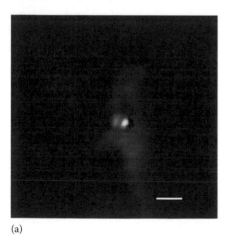
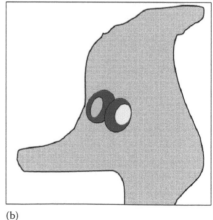

(a)   (b)

Figure 13.4 Real-time image of mitotic cancer cells, expressing GFP in the nucleus and RFP in the cytoplasm in ear of a live mouse captured 12 h after cell injection. (a) High-magnification image. Bar = 50 μm. (b) Schema of (a). (From Yamamoto, N. et al., *Cancer Res.*, 64, 4251, 2004.)

The heterogeneous and structurally complex nature of the interactive TME is little understood. The relative amount of stroma and its composition vary considerably from tumor to tumor and vary within a tumor over the course of tumor progression. The interaction between cancer cells and stromal cells largely determines the phenotype of the tumor. For example, studies have shown that the growth, invasiveness, and angiogenesis of human breast cancer xenografts in mice depend on the presence of stromal fibroblasts (Orimo et al. 2005) and that splenocytes are obligate passengers of cancer cells that metastasize to the liver after injection in the spleen (Bouvet et al. 2006).

To noninvasively visualize cellular and subcellular events in the TME in real time in the live mouse, we used a laser scanning microscope with a 0.3 mm diameter stick objective up to 2 cm in length (IV100 laser scanning microscope, Olympus Corp., Tokyo, Japan). This novel imaging system, coupled with the use of the dual-color cancer cells and transgenic GFP mouse, has enabled noninvasive *in vivo* imaging of the cancer and stromal cells in the TME at the subcellular level (Figure 13.5) (Yang et al. 2007).

The model consists of transgenic GFP-expressing nude mice (please see above) transplanted with dual-color cancer cells labeled with GFP in the nucleus and RFP in the cytoplasm as described earlier. The GFP-expressing stroma interacting with the dual-color cancer cells could be imaged noninvasively. In this model, drug response of both cancer and stromal cells in the intact live animal was also imaged in real time. Mitotic and apoptotic tumor cells, stromal cells interacting with the cancer cells, tumor vasculature, and tumor blood flow were all imaged noninvasively. This model system enabled the first cellular and subcellular images of unperturbed tumors in the live intact animal (Yang et al. 2007).

13.6 COLOR-CODED IMAGING OF THE CELL CYCLE

Miyawaki's groups Fucci probes label individual G1 phase nuclei red and those in S/G2/M phases green, thereby identifying the phase of the cell cycle of each cell by the color of their nuclei (Sakaue-Sawano et al. 2008).

13.7 NONINVASIVE IMAGING OF TUMOR ANGIOGENESIS WITH GFP

The nonluminous angiogenic blood vessels appear as sharply-defined dark networks against this bright background of a GFP-expressing tumor. Whole-body optical imaging of tumor angiogenesis was demonstrated by injecting GFP-expressing Lewis lung carcinoma cells into the s.c. site of the footpad of nude mice. The footpad is relatively transparent, with comparatively few resident blood vessels, allowing quantitative imaging of tumor angiogenesis in the intact animal. Capillary density increased linearly over a 10-day period as determined by whole-body imaging. Similarly, the GFP-expressing human breast tumor MDA-MB-435 was orthotopically transplanted to the mouse fat pad, where whole-body optical imaging showed that blood vessel density increased linearly over a 20-week period. These powerful and clinically-relevant angiogenesis mouse models can be used for real-time *in vivo* evaluation of agents inhibiting or promoting tumor angiogenesis in physiological microenvironments (Yang et al. 2001).

Applications

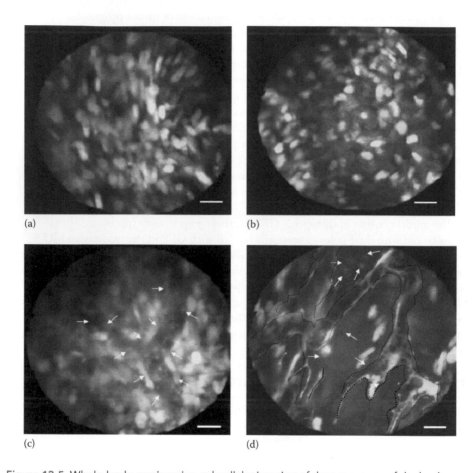

(a)

(b)

(c)

(d)

Figure 13.5 Whole-body, noninvasive, subcellular imaging of drug response of dual-color mouse mammary tumor (MMT) cells, expressing GFP in the nucleus and RFP in the cytoplasm, and GFP stromal cells in the live GFP nude mouse with and without doxorubicin. Dual-color MMT cells were injected in the footpad of GFP transgenic nude mice. (a) Whole-body image of untreated dual-color MMT cells in the footpad of a live GFP mouse. Note the numerous spindle-shaped dual-color MMT cells interdispersed among the GFP host cells. (b) Whole-body image of MMT dual-color cancer cells in a live GFP nude mouse 12 h after treatment with doxorubicin (10 mg kg^{-1}). The cancer cells lost their spindle shape, and the nuclei appear contracted. (c) Whole-body image of dual-color MMT tumor. Numerous dual-color spindle-shaped MMT cells interacted with GFP-expressing host cells. Well-developed tumor blood vessels and real-time blood flow were visualized by whole-body imaging (arrows). (d) *In vivo* drug response of dual-color MMT tumor 12 h after i.v. injection of 10 mg kg^{-1} doxorubicin. All of the visible MMT cells lost their spindle shape. Many of the cancer cells fragmented (arrows). Tumor blood vessels were damaged (dashed black lines), and the number of cancer cells was dramatically reduced 12 h after chemotherapy. Bar = 20 μm. (From Yang, M. et al., *Cancer Res.*, 67, 5195, 2007.)

13.8 NONINVASIVE IMAGING OF GENE EXPRESSION WITH GFP

Mice were labeled by directly injecting adenoviral GFP into either the brain, liver, pancreas, prostate, or bone marrow. Within 5–8 h after adenoviral GFP injection, the fluorescence of the expressed GFP in brain and liver became visible, and noninvasive, whole-body images were recorded at video rates. The GFP fluorescence continued to increase for at least 12 h and remained detectable in liver for up to 4 months. The rapidity of image acquisition makes possible real-time recording. The method requires only that the expressed gene or promoter be fused or operatively linked to GFP (Yang et al. 2000b).

13.9 NONINVASIVE IMAGING OF BACTERIAL INFECTION WITH GFP

Escherichia coli, expressing GFP, are sufficiently bright as to be clearly visible noninvasively from outside the infected animal. GFP-expressing bacteria were observed noninvasively in several mouse organs including the peritoneal cavity, stomach, small intestine, and colon. The progress of *E. coli*-GFP through the mouse gastrointestinal tract after gavage was followed in real time by noninvasive, whole-body imaging. Bacteria, seen first in the stomach, migrated into the small intestine and subsequently into the colon—an observation confirmed by intravital direct imaging. Infection was established by intraperitoneal (i.p.) injection of *E. coli*-GFP. The development of infection over 6 h and its regression after kanamycin treatment were visualized by whole-body imaging. This imaging technology affords a powerful approach to visualizing the infection process, determining the tissue specificity of infection, and the spatial migration of the infectious agents (Zhao et al. 2001).

13.10 NONINVASIVE IMAGING OF STEM CELLS WITH GFP

Multiphoton tomography is suitable for nondestructive long-term analysis of living cells due to the absence of out-of-focus absorption, out-of focus photobleaching, and out-of-focus phototoxicity. Multicolor imaging is possible with a single NIR laser excitation. In addition to two-photon excited fluorescence, multiphoton imaging enables second harmonic generation (SHG) imaging that can visualize collagen (Denk et al. 1990, Koenig 2000, Uchugonova et al. 2011).

A femtosecond-laser-based multiphoton tomograph with a mechano-optical articulated arm and an active beam stabilizer in combination with a compact scan head for flexible 3D imaging was used for imaging of multipotent hair follicle stem cells in the living mouse to noninvasively visualize nestin-expressing multipotent stem cells, which also express GFP, originating in the bulge area of the hair follicle in their native niche as well as when they migrate out of the niche (Amoh et al. 2005a,b, Li et al. 2003, Liu et al. 2011, Mii et al. 2013, Uchugonova et al. 2011).

Applications

13.11 CONCLUSIONS

The use of fluorescent proteins for *in vivo* imaging is perhaps their most important revolution. Not since the discovery of x-rays by Röntgen has there been a development in imaging with such great potential: the possibility of cellular and subcellular imaging *in vivo*. Fluorescent proteins are genetic reporters and thus can be used to permanently label essentially any cellular process. Fluorescent proteins come in many colors; therefore, many cellular processes can be labeled, enabling their functions to be imaged simultaneously. The host mouse can also express a fluorescent protein; therefore, tumor–host interaction can be followed by color-coded imaging. The next step in the *in vivo* revolution with fluorescent proteins will be their direct application in the clinic.

REFERENCES

Amoh, Y., Li, L., Campillo, R. et al. 2005b. Implanted hair follicle stem cells form Schwann cells that support repair of severed peripheral nerves. *Proc Natl Acad Sci USA* 102: 17734–17738.

Amoh, Y., Li, L., Katsuoka, K., Penman, S., Hoffman, R.M. 2005a. Multipotent nestin-positive, keratin-negative hair-follicle-bulge stem cells can form neurons. *Proc Natl Acad Sci USA* 102: 5530–5534 .

Ando, R., Hama, H., Yamamoto-Hino, M., Mizuno, H., Miyawaki, A. 2002. An optical marker based on the UV-induced green-to-red photoconversion of a fluorescent protein. *Proc Natl Acad Sci USA* 99: 12651–12656.

Betzig, E., Patterson, G.H., Sougrat, R. et al. 2006. Imaging intracellular fluorescent proteins at nanometer resolution. *Science* 313: 1642–1645.

Bouvet, M., Tsuji, K., Yang M., Jiang, P., Moossa, A.R., Hoffman, R.M., 2006. *In vivo* color-coded imaging of the interaction of colon cancer cells and splenocytes in the formation of liver metastases. *Cancer Res* 66: 11293–11297.

Chalfie, M. 2009. GFP: Lighting up life (Nobel Lecture). *Angew Chem Int Ed Engl* 48: 5603–5611.

Chambers, A.F., MacDonald, I.C., Schmidt, E.E. et al. 1995. Steps in tumor metastasis: New concepts from intravital videomicroscopy. *Cancer Metastasis Rev* 14: 279–301 .

Cheng, L., Fu, J., Tsukamoto, A., Hawley, R.G. 1996. Use of green fluorescent protein variants to monitor gene transfer and expression in mammalian cells. *Nat Biotechnol* 14: 606–609.

Chishima, T., Miyagi, Y., Li, L. et al. 1997e. Use of histoculture and green fluorescent protein to visualize tumor cell host interaction. *In Vitro Cell Dev Biol* 33: 745–747.

Chishima, T., Miyagi, Y., Wang, X. et al. 1997a. Cancer invasion and micrometastasis visualized in live tissue by green fluorescent protein expression. *Cancer Res* 57: 2042–2047.

Chishima, T., Miyagi, Y., Wang, X. et al. 1997b. Metastatic patterns of lung cancer visualized live and in process by green fluorescent protein expression. *Clin Exp Metastasis* 15: 547–552.

Chishima, T., Miyagi, Y., Wang, X. et al. 1997c. Visualization of the metastatic process by green fluorescent protein expression. *Anticancer Res* 17: 2377–2384.

Chishima, T., Yang, M., Miyagi, Y. et al. 1997d. Governing step of metastasis visualized in vitro. *Proc Natl Acad Sci USA* 94: 11573–11576.

Cody, C.W., Prasher, D.C., Welstler, V.M., Prendergast, F.G., Ward, W.W. 1993. Chemical structure of the hexapeptide chromophore of the *Aequorea* green fluorescent protein. *Biochemistry* 32: 1212–1218.

Condeelis, J., Segall, J.E. 2003. Intravital imaging of cell movement in tumours. *Nat Rev Cancer* 3: 921–930 .

Cormack, B., Valdivia, R., Falkow, S. 1996. FACS-optimized mutants of the green fluorescent protein (GFP). *Gene* 173: 33–38.

Crameri, A., Whitehorn, E.A., Tate, E., Stemmer, W.P.C. 1996. Improved green fluorescent protein by molecular evolution using DNA shuffling. *Nat Biotechnol* 14: 315–319.

Delagrave, S., Hawtin, R.E., Silva, C.M., Yang, M.M., Youvan, D.C. 1995. Red-shifted excitation mutants of the green fluorescent protein. *Biotechnology* 13: 151–154.

Denk, W., Strickler, J.H., Webb, W.W. 1990. Two-photon laser scanning fluorescence microscopy. *Science* 248: 73–76.

Gross, S., Piwnica-Worms, D. 2005. Spying on cancer: Molecular imaging in vivo with genetically encoded reporters. *Cancer Cell* 7:5–15.

Hadjantonakis, A.K., Macmaster, S., Nagy, A. 2002. Embryonic stem cells and mice expressing different GFP variants for multiple non-invasive reporter usage within a single animal. *BMC Biotechnol* 2: 11.

Hayashi, K., Jiang, P., Yamauchi, K. et al. 2007. Real-time imaging of tumor-cell shedding and trafficking in lymphatic channels. *Cancer Res* 67: 8223–8228.

Heim, R., Cubitt, A.B., Tsien, R.Y. 1995. Improved green fluorescence. *Nature* 373: 663–664.

Hoffman, R.M. 2005. The multiple uses of fluorescent proteins to visualize cancer in vivo. *Nat Rev Cancer* 5: 796–806.

Hoffman, R.M. 2008. A better fluorescent protein for whole-body imaging. *Trends Biotechnol* 26: 1–4.

Hoffman, R.M., Yang, M. 2006a. Subcellular imaging in the live mouse. *Nature Protoc* 1: 775–782.

Hoffman, R.M., Yang, M. 2006b. Whole-body imaging with fluorescent proteins. *Nat Protoc* 1: 1429–1438.

Hoffman, R.M., Yang, M. 2006c. Color-coded fluorescence imaging of tumor-host interactions. *Nature Protoc* 1: 928–935.

Katz, M., Takimoto, S., Spivac, D. et al. 2003. A novel red fluorescent protein orthotopic pancreatic cancer model for the preclinical evaluation of chemotherapeutics. *J Surg Res* 113: 151–160.

Kaushal, S., McElroy, M.K., Luiken, G.A. et al. 2008. Fluorophore-conjugated anti-CEA antibody for the intraoperative imaging of pancreatic and colorectal cancer. *J Gastrointest Surg* 12: 1938–1950.

Kishimoto, H., Zhao, M., Hayashi, K., Urata, Y., Tanaka, N., Fujiwara, T., Penman, S., Hoffman, R.M. 2009. *In vivo* internal tumor illumination by telomerase-dependent adenoviral GFP for precise surgical navigation. *Proc Natl Acad Sci USA* 106: 14514–14517.

Kocher, B., Piwnica-Worms, D. 2013. Illuminating cancer systems with genetically engineered mouse models and coupled luciferase reporters in vivo. *Cancer Discov* 3: 616–629.

Koenig, K. 2000. Multiphoton microscopy in life sciences. *J Microsc* 200: 83–104.

Li, L., Mignone, J., Yang, M. et al. 2003. Nestin expression in hair follicle sheath progenitor cells. *Proc Natl Acad Sci USA* 100: 9958–9961.

Liu, F., Uchugonova, A., Kimura, H. et al. 2011. The bulge area is the major hair follicle source of nestin-expressing pluripotent stem cells which can repair the spinal cord compared to the dermal papilla. *Cell Cycle* 10:830–839.

Lukyanov, K.A., Chudakov, D.M., Lukyanov, S., Verkhusha, V.V. 2005. Innovation: Photoactivatable fluorescent proteins. *Nat Rev Mol Cell Biol* 6: 885–891.

Matz, M.V., Fradkov, A.F., Labas, Y.A. et al. 1999. Fluorescent proteins from nonbioluminescent Anthozoa species. *Nat Biotechnol* 17: 969–973.

McElroy, M., Bouvet, M., Hoffman, R.M. 2008b. Color-coded fluorescent mouse models of cancer cell interactions with blood vessels and lymphatics. *Methods Enzymol* 445: 27–52.

McElroy, M., Hayashi, K., Garmy-Susini, B. et al. 2009. Fluorescent LYVE-1 antibody to image dynamically lymphatic trafficking of cancer cells in vivo. *J Surg Res* 151: 68–73.

McElroy, M., Kaushal, S., Luiken, G., Moossa, A.R., Hoffman, R.M., Bouvet, M. 2008a. Imaging of primary and metastatic pancreatic cancer using a fluorophore-conjugated anti-CA19-9 antibody for surgical navigation. *World J Surg* 32: 1057–1066.

Metildi, C.A., Kaushal, S., Lee, C., Hardamon, C.R., Snyder, C.S., Luiken, G.A., Talamini, M.A., Hoffman, R.M., Bouvet, M. 2012a. An LED light source and novel fluorophore combinations improve fluorescence laparoscopic detection of metastatic pancreatic cancer in orthotopic mouse models. *J Am Coll Surg* 214: 997–1007.e2.

Metildi, C.A., Kaushal, S., Hardamon, C.R., Snyder, C.S., Pu, M., Messer, K.S., Talamini, M.A., Hoffman, R.M., Bouvet, M. 2012b. Fluorescence-guided surgery allows for more complete resection of pancreatic cancer, resulting in longer disease-free survival compared with standard surgery in orthotopic mouse models. *J Am Coll Surg* 215: 126–136.

Metildi, C.A., Hoffman, R.M., Bouvet, M. 2013a. Fluorescence-guided surgery and fluorescence laporascopy for gastrointestinal cancers in clinically-relevant mouse models. *Gastroenterol Res Pract*: Article ID 290634, 8 pages.

Metildi, C.A., Kaushal, S., Luiken, G.A., Talamini, M.A., Hoffman, R.M., Bouvet, M. 2013b. Fluorescently labeled chimeric anti-CEA antibody improves detection and resection of human colon cancer in a patient-derived orthotopic xenograft (PDOX) nude mouse model. *J Surg Oncol* (in press). doi:10.1002/iso.23507.

Mii, S., Duong, J., Tome, Y., et al. 2013. The role of hair follicle nestin-expressing stem cells during whisker sensory-nerve growth in long-term 3D culture. *J Cell Biochem* 114: 1674–1684.

Morin, J., Hastings, J. 1971. Energy transfer in a bioluminescent system. *J Cell Physiol* 77: 313–318.

Naumov, G.N., Wilson, S.M., MacDonald, I.C. et al. 1999. Cellular expression of green fluorescent protein, coupled with high-resolution in vivo videomicroscopy, to monitor steps in tumor metastasis. *J Cell Sci* 112: 1835–1842.

Ntziachristos, V., Ripoll, J., Wang, L.V., Weissleder, R. 2005. Looking and listening to light: The evolution of whole-body photonic imaging. *Nat Biotechnol* 23: 313–320.

Okabe, M., Ikawa, M., Kominami, K., Nakanishi, T., Nishimune, Y. 1997. "Green mice" as a source of ubiquitous green cells. *FEBS Lett* 407: 313–319.

Orimo, A., Gupta, P.B., Sgroi, D.C. et al. 2005. Stromal fibroblasts present in invasive human breast carcinomas promote tumor growth and angiogenesis through elevated SDF-1/CXCL12 secretion. *Cell* 121: 335–348.

Prasher, D.C., Eckenrode, V.K., Ward, W.W., Prendergast, F.G., Cormier, M.J. 1992. Primary structure of the *Aequorea victoria* green-fluorescent protein. *Gene* 111: 229–233.

Ray, P., De, A., Min, J.J., Tsien, R.Y., Gambhir, S.S. 2004. Imaging tri-fusion multimodality reporter gene expression in living subjects. *Cancer Res* 64: 1323–1330.

Sakaue-Sawano, A., Kurokawa, H., Morimura, T. et al. 2008. Visualizing spatiotemporal dynamics of multicellular cell-cycle progression. *Cell* 132: 487–498.

Shaner, N.C., Campbell, R.E., Steinbach, P.A. et al. 2004. Improved monomeric red, orange and yellow fluorescent proteins derived from *Discosoma* sp. red fluorescent protein. *Nat Biotechnol* 22: 1567–1572.

Shcherbo, D., Merzlyak, E.M., Chepurnykh, T.V. et al. 2007. Bright far-red fluorescent protein for whole body imaging. *Nat Methods* 4: 741–746.

Shimomura, O. 2009. Discovery of green fluorescent protein (GFP) (Nobel Lecture). *Angew Chem Int Ed Engl* 48: 5590–5602.

Sweeney, T.J., Mailander, V., Tucker, A.A. et al. 1999. Visualizing the kinetics of tumor-cell clearance in living animals. *Proc Natl Acad Sci USA* 96: 12044–12049.

Taubes, G. 1997. Play of light opens a new window into the body. *Science* 276: 1991–1993.

Tearney, G.J., Brezinski, M.E., Bouma, B.E. et al. 1997. In vivo endoscopic optical biopsy with optical coherence tomography. *Science* 276: 2037–2939.

Tran Cao, H.S., Reynoso, J., Yang, M. et al. 2009. Development of the transgenic cyan fluorescent protein (CFP)-expressing nude mouse for "Technicolor" cancer imaging. *J Cell Biochem* 107: 328–334.

Tsien, R.Y. 2009. Constructing and exploiting the fluorescent protein paintbox (Nobel Lecture). *Angew Chem Int Ed Engl* 48: 5612–5626.

Uchugonova, A., Hoffman, R.M., Weinigel, M., Koenig, K. 2011. Watching stem cells in the skin of living mice noninvasively. *Cell Cycle* 10: 2017–2020.

Verkhusha, V.V., Lukyanov, K.A. 2004. The molecular properties and applications of Anthozoa fluorescent proteins and chromoproteins. *Nat Biotechnol* 22: 289–296.

Applications

Vintersten, K., Monetti, C., Gertsenstein, M. et al. 2004. Mouse in red: Red fluorescent protein expression in mouse ES cells, embryos, and adult animals. *Genesis* 40: 241–246.

Weissleder, R., Ntziachristos, V. 2003. Shedding light onto live molecular targets. *Nat Med* 9:123–128.

Weissleder, R., Tung, C. H., Mahmood, U., and Bogdanov, A., Jr. 1999. In vivo imaging of tumors with protease-activated near-infrared fluorescent probes. *Nat Biotechnol* 17: 375–378.

Yamamoto, N., Jiang, P., Yang, M. et al. 2004. Cellular dynamics visualized in live cells in vitro and in vivo by differential dual-color nuclear-cytoplasmic fluorescent-protein expression. *Cancer Res* 64: 4251–4256.

Yamauchi, K., Yang, M., Jiang, P., et al. 2005. Real-time *in vivo* dual-color imaging of intracapillary cancer cell and nucleus deformation and migration. *Cancer Res* 65: 4246–4252.

Yamauchi, K., Yang, M., Jiang, P. et al. 2006. Development of real-time subcellular dynamic multicolor imaging of cancer cell-trafficking in live mice with a variable-magnification whole-mouse imaging system. *Cancer Res* 66: 4208–4214.

Yang, F., Moss, L.G., Phillips, G.N. Jr. 1996. The molecular structure of green fluorescent protein. *Nat Biotechnol* 14: 1246–1251.

Yang, M., Baranov, E., Jiang, P. et al. 2000a. Whole-body optical imaging of green fluorescent protein-expressing tumors and metastases. *Proc Natl Acad Sci USA* 97: 1206–1211.

Yang, M., Baranov, E., Li, X.-M. et al. 2001. Whole-body and intravital optical imaging of angiogenesis in orthotopically implanted tumors. *Proc Natl Acad Sci USA* 98: 2616–2621.

Yang, M., Baranov, E., Moossa, A.R., Penman, S., Hoffman, R.M. 2000b. Visualizing gene expression by whole-body fluorescence imaging. *Proc Natl Acad Sci USA* 97: 12278–12282.

Yang, M., Hasegawa, S., Jiang, P. et al. 1998. Widespread skeletal metastatic potential of human lung cancer revealed by green fluorescent protein expression. *Cancer Res* 58: 4217–4221.

Yang, M., Jiang, P., An, Z. et al. 1999b. Genetically fluorescent melanoma bone and organ metastasis models. *Clinical Cancer Res* 5: 3549–3559.

Yang, M., Jiang, P., Hoffman, R.M. 2007. Whole-body subcellular multicolor imaging of tumor-host interaction and drug response in real time. *Cancer Res* 67: 5195–5200.

Yang, M., Jiang, P., Sun, F.X. et al. 1999a. A fluorescent orthotopic bone metastasis model of human prostate cancer. *Cancer Res* 59: 781–786.

Yang, M., Li, L., Jiang, P. et al. 2003. Dual-color fluorescence imaging distinguishes tumor cells from induced host angiogenic vessels and stromal cells. *Proc Natl Acad Sci USA* 100: 14259–14262.

Yang, M., Luiken, G., Baranov, E., and Hoffman, R.M. 2005. Facile whole-body imaging of internal fluorescent tumors in mice with an LED flashlight. *BioTechniques* 39: 170–172.

Yang, M., Reynoso, J., Bouvet, M., Hoffman, R.M. 2009. A transgenic red fluorescent protein-expressing nude mouse for color-coded imaging of the tumor microenvironment. *J Cell Biochem* 106: 279–284.

Yang, M., Reynoso, J., Jiang, P. et al. 2004. Transgenic nude mouse with ubiquitous green fluorescent protein expression as a host for human tumors. *Cancer Res* 64: 8651–8656.

Zhao, H., Doyle, T.C., Coquoz, O., Kalish, F., Rice, B.W., Contag, C.H. 2005. Emission spectra of bioluminescent reporters and interaction with mammalian tissue determine the sensitivity of detection in vivo. *J Biomed Opt* 10: 41210.

Zhao, M., Yang, M., Baranov, E. et al. 2001. Spatial-temporal imaging of bacterial infection and antibiotic response in intact animals. *Proc Natl Acad Sci USA* 98: 9814–9818.

Zolotukhin, S., Potter, M., Hauswirth, W.W., Guy, J., Muzycka, N. 1996. A "humanized" green fluorescent protein cDNA adapted for high-level expression in mammalian cells. *J Virol* 70: 4646–4654.

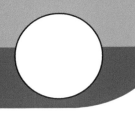

Index